Nordic Tourism

ASPECTS OF TOURISM
Series Editors: Chris Cooper, *Nottingham University Business School, UK*; C. Michael Hall, *University of Canterbury, New Zealand* and Dallen J. Timothy, *Brigham Young University, USA*

Aspects of Tourism is an innovative, multifaceted series, which comprises authoritative reference handbooks on global tourism regions, research volumes, texts and monographs. It is designed to provide readers with the latest thinking on tourism worldwide and push back the frontiers of tourism knowledge. The volumes are authoritative, readable and user-friendly, providing accessible sources for further research. Books in the series are commissioned to probe the relationship between tourism and cognate subject areas such as strategy, development, retailing, sport and environmental studies.

Full details of all the books in this series and of all our other publications can be found on http://www.channelviewpublications.com, or by writing to Channel View Publications, St Nicholas House, 31–34 High Street, Bristol, BS1 2AW, UK.

ASPECTS OF TOURISM
Series Editors: Chris Cooper, C. Michael Hall
and Dallen J. Timothy

Nordic Tourism

Issues and Cases

C. Michael Hall, Dieter K. Müller and
Jarkko Saarinen

CHANNEL VIEW PUBLICATIONS
Bristol • Buffalo • Toronto

Library of Congress Cataloging in Publication Data
A catalog record for this book is available from the Library of Congress.

Hall, Colin Michael, 1961-
Nordic Tourism: Issues and Cases
C. Michael Hall, Dieter K. Müller and Jarkko Saarinen.
Aspects of tourism: 36
Includes bibliographical references and index.
1. Tourism--Scandinavia--Management. 2. Travel--Scandinavia. I. Müller, Dieter K.
II. Saarinen, Jarkko, 1968- III. Title.
G155.S35H35 2008
914.8068--dc22 2008026659

British Library Cataloguing in Publication Data
A catalogue entry for this book is available from the British Library.

ISBN-13: 978-1-84541-094-0 (hbk)
ISBN-13: 978-1-84541-093-3 (pbk)

Channel View Publications
UK: St Nicholas House, 31–34 High Street, Bristol, BS1 2AW.
USA: UTP, 2250 Military Road, Tonawanda, NY 14150, USA.
Canada: UTP, 5201 Dufferin Street, North York, Ontario M3H 5T8, Canada.

Copyright © 2009 C. Michael Hall, Dieter Müller and Jarkko Saarinen

All rights reserved. No part of this work may be reproduced in any form or by any means without permission in writing from the publisher.

The policy of Multilingual Matters/Channel View Publications is to use papers that are natural, renewable and recyclable products, made from wood grown in sustainable forests. In the manufacturing process of our books, and to further support our policy, preference is given to printers that have FSC and PEFC Chain of Custody certification. The FSC and/or PEFC logos will appear on those books where full certification has been granted to the printer concerned.

Typeset by Techset Composition Ltd., Salisbury, UK
Printed and bound in Great Britain by MPG Books Ltd

Contents

Contributors.. vi
Cases and Issues ix
Figures, Tables and Plates xii
Abbreviations xvi
Preface... xvii

1 Nordic Tourism: Introduction to Key Concepts 1
2 Tourism Marketing 25
3 Nordic Tourism Governance and Planning Issues 52
4 Urban Tourism 83
5 Rural Tourism: Tourism as the Last Resort? 109
6 Nature-based Tourism in Northern Wildernesses 130
7 Coastal, Marine and Ocean Tourism............................. 153
8 Second Homes in the Nordic Countries.......................... 172
9 Culture and Tourism 197
10 Winter Tourism: Changing 'Snow Business' 224
11 The Future of Nordic Tourism: Regional and Environmental Change 243

References.. 266
Index 291

Contributors

Tor Arnesen, Eastern Norway Research Institute, postboks 223, 2601 Lillehammer, Norway; Tor.Arnesen.II@ostforsk.no

Trude Borch, Norut Social Science Research, Forskningsparken, NOR-9294 Tromsø, Norway; trude.borch@samf.norut.no

Lars Emmelin, Blekinge Institute of Technology, SE-371 79 Karlskrona, Sweden; lars.emmelin@bth.se

Birgitta Ericsson, Eastern Norway Research Institute. NO-2626 Lillehammer, Norway; be@ostforsk.no

Thor Flognfeldt Jr, Faculty of Tourism and Applied Social Sciences, Lillehammer University College, NOR-2626 Lillehammer, Norway; thor.flognfeldt@hil.no

Guðrún Þóra Gunnarsdóttir, Holar Agricultural College, IS-551 Sauðárkrókur, Iceland; ggunn@holar.is

Stefan Gössling, Centre for Sustainable- and Geotourism, Western Norway Research Institute, Sogndal, Norway; stefan.gossling@vestforsk.no

Szilvia Gyimóthy, Department of Service Management, Lund University, Box 882, SE-251 08 Helsingborg, Sweden; szilvia.gyimothy@msm.lu.se

Michel Haldrup, Department of Geography and International Development Studies, University of Roskilde, POB 260, DK-4000 Roskilde, Denmark; mhp@ruc.dk

C. Michael Hall, Department of Management, University of Canterbury, Christchurch, New Zealand; and Docent, Department of Geography, University of Oulu, Oulu, Finland; michael.hall@canterbury.ac.nz

Laufey Haraldsdóttir, Holar Agricultural College, IS-551 Sauðárkrókur, Iceland; laufey@holar.is

Mervi Hiltunen, Centre for Tourism Studies, University of Joensuu, POB 78, FIN-57101 Savonlinna, Finland; mervi.hiltunen@joensuu.fi

Eva Holmberg-Anttila, Supply Chain Management and Corporate Geography, HANKEN – Swedish School of Economics and Business Administration, PB 479 (Arkadiankatu 22), 00101 Helsinki, Finland; eva.holmberg@hanken.fi

Øystein Jensen, University of Stavanger, The Norwegian School of Hotel Management, N-4036 Stavanger, Norway; oystein.jensen@uis.no

Gunnar Thór Jóhannesson, Institut III for Geografi og Internationale Udviklingsstudier, Roskilde Universitets Center, Bygning 02, Postboks 260, DK-4000 Roskilde, Denmark; gunnarjo@ruc.dk

Anna Karlsdóttir, Department of Geography and Tourism, University of Iceland, 107 Reykjavík, Iceland; annakar@hi.is

Björn Kaltenborn, NINA, Fakkelgården, NOR-2624 Lillehammer, Norway; bjorn.kaltenborn@nina.no

Pekka Kauppila, Department of Geography, University of Oulu, PL 3000, FIN-90014 Oulu, Finland; pekka.kauppila@oulu.fi

Raija Komppula, Department of Business and Economics, University of Joensuu, POB 111, FIN-80101 Joensuu, Finland; raija.komppula@joensuu.fi

Peter Kvistgaard, Tourism Research Unit, Aalborg University, Fibigerstraede 2, DK-9220 Aalborg East, Denmark; kvist@ihis.aau.dk or kvist@solutionsinmotion.dk

Roger Marjavaara, Department of Social and Economic Geography, Umeå University, SE-901 87 Umeå, Sweden; roger.marjavaara@geography.umu.se

Samu Mäkelä, Hospitality, Åland Polytechnic, PB 1010 AX-22111 Mariehamn, Åland; sma@ha.ax

Dieter Müller, Department of Social and Economic Geography, Umeå University, 90187 Umeå, Sweden; dieter.muller@geography.umu.se

Per-Åke Nilsson, Department of Social Sciences, Mid Sweden University, Östersund, Sweden; per-ake.nilsson@miun.se; and Hammerdal Förlag & Reportage, Häradsvägen 11, 830 70 Hammerdal, Sweden; pan@hammerdalforlag.se

Kjell Overvåg, Eastern Norway Research Institute, NO-2626 Lillehammer, Norway; ko@ostforsk.no

Anna Dóra Sæþórsdóttir, Department of Geography and Tourism, University of Iceland, 107 Reykjavík, Iceland; annadora@hi.is

Jarkko Saarinen, Department of Geography, PO Box 3000, FIN-90014 University of Oulu, Oulu, Finland; jarkko.saarinen@oulu.fi

Ola Sletvold, Department of Tourism and Hotel Management, Finnmark University College, Follums vej 31, NOR-9509 Alta, Norway; ola.sletvold@hifm.no

Jens Kr. Steen Jacobsen, University of Stavanger, N-4036 Stavanger, Norway & Institute of Transport Economics, N-0349 Oslo, Norway; jsj@toi.no

Jaakko Suvantola, Department of Tourism Studies and Research, University of Joensuu, P.O.B. 126, FIN-57101 Savonlinna, Finland; jaakko.suvantola@joensuu.fi

Jan Åge Riseth, NORUT, PO Box 250, N-8504 Narvik, Norway; janar@samf.norut.no

Anette Therkelsen, Tourism Research Unit, Aalborg University, Fibigerstraede 2, DK-9220 Aalborg East, Denmark; at@ihis.aau.dk

Matti Vaara, Faculty of Forest Sciences, University of Joensuu, P.O. Box 111, FI-80101 Joensuu, Finland; Matti.Vaara@joensuu.fi

Arvid Viken, Department of Tourism and Hotel Management, Finnmark University College, Follums vej 31, NOR-9509 Alta, Norway; arvid.viken@hifm.no

Tuomas Vuorio, 4Event, Käpytie 6 B 19, 33180 Tampere, Finland; tuomas.vuorio@4event.fi

Cases and Issues

1.1	Nordic is not Scandinavia!	3
1.2	Tourism satellite accounts	12
1.3	The competitiveness of the Nordic states	17
2.1	Prominent promontory: The social construction of North Cape *Jens Kr. Steen Jacobsen*	30
2.2	The Hans Christian Andersen bicentenary	36
2.3	Marketing the 'natural' in Iceland *Anna Dóra Sæþórsdóttir and Anna Karlsdóttir*	39
2.4	Branding Denmark: Present strategies and future options *Anette Therkelsen*	41
2.5	Marketing Norway *Øystein Jensen*	46
3.1	Evaluating an EU-project on improving sustainable competences in micro, small and medium sized Danish tourism enterprises *Peter Kvistgaard*	71
3.2	Protection and equity?: Local and indigenous encounters with the grand scheme of area protection in Norway *Tor Arnesen and Jan Åge Riseth*	77
4.1	Place competition: 'Cities of the Future'	89
4.2	Development of hotels in Nordic urban centres: Expansion, diversification and yield maximising strategies *Szilvia Gyimóthy*	101
4.3	The status of an Olympic town: A dozen years thereafter *Thor Flognfeldt Jr*	104
5.1	Tourism in the Åland Islands: An overview *Eva Holmberg-Anttila*	111

5.2	Culinary tourism project in Northern Iceland *Guðrún Gunnarsdóttir and Laufey Haraldsdóttir*	117
5.3	Rural development and tourism entrepreneurship in Finland *Raija Komppula*	122
6.1	Svalbard: Wilderness tourism in the High North *Bjørn P. Kaltenborn*	131
6.2	Wilderness tourism in Iceland: Threats and opportunities *Anna Dóra Sæþórsdóttir*	139
6.3	Nature-based tourism in mountain areas: Hikers in Södra Jämtlandsfjällen, Sweden *Tuomas Vuorio and Lars Emmelin*	146
6.4	Tourism development, amenity values and conflicting interests in Pyhätunturi National Park, Finland *Jarkko Saarinen and Matti Vaara*	148
7.1	Hurtigruten *Ola Sletvold*	156
7.2	Marine fishing tourism in Norway *Trude Borch*	163
8.1	Inhabiting the second-home *Michael Haldrup*	176
8.2	Second home tourism in Norway *Birgitta Ericsson and Kjell Overvåg*	183
8.3	Second homes in Finland *Mervi Hiltunen*	185
8.4	Cottage holidays in Åland *Samu Mäkelä*	190
8.5	Second home tourism and displacement in the Stockholm archipelago *Roger Marjavaara*	192
9.1	Interpretation in Finland *Jaakko Suvantola*	202
9.2	The controversies of a heritage tourism site *Arvid Viken*	204
9.3	The Maihaugen folk museum at Lillehammer *Thor Flognfeldt Jr*	212
9.4	Emergent Vikings: Changing character of a micro-festival in Iceland *Gunnar Thór Jóhannesson*	213
9.5	The Sami winter festival in Jokkmokk, Sweden *Robert Pettersson*	216

10.1	The development process of Ruka tourist resort, Finland *Pekka Kauppila*	226
10.2	Mountain resort labour market: The case of the Swedish mountain range *Linda Lundmark*	231
11.1	Lordi and Finnish tourism	244
11.2	Tourism mobility and climate change: The case of Sweden *Stefan Gössling*	251
11.3	Climate change responses of Swedish tourism actors: An analysis of actor websites *Stefan Gössling and C. Michael Hall*	256

Figures, Tables and Plates

FIGURES

1.1	Nordic tourism space	3
1.2	Political history of the Nordic countries from the 11th century to the present-day	4
1.3	Popular and academic conceptions of tourism	5
1.4	Temporary mobility in space and time (after Hall, 2003)	6
1.5	Geographical elements of a tourism system	8
1.6	Tourism system with multiple destinations in a seven day trip	8
1.7	Extent of mobility in time and space (after Hall, 2003)	17
1.8	The Nordic countries	20
1.9	The Nordic countries and their autonomous territories	21
2.1	The tourism experience at the intersection of consumption and production	29
2.2	The embedded nature of the tourism product (after Hall, 2005a, 2007b)	34
3.1	The Norwegian paradox of protection need and policy	79
5.1	The impact of tourism on the economy of Åland in 2003 (%)	112
5.2	The location of the study area, Skagafjörður, northern Iceland	119
5.3	Logo for the culinary project. It references one of the isands on the fjord, which for centuries was called the food chest of Skagafjörður due to its abundant bird life	120
6.1	National parks in the Nordic countries	144

7.1	Selected ferry connections and transport corridors in the Nordic countries	158
7.2	Selected cruise ship routes	161
7.3	MV Polar Star's Svalbard route	162
8.1	Second homes in the Nordic countries	177
8.2	Second home landscapes and their characteristics	181
8.3	Density of second homes. Number of second homes per km² land per municipality, January 2005 (statistics: SSB)	182
8.4	Change in number of second homes 1970–2005 by municipality (statistics: SSB)	184
8.5	Second home development in Finland 1920–2005	186
8.6	Distribution of second homes in Finland, municipal level	187
8.7	Time distance between permanent home and second home for a selection of destinations in Sweden and Finland	189
8.8	County of Stockholm	192
9.1	Demand for heritage tourism	203
9.2	Nordic world heritage sites	210
9.3	The location of Jokkmokk	216
10.1	A selection of ski resorts in the Swedish mountain range	232
10.2	Age profile and number employed in tourism in the mountain municipalities	234
11.1	The awareness-responsibility spectrum of key Swedish tourism actors	263

TABLES

1.1	Key characteristics of the Nordic states	2
1.2	Total tourism consumption in Norway, 2004 and 2005. NOK million, current prices (includes tourism consumption in Norway by resident households, resident industries and non-residents)	13
1.3	Key TSA figures for tourism demand and supply for Finnish regions with more than 2000 tourism generated employees	14
1.4	The Nordic countries in the World Economic Forum's Global Competitiveness index rankings 2000–2008	18
1.5	Rankings of the Nordic countries with respect to the World Economic Forum tourism and travel index	19
1.6	Nordic government and other relevant websites	23
2.1	Consumption, market, product and production characteristics of services	27

2.2	The applicability of the characteristics of services to tourism	28
2.3	Denmark: The brand	41
3.1	Regulatory regimes with respect to coordination of central–local relations in Denmark, Finland, Norway and Sweden	55
3.2	Supranational institutions of the Nordic region	61
3.3	Nordic Interreg IIIA and B programmes 2000–2006	64
3.4	Evaluation model for evaluating sustainable competences focusing on process and users	73
3.5	National park processes in mid- and north Norway	78
4.1	Categories and criteria for 'cities of the future'	90
4.2	Copenhagen and place marketing	93
4.3	Urban world heritage in Nordic countries	97
4.4	Ranking of Nordic capitals for hosting of international meetings	101
4.5	Key indicators of the Swedish hotel market in 2006	102
5.1.	Employment in agriculture, forestry and fishery (persons in thousands) in 1990–2005	110
5.2	The range of tourist leisure activities related in rural areas listed by the Council of Europe	116
6.1	Four attitudes to wilderness areas	136
6.2	The visitor numbers of selected National Parks in Finland in 2000–2006	145
6.3	Forms of tourist use and activities regarded by second home owners or residents as suitable, suitable with reservations, or unsuitable for the Pyhätunturi National Park, Finland	149
8.1	Population and second homes in the Nordic countries	178
8.2	Some average values of the 2004 Lakeland study compared to national means of second home tourism	188
9.1	Museums and tourism in the Nordic countries, 2006	211
11.1	Climate change and key actors in Swedish tourism	257
11.2	Balance of tourism trade (tourism receipts and expenditure) 2005 (million euro)	264

PLATES

2.1	The statue of the 'Little Mermaid' in Copenhagen	36
4.1	Waterfront development in Helsingborg, Sweden	84
4.2	Waterfront development in Copenhagen, Denmark	84

4.3	Waterfront development in Helsinki, Finland	85
4.4	Historic waterfront in Oulu, Finland	85
4.5	Historic waterfront in Stockholm, Sweden	86
5.1	Rural tourism is traditionally a small scale and family-owned business in Nordic countries. Farm tourism businesses in Norway: (a) showing a property where horse riding is available; (b) showing a farm that offers artistic space and events in summer	118
7.1	The fishing village on Ulvön within the High Coast World Heritage area in northern Sweden	154
8.1	Second homes in the Swedish mountain resort of Tärnaby	173
8.2	Second homes in Iceland	174
9.1	Guided tour on Norrbyskär Island	213
9.2	Two Sami attributes at the winter festival: The Sami dress and the reindeer	219
9.3	A Buddha-statue in Fredrika, northern Sweden	223
10.1	Olympic ski jump, Lillehammer, Norway	230
10.2	Santa Park, Rovaniemi, Finland	240

Abbreviations

ACIA	Arctic Climate Impact Assessment
AT	Alternative Tourism
ÅSUB	Ålands Statistik och Utredningsbyrå
ÅTF	Ålands Turistförbund
CBD	Central Business District
CBSS	Council of Baltic Sea States
CLIA	Cruise Lines International Association
DMO	Destination Marketing Organization
EEA	European Economic Area
EU	European Union
EUR	Euros
GoÅ	Government of Åland
IATA	International Air Transport Association
MNT	Midt-Nord Turisme
NORTRA	Nortravel Marketing
NP	National Park
OECD	Organization of Economically Developed Countries
PRF	Passagerarrederiernas förening
RevPAR	Revenue per Available Room
SLAO	Svenska Liftanläggningars Organisation
SME	Small and Medium Enterprise
SND	Norwegian Industry Development Fund
SSB	Statistisk sentralbyrå (Statistics Norway)
TSA	Tourism Satellite Account
TTCI	Travel and Tourism Competitiveness Index
UIA	The Union of International Associations
UNWTO	United Nations World Tourism Organization
VFR	Visiting Friends and Relations
WEF	World Economic Forum
WTTC	World Travel & Tourism Council

Preface

Writing a book on tourism in the Nordic countries appeared to be a great idea and a good way to fill a gap in the international literature on tourism. That was at least what we thought in 2005 when a book proposal to Channel View was put on paper. As it turned out, the task was not an easy one at all. Instead, it was a good way to get to know your own limits to knowledge despite being the so-called 'experts' teaching tourism classes on issues relating to tourism in the Nordic countries. The problem has proved to be twofold. In certain areas, such as second homes or urban and regional development, there is a great amount of research available, whereas others have hardly been addressed at all. Nordic research has been relatively lacking in international tourism journals although the *Scandinavian Journal of Hospitality and Tourism* has meant a great deal as an English-language outlet of Nordic tourism research. This made research for this book a jigsaw puzzle work. However, with the help of many colleagues and friends all around the Nordic realm we were able to finalise this volume to highlight the main traits in Nordic tourism.

For this book in particular, many researchers, colleagues and friends deserve our thankfulness. Dieter would first of all like to thank his co-authors Michael Hall and Jarkko Saarinen for getting involved in yet another book and a first common book, respectively. Poor Michael had the 'rewarding' task to convert what Dieter thought was English into what is commonly accepted as English. However, we hope that supplying him with Västerbotten cheese is a suitable appreciation of his engagement – but to be honest we would supply it anyway since Michael is the favourite guest in Dieter's and Åsa's apartment in Umeå. Besides, the Department of Social and Economic Geography, Umeå University, provided a creative and supportive environment. In particular, discussions with Linda Lundmark, Roger Marjavaara, Malin Zillinger and Bruno Jansson made an impact on this book. In an interconnected world even people outside Umeå contributed in various ways to this book. Mervi Hiltunen, Kati Pitkänen, Mia Vepsäläinen, Jaakko Suvantola, Petri Hottola, Kjell Overvåg, Tor Arnesen, Terje Sjeggedal, Thor Flognfeldt Jr, Stefan Gössling, Klas Sandell, Göran Ericsson, Peter Fredman, Robert Pettersson and all other people at Etour are important friends within the Nordic realm.

Meanwhile, Gustav Visser, Alan Lew, David Duval, Pat Maher and Dallen Timothy are some of the people in the wide world who mattered in various ways for this book. Most deserving of gratitude however, is, Åsa who has to stand with Dieter's compassion for research and tourism. Luckily she enjoys Dieter's recent taste for country music more than Michael, who it is hoped will not be stopped from returning regularly to Umeå. I guess Jarkko is more a country guy (although from the country where heavy metal has gone for hibernation) and we hope he will visit Umeå more often in the future.

Michael would similarly like to thank his co-authors for their efforts in working on the book and for the opportunity to share with them the enthusiasm for research on a part of the tourism world that deserves far greater recognition in the English language literature for its innovation, sophistication and interest. Although he has been fortunate that the book has been written in English so that it can be more easily shared both in the Nordic and international context, his reading of Swedish and Finnish has improved though his speaking has not. He would like to thank colleagues in the Department of Social and Economic Geography at Umeå, the Department of Geography at Oulu, Tourism Studies at the Savonlinna campus of the University of Joensuu, and the Department of Service Management at Lund Helsingborg for their hospitality and warm welcome when in Nordland. He would also like to acknowledge all those on Dieter's list plus Tim Coles, Fiona Crawford, Anna Dóra Sæþórsdóttir, Stephen Page and Sandra Wilson. While not sharing Dieter's taste in country music he has certainly been exposed to some excellent Swedish music, especially Ebba Forsberg, and developed a taste for Västerbotten cheese along with other Nordland specialties such as cloudberry, reindeer, moose, salmon and mushrooms. He would especially like to thank Åsa and Satu for coping with his visits and allowing him to return. Finally, he would like to express gratitude to Jody for making it possible to spend time in Nordland, sometimes even together, as well as retaining some Nordland sensibilities when in New Zealand or wherever in the world.

Jarkko is pleased to be working with Dieter and Michael and being a part of the interesting but challenging writing process opening new avenues to Nordic tourism studies. As he expected, the place of being the only 'indigenous' Nordic – although not strictly a Scandinavian – member of the group did not give any advantage on the knowledge base of tourism and tourism research in the wider region. On the contrary, he noted his co-authors' ability to correct misspellings not only from Finglish but also increasingly from Finnish place names, citations and references. Dieter and Michael have always been waited and wanted visitors at the University of Oulu, and without Michael's visits, many berries in the garden and forests near the house of Metsänreuna would have been wasted. Michael's visits have upgraded the quality of wine in the house and only during the visits there was also some dessert served – after he has found suitable means and not too old best before dated 'materials' to prepare it with. True Finnish hospitality: come, bring something to drink and cook! All we would need is an additional ingredient of Dieter's country music or something potentially stronger and darker. In addition to Dieter's and Michael's lists Jarkko would like to thank colleagues at the Department of

Geography and especially the Tourism Geographies research group, David Fennell, Timo Helle, Antti Honkanen, Jari Järviluoma, Arto Naskali, Maano Ramutsindela, Soile Veijola and OYUS Rugby Club. Finally, thanks are due to Satu, Mira and Katlego who have been relatively patient during the periodically time-consuming writing process.

All the authors would like to express their thanks to the contributors of the cases, especially those who did so at relatively short notice when some of the originally promised contributions failed to appear or requests were never answered. Finally, we would like to thank the staff of Channel View for supporting this project and for their broader support of academic publishing in tourism in general.

Chapter 1
Nordic Tourism: Introduction to Key Concepts

Learning Objectives

After reading this chapter, you should be able to:

- Understand the concept of the Nordic countries and region.
- Understand key concepts of tourism and mobility.
- Identify the key elements in the tourism system.

Introduction

This first chapter will provide a broad introduction to the Nordic countries and their autonomous regions: Denmark, Faeroe Islands, Finland, Åland, Greenland, Norway, Iceland and Sweden. The chapter also provides an introduction to some of the key concepts that will be used in the book such as domestic and international tourism, trip, and tourism industry.

The Nordic Concept

> In Asia there are hundreds of millions whose image of the Nordic Region is still unformed – a canvas with only a few dots, if any. Therefore, the Nordic Region, finally encompassing all shores of the Baltic Sea, has what it takes to satisfy an intelligent traveller from the polluted industrial centres of China and Japan: Nature and wilderness, culture and history. Luckily, mass tourism can never be our forte. We Nordics have a far more satisfying challenge: To meet and surpass the expectations of the truly demanding traveller. It can be done by exploiting our image to the full – prejudices and misunderstandings included. (Toivanen, 2006: 354)

2 Nordic Tourism

The term 'nordic' refers to the countries of northern Europe: Denmark, Finland, Iceland, Norway and Sweden and the associated territories of Greenland, Faroe Islands and the Åland islands (Table 1.1). The term is derived from the Scandinavian language equivalent of *Norden*, which is *Pohjola* or *Pohjoismaat* in the Finnish language, meaning northern or northern countries. However, the term does not just refer to a geographical space but is also a reference point to a political

Table 1.1 Key characteristics of the Nordic states

State	Area km²	Population	Form of government	Capital	Capital population
Denmark	43,376	5,411,405	Constitutional monarchy	Copenhagen	503,699 (municipality)* 1,825,814 (capital region)*
Finland	338,145	5,236,611	Republic	Helsinki	568,146 (municipality) 1,293,262 (metropolitan area)
Iceland	103,300	293,577	Republic	Reykjavik	117,721 (city council region)* approx. 200,000 (metropolitan area)*
Norway	323,802	4,606,363	Constitutional monarchy	Oslo	548,617 (city council area)* 1,121,020 (metropolitan area)
Sweden	449,964	9,011,392	Constitutional monarchy	Stockholm	792,593 (municipality)* 1,942,233 (metropolitan area)*
Faroe Islands	1393	48,379	Home rule within the Kingdom of Denmark	Tórshavn	17,447
Greenland	2,166,086	56,969	Home rule within the Kingdom of Denmark	Nuuk	14,874
Åland	1552	26,530	Home rule within the Republic of Finland	Mariehamn	10,712

Note: All information 2005 unless otherwise stated. Population of Finland includes population of Åland.
*2007

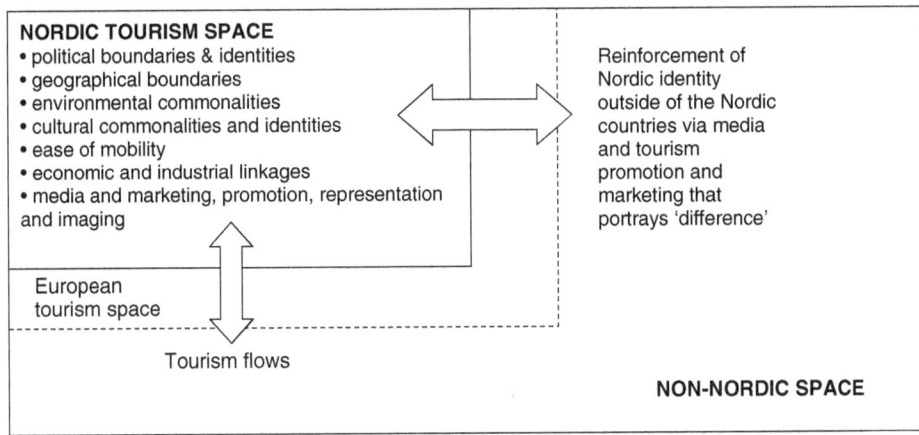

Figure 1.1 Nordic tourism space

space, primarily occupied in institutional terms by the Nordic Council; a cultural space, in terms of common elements of a northern identity, as well as the related historical linkages and relationships between the various territories; and an economic space, with respect to the economic agreements that link the countries as well as the substantial intra-regional trade that occurs. Finally, related to the various ways in which the Nordic idea is understood, as well as the openness of movement between the Nordic countries, is the tourism space of the Nordic countries.

The notion of a distinct Nordic tourism space is obviously a major focus for this book with such a space being both internally and externally defined (Figure 1.1). Internally such a space has been determined by a number of political, cultural and economic factors that have led to the development of substantial intra-regional and cross-border travel, while there is also a substantial commonality of approach with respect to state involvement in tourism development, especially in peripheral areas, that reflects a common social democratic tradition. Externally the various elements of Nordic identity are reinforced through the media as well as via tourism specific promotion and marketing. In addition, from a global perspective the countries of the Nordic region are an important inbound and outbound market in their own right with several significant international tourism businesses including SAS Airlines and Finnair. Furthermore, the Nordic region has also contributed to innovations in tourism business, education and research which also reinforce the importance of a specific examination of Nordic tourism.

Cases and issues 1.1: Nordic is not Scandinavia!

In English the term 'Scandinavia' is often incorrectly used with reference to the Nordic countries. The term Scandinavia was originally applied in the 16th century to the countries that then occupied the Scandinavian peninsula: Norway, Sweden and Denmark. However, Denmark ceased having territory on the

Peninsula from 1658, while Finland ceased being a part of Sweden in 1809. The Scandinavian peninsula, Denmark, Iceland, Scandinavian Greenland and the Faroes (along with the Shetland and Orkney Islands, and the Isle of Man and other parts of Ireland and Britain) did share a common linguistic, cultural and religious environment as late at the high Middle Ages, and many commonalities remain, but they are geographically separate. Indeed, geologically Greenland is part of North America! Present-day Finland is also not part of geographic Scandinavia and although it was part of Sweden from the 15th to the early 19th century. In terms of language, Finnish, is also completely separate from the Scandinavian languages belonging instead to the Finno-Ugric/Uralic languages together with the languages of the Sami, the indigenous people of far northern Europe. However, the extent to which the region shares a common history can be indicated by there being a significant Swedish speaking minority in Finland (approximately 6%) and Finnish minorities in Sweden and Norway.

The Nordic idea is one that has emerged over time as a result of political and cultural interrelationships (Figure 1.2). At various times the countries have been linked under a common government or crown although it was only in the 15th century at the time of the Kalmar Union that that the territories were united. For much of the last thousand years the Nordic region has been marked by substantial political rivalry between the various countries and especially between Denmark and Sweden. External political influence in the region has also been prominent with respect to Russia, which controlled Finland for most of the 19th century and early 20th century and, to a lesser extent, the UK and Germany. However, since Finnish

Century	Denmark	Faroes	Greenland	Iceland	Norway	Sweden	Åland	Finland
21st								
20th Nordic Council membership	Joined EU 1973 Joined 1953	Associate membership FAROES Home rule under Danish sovereignty	Associate membership GREENLAND Home rule under Danish sovereignty	Joined ICELAND Independence declared 1944	Joined 1953 NORWAY Independence declared 1905	Joined EU 1995 Joined 1953	Associate membership ÅLAND Special status under Finnish sovereignty	Joined EU 1995 Joined 1956 FINLAND Independence declared in 1917
19th	DENMARK					SWEDEN AND NORWAY		GRAND DUCHY OF FINLAND (part of Russia from 1809)
18th 17th 16th	DENMARK AND NORWAY					SWEDEN		
15th	KALMAR UNION							
14th 13th				NORWAY		SWEDEN		Primarily western Finland under Swedish influence
12th 11th		FAROES	GREENLAND	ICELANDIC COMMONWEALTH	NORWAY			

Figure 1.2 Political history of the Nordic countries from the 11th century to the present-day

independence, and especially in the post Second World War period, the region has been marked by peace and substantial cooperation. In fact, the region has some of the most open borders in the world, further reinforcing its tourism identity.

Under the Nordic Passport Union, any citizen can travel between the Nordic countries without having passports checked, with an identity card being sufficient. Other citizens can also travel between the Nordic countries' borders without having their passport checked, but they still have to carry a passport or other kinds of approved travel identification papers. Hence, the membership of Denmark, Sweden and Finland in the European Union did not provide any major changes, particularly since even Norway and Iceland signed the Schengen agreement abolishing systematic border controls within Europe. Such relative openness obviously raises questions as to how international and domestic tourism can be defined, and it is to these issues that the chapter will now turn.

What is Tourism? Definitions of Key Concepts

> Conceptualisation is critical for a culture of high-quality information and for creating knowledge. It is hard to develop and work with tourism statistics and statistics-based information in the absence of distinct conceptualisation which has been firmly grounded with both the producer and the user of the information. (Nortek, 2007: 4)

Tourism is a concept that while initially looks very easy to define is actually quite complicated. Much of the problem with considering the concept of tourism is that most people think of tourism in terms of vacation or leisure oriented travel. However, in academic terms the concept is much wider than that (Figure 1.3) and

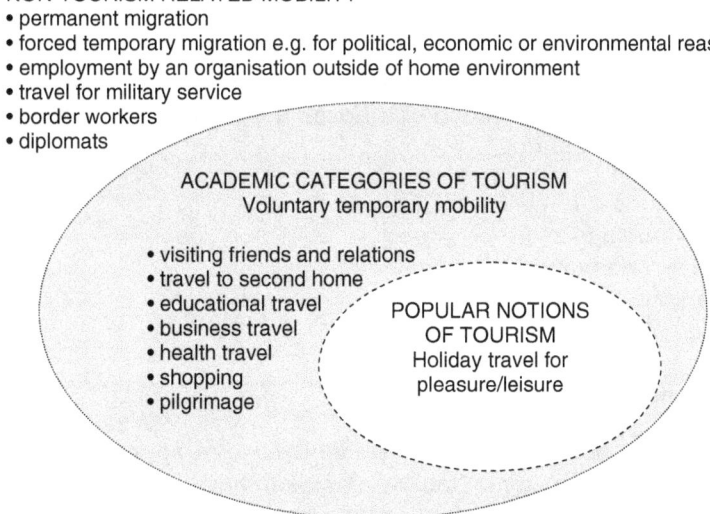

Figure 1.3 Popular and academic conceptions of tourism

Figure 1.4 Temporary mobility in space and time (after Hall, 2003)

includes consideration of a range of forms of voluntary travel in which people travel from their usual home environment to another location and then return, with the time and space over which they travel being an important influence on the definition of different forms of mobility (Figure 1.4). Therefore, within academic and research oriented consideration of tourism, there is a range of other types of travel that we study, including

- visiting friends and relations (VFR);
- business tourism;
- travel to second homes;
- health and medical related travel;
- education related travel;
- religious travel and pilgrimage;
- travel to shopping and retail.

One of the major problems that many students have in approaching tourism is the confusion among the terms 'tourist', 'tourism' and 'tourism industry'. This difficulty is a product of the definitions of the terms, and the uses for which these definitions were designed (Smith, 2004), as well as the inherent characteristics of tourism itself:

- visitors consume both tourism and non-tourism commodities;
- locals (non-visitors) consume both tourism and non-tourism commodities;
- tourism industries produce (and often consume) both tourism and non-tourism commodities;
- non-tourism industries produce (and often consume) tourism and non-tourism commodities.

Definitions are fundamental to any subject. Each area of scholarship and research has, as one of its first tasks, the identification of the things that comprise the foci of study. In tourism studies we are faced with four interrelated concepts – tourism, tourist, tourism industry and tourism resources – which provide the basis, in one form or another, for the subject that we study. By defining terms we give meaning to what we are doing. Just as important, we are able to give each term a specific, technical basis that can be used to help communicate more effectively and to improve the quality of our research, and business and management practices.

Definitions of tourism tend to share a range of common elements:

- tourism is the temporary, short-term travel of people (non-residents) along transit routes to and from a destination that is outside of their normal home environment;
- it can have a wide variety of impacts on the destination, the transit route and the source point of tourists;
- it can influence the attitudes and behaviours of the tourist as well as the people that provide tourism experiences;
- it is voluntary; and
- it is primarily for leisure or recreation, although business is also important.

The *concept of the home or usual environment of an individual* is an important concept of tourism statistics. It refers to the geographical boundaries within which an individual moves within his/her regular routine of life. According to the UN and UNWTO (2007: 16): 'the usual environment of an individual includes the place of usual residence of the household to which he/she belongs, his/her own place of work or study and any other place that he/she visits regularly and frequently within his/her current routine of life, even when this place is located far away from the place of usual residence'. However, it should be noted that the notion of home environment in tourism statistical terms is being increasingly challenged because of the increasing regularity of mobility in society (Hall, 2005a). For example, many Nordic people have access to second homes that they use on an extremely regular basis which also constitute a form of usual or routine environment. Because of the availability of transport connections and cross-border access, such second homes may even be international in scope (Hall & Müller, 2004a).

The term *trip* is also used extensively in tourism research and statistics and refers to the movement of an individual outside their home environment until they return.

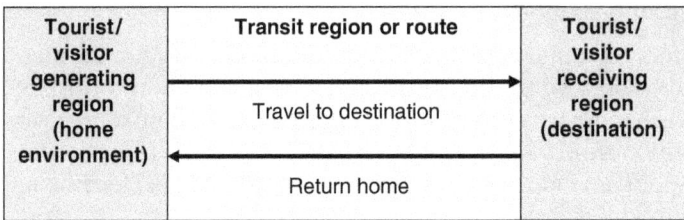

Figure 1.5 Geographical elements of a tourism system

It therefore refers to a roundtrip. The trip idea has served as the basis for identifying a tourism system which includes the various elements that make up a trip: the generation region, the transit region, the destination and the environment (Figure 1.5). A trip may also be made up of various visits to different places although it is usually characterised by its main destination which is the location outside of the home environment in which most time was spent or the place which most influenced the decision to take the trip (Figure 1.6). If the same amount of time was spent in two or more places during the trip, then the main destination is usually defined as the one that is the farthest from the place of usual residence.

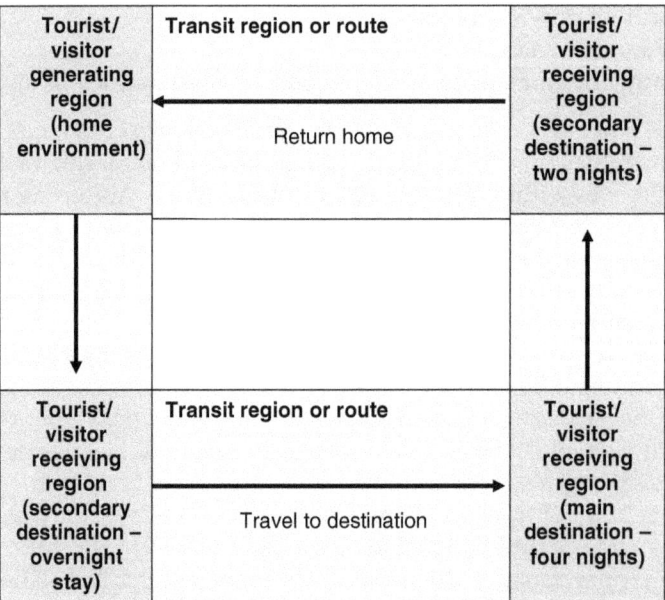

Figure 1.6 Tourism system with multiple destinations in a seven day trip

An international trip is therefore one in which the main destination is outside the country of residence of the traveller, whereas a domestic trip is one in which the main destination is within the country of residence of the traveller. However, an international trip might include visits to places within the country of residence in the same way as a domestic trip might include the crossing of international borders and visits outside the country of residence of the traveller. In international and national tourism statistics, the term tourism trip usually refers to a trip of not more than 12 months, and for a main purpose other than being employed in the destination. However, there are national differences in how this is applied and therefore how national and international mobility is classified.

The term *visit* refers to the stay (overnight or same-day) in a place visited during a trip. The stay need not be overnight to qualify as a visit. Nevertheless, the notion of stay supposes that there is a stop. Entering a geographical area without stopping therefore usually does not qualify as a visit to that area (UN & UNWTO, 2007).

Nordic and other national definitions of tourism have been strongly influenced by international definitions of tourism from such organisations as the Statistical Division of the United Nations (UN) and the United Nations World Tourism Organization (UNWTO) (see also Lennon, 2003) that have been developed as a guide for national statistical collection and research. In order to improve statistical collection and improve understanding of tourism, the UN and the UNWTO have recommended differentiating between visitors, tourists and excursionists.

Usually three types of tourism are recognised: (1) *domestic tourism*, which includes the activities of resident visitors within the country or economy of reference either as part of a domestic or an international trip; (2) *inbound tourism*, which includes the activities of non-resident visitors within the country or economy of reference either as part of a domestic or an international trip (from the perspective of his/her country of residence); and (3) *outbound tourism*, which includes the activities or resident visitors outside the country or economy of reference, either as part of a domestic or an international trip.

However, for statistical purposes the UN and UNWTO (2007: 21–22) have recommended the adoption of the following concepts: *internal tourism*, which comprises domestic tourism and inbound tourism, that is, the activities of resident and non-resident visitors within the economy of reference as part of a domestic or an international trip; *national tourism*, which comprises domestic tourism and outbound tourism, that is, the activities of resident visitors within and outside the economy of reference either as part of a domestic or an international trip; and *international tourism*, which comprises inbound tourism and outbound tourism, that is, the activities of resident visitors outside the economy of reference either as part of a domestic or an international trip and the activities of non-resident visitors within the economy of reference as part of a domestic or an international trip (from the perspective of their country of residence).

The UNWTO has recommended that an international tourist be defined as: 'a visitor who travels to a country other than that in which he/she has his/her usual residence for at least one night but not more than one year, and whose main purpose of visit is other than the exercise of an activity remunerated from within the country visited'; and that an international excursionist (e.g. a cruise-ship visitor) be defined as '[a] visitor residing in a country who travels the same day to a country other than which he/she has his/her usual environment for less than 24 hours without spending the night in the country visited and whose main purpose of visit is other than the exercise of an activity remunerated from within the country visited' (WTO, 1991). Similar definitions have also been developed for domestic tourists, with a domestic tourists having a time limit of 'not more than six months' (UN, 1994; WTO, 1991).

More recent recommendations from the UN and UNWTO have focused more on the category of 'visitor' rather than 'tourist' per se, with a number of criteria needing to be satisfied for an international traveller to qualify as an international visitor:

(1) The place of destination within the country visited is outside the traveller's usual environment.
(2) The stay, or intended stay, in the country visited should last no more than 12 months, beyond which this place in the country visited would become part of his/her usual environment. At which point this would lead to a classification as migrant or permanent resident. The UN and UNWTO recommend that this criterion should be applied to also cover long-term students and patients, even though their stay might be interrupted by short stays in their country of origin or elsewhere.
(3) The main purpose of the trip is other than being employed by an organisation or person in the country visited.
(4) The traveller is not engaged in travel for military service nor is a member of the diplomatic services.
(5) The traveller is not a nomad or refugee. According to the UN and UNWTO (2007: 21), 'For nomads, by convention, all places they visit are part of their usual environment so that beyond the difficulty in certain cases to determine their country of residence ... For refugees or displaced persons, they have no longer any place of usual residence to which to refer, so that their place of stay is considered to be their usual environment.'

Domestic visitors can also be similarly classified. Therefore, for any traveller to be considered a domestic visitor to a place in the country he or she is resident, the following conditions should be met:

(1) The place (or region) visited should be outside the visitor's usual environment which would exclude frequent trips, although the UN and UNWTO (2007) recommend that trips to vacation homes should always be considered as tourism trips.
(2) The stay, or intended stay, in the place (or region) visited should last no more than 12 months, beyond which this place would become part of his/her usual environment. As with the international visitor classification the UN and UNWTO recommend that this criterion should be applied to also cover long-term students and patients, even though their stay might be interrupted by short stays in their place of origin or elsewhere.
(3) The main purpose of the visit should be other than being employed by an organisation or person in the place visited.

For many people the notion of 'tourism' is inseparable from that of 'the tourism industry' and has long been a source of debate over an appropriate definition (Smith, 2004). Defining services industries, such as tourism, can be an extremely difficult exercise. The essential characteristics of services are that they cannot be produced without the agreement and cooperation of the consumer, and that the outputs produced are not separate entities that exist independently of the producers or consumers. In the case of tourism, such consumption and production usually occurs outside of the home environment of the consumer. However, not entirely, as it is possible to purchase tourism products from a home computer over the internet. Although of course to be able to completely experience them it is still necessary to travel to the destination for the tourist experience.

The consumer service nature of tourism has led to a situation in which tourism has not usually been identified as a separate industry classification in the official statistics of many countries. Nevertheless, in recent years there has been an attempt to develop a production or supply-side approach so as to compare the economic dimensions of tourism with other sectors and industries, particularly through the development of tourism satellite accounts (TSAs). From a supply-side perspective, the tourism industry can be defined as the aggregate of all businesses that directly provide goods and/or services to facilitate business, pleasure, and leisure activities to people who are voluntarily away from their home environment. Therefore, for Smith (2004: 31), when utilising a supply-side approach, 'A tourism industry is any industry that produces a tourism commodity.' Three key features emerge from this definition.

(1) The tourism industry produces tourism commodities that can be defined as 'any good or service for which a significant portion of demand comes from persons engaged in tourism as consumers' (Smith, 2004: 30).
(2) The inclusion of business, pleasure, and leisure activities emphasises the nature of the goods and services a traveller requires to undertake the trip.
(3) The definition includes the notion of a 'home environment', which refers to the delineation of a distance threshold or period of overnight stay (Hall, 2005a; Hall, 2006b). These three elements of the definition can be combined to conceptualise and measure tourism in a manner which is consistent with that of other industries.

Business elements that would usually be recognised as contributing to the tourism industry therefore include:

- international and domestic transport operators and carriers;
- accommodation operators, including hotels, motels, caravan parks and camping grounds;
- restaurants and other catering establishments;
- tour operators, wholesalers, travel and booking agents;
- attraction, entertainment and event facility operators;
- national parks;
- manufacturers of souvenirs;
- specialist travel information suppliers;
- specialist event, convention and meeting centre operators; and
- specialist retailers, such as souvenir shops.

Cases and issues 1.2: Tourism satellite accounts

The concept of a satellite account is a relatively new idea that originated in France as a way of addressing the problem of new and growing economic activities not recognised in the official measures of certain industries, such as tourism, energy and information technology. Such an account is a model that is separate from, but linked to, the national accounts, and uses the same concepts and structure as the core accounts. The system of national accounts is a set of guidelines for organising information about an economy in a useful way. It provides concepts, definitions, classifications, accounting rules, accounts and tables to provide a comprehensive, integrated framework for production, consumption, capital investment, income, stocks, flows of financial and non-financial wealth, and related economic variables.

Tourism satellite accounts (TSAs) have been developed in a number of countries including Canada, New Zealand, Norway and Australia. The World Tourism Organization, the OECD and the World Travel & Tourism Council (WTTC) (1997) encourage the development of an agreed international framework for TSAs to provide for better international comparisons. Although the accounts produced in international TSAs have not yet been standardised, they usually contain one or more of the following:

- tourism consumption by commodity;
- tourism consumption impact on supply;
- production accounts of the tourism industry;
- tourism-related gross fixed capital formation;
- employment related to tourism;
- stocks and flows of fixed assets related to tourism;
- imports and exports of goods and services generated by tourism;
- tourism balance of payments; and
- tourism value added (contribution to GDP) by commodity and activity.

According to the Norwegian TSA, Norwegian tourists increased their tourist consumption from about NOK 41 billion in 2004 to almost NOK 43 billion in 2005. Foreign tourists spent NOK 26.4 billion in Norway, while business travellers spent close to NOK 18 billion (Statistics Norway, 2006a). According to preliminary figures, the value of the total production of tourism industries was NOK 7.4 billion higher in 2005 than in 2004. In volume this is a 5% increase. Hotels and restaurants are the largest industry and account for 33% of the total production. Tourism industries accounted for 3.1% of Norwegian gross domestic product (GDP) in 2005 (Table 1.2).

In Finland a TSA was first introduced in 1999 and is updated on a regular basis. The TSA was partly regionalised in June 2005 with research being carried out by Statistics Finland and financed by Ministry of Trade and Industry, and Ministry of the Interior. Table 1.3 indicates some of the TSA figures for Finnish regions with relatively high levels of tourism generated employment. In terms of a national figure Konttinen (2005) reported that the total tourism demand in Finland in 2002 was 6.54 billion EUR, with other domestic tourism demand (compensated business trips, own free-time residences) being valued at 1.7 billion EUR leading to a combined figure of 8.3 billion EUR. The total direct employment effects were estimated to be 127,000 persons (number of employed persons).

See Statistics Norway TSA website: http://www.ssb.no/turismesat_en/main.html.

Table 1.2 Total tourism consumption in Norway, 2004 and 2005. NOK million, current prices (includes tourism consumption in Norway by resident households, resident industries and non-residents)

Characteristic tourism products	2004	2005
Accommodation services	8418	8809
Hotel services	6942	7324
Food and beverage services	11,942	12,741
Passenger transport services	24,710	26,122
Transport by railway, tramway and suburban transport	1556	1617
Transport by scheduled motor bus transportation and taxi operation	3203	3572
Transport by inland water transport	1322	1156
Transport by ocean and coastal water passenger transport abroad	3419	3548
Passenger transportation by air	15,210	16,230
Package tours and car rental services	9231	9844
Museum, sporting activities, etc.	2178	2374
Total tourism consumption of tourism products	56,479	59,890
Other products		
Food, beverages and tobacco	4671	4973
Clothing and footwear	1015	1079
Souvenirs, maps, etc.	1286	1355
Other transportation costs	7502	7506
Other commodities and services	11,905	12,286
Total tourism consumption of other products	26,379	27,199
Total consumption expenditures by tourists	82,858	87,088

Source: Statistics Norway (2006b)

Table 1.3 Key TSA figures for tourism demand and supply for Finnish regions with more than 2000 tourism generated employees

Figures	Uusimaa/ southern Finland (Helsinki/ Vantaa region)	Varsinais-Suomi/ southwest Finland (Turku region)	Pirkanmaa/ Tampere region	Keski-Suomi/ central Finland (Jyväskylä region)	Pohjois-Pohjanmaa/ northern Ostrobothnia (Oulu/Kuusamo region)	Lappi/ Lapland	Ahvenanmaa/ Åland
Inbound tourism demand total, EUR million	1469	114	78	45	83	156	60
Inbound tourism demand total, share of total tourism demand %	40%	22%	14%	15%	20%	40%	36%
Domestic leisure tourism demand, EUR million	1353	322	361	195	255	176	98
Domestic leisure tourism demand, share of total tourism demand %	37%	61%	66%	63%	62%	45%	59%
Other domestic tourism demand (compensated business trips, own free-time residences), EUR million	831	92	109	69	75	56	7
Other domestic tourism demand (compensated business trips, own free-time residences), share of total tourism demand %	23%	17%	20%	22%	18%	14%	4%

Table 1.3 Continued

Figures	Uusimaa/ southern Finland (Helsinki/ Vantaa region)	Varsinais-Suomi/ southwest Finland (Turku region)	Pirkanmaa/ Tampere region	Keski-Suomi/ central Finland (Jyväskyla region)	Pohjois-Pohjanmaa/ northern Ostrobothnia (Oulu/Kuusamo region)	Lappi/ Lapland	Ahvenanmaa/ Åland
Total tourism demand in Finland, EUR million	3653	528	169	309	413	388	165
Value added generated by tourism demand, EUR million, incl. employers' expenses	1353	179	56	106	141	123	53
Value added generated by tourism demand, EUR million, excl. employers' expenses	1045	148	45	82	115	105	51
GDP at basic prices, EUR million	44,712	10,980	3091	5463	7898	3698	838
Tourism value added as a share of GDP at basic prices, %, incl. employers' expenses	3.0%	1.6%	1.8%	1.9%	1.8%	3.3%	6.4%
Tourism value added as a share of GDP at basic prices, %, excl. employers' expenses	2.3%	1.3%	1.4%	1.5%	1.5%	2.8%	6.1%
Imputed employment generated by tourism (number of employed persons)	20,150	3842	1365	2132	2815	3381	2201

Source: Derived from Ministry of Trade and Industry (2006: 60–62)

Tourism and Mobility

As noted above, tourism can be conceptualised as one form of human mobility among many others, including migration, refugees and cross-border and expatriate workers. In order to understand the wide range of temporary human movement, which is central to an academic notion of tourism, an appropriate framework therefore requires an approach that involves the relationships between tourism, leisure and other activities, and behaviours related to human movement, for example, migration, retirement and sojourning. Such an assessment also needs to consider the extent to which improvements in transport technology have made it easier for those with sufficient time and money to travel further and more quickly than ever before. What was the routine or home environment 50 years ago is not the same as the routine environment of today. Travel which once took two or three days to accomplish may now be completed as a daytrip – a factor which obviously challenges how we conceive of the notion of a home or routine environment. In addition, the technological capacity to travel is complemented by advances in information and communications technology that extend the 'reach' of people with access to such technology further than ever before. Clearly, such changes in mobility not only have implications for tourism but also for a wide range of human activities, as well as ideas of accessibility, extensibility, distance, identity, networks and proximity (e.g. Coles & Timothy, 2004; Coles et al., 2004, 2005; Frändberg & Vilhelmson, 2003; Hall, 2005a).

Figure 1.7 presents a model for describing different forms of temporary mobility in terms of three dimensions of space, time and number of trips. Figure 1.7 illustrates the decline in the overall number of trips or movements with time and distance away from a central generating point which would often be termed as 'home'. The fact that the number of movements declines the further one travels in time and space away from the point of origin is well recognised in the study of travel behaviour. However, it has not been utilised as a means to illustrate the totality of trips that are undertaken by individuals. The relationship represented in Figure 1.7 holds whether one is describing the totality of movements of an individual over their life span from a central point (home), or whether one is describing the total characteristics of a population (Hall, 2005a, 2006a).

Such distance decay effects with respect to travel frequency and trip characteristics have been well documented (e.g. Hall, 2005a). In addition, Figure 1.7 illustrates the relationship between tourism and other forms of temporary mobility including various forms of what is often regarded as migration or temporary migration. Such activities, which have increasingly come to be discussed in the tourism literature, include travel for work and international or 'overseas' experiences, education, health, as well as travel to second homes, return migration and travel among diasporas. The inclusion of diasporic travel and return migration issue are important in the Nordic context because of the extent to which Norwegians, Swedes, Icelanders and Finns have migrated internationally over the past 150 years only to return later in life or have their descendents return to the Nordic countries as visitors. In addition, pockets of Nordic migrants and their descendents in the United States have provided a focal point for outbound Nordic tourism, as Timothy (2002) has demonstrated with respect to Finnish international tourism to some parts of the United States.

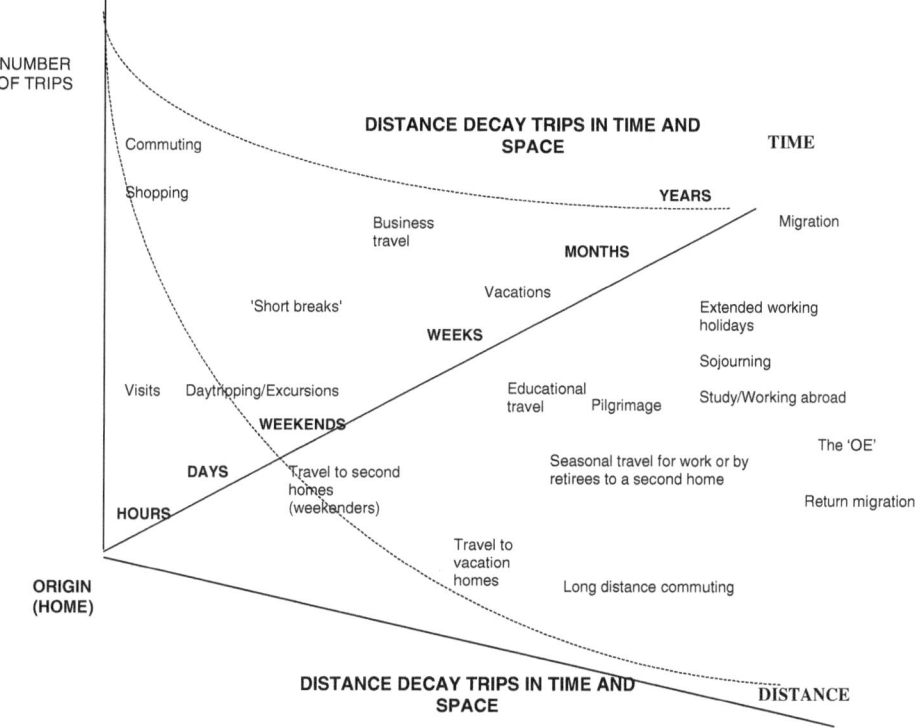

Figure 1.7 Extent of mobility in time and space (after Hall, 2003)

Cases and issues 1.3: The competitiveness of the Nordic states

The positive economic and political images of the Nordic states are reflected in the rankings of the countries in international competitiveness studies. Although such studies can be criticised for their superficiality and their failure to take into account numerous aspects of place development (Karppi, 2001) they 'nonetheless, due to their extensive media coverage, ... at least to a certain degree most likely do influence international business behaviour regarding trade and investment decisions, and as such they cannot be ignored or considered as merely of "entertainment value" only' (Hanell *et al.*, 2002: 3).

One of the most well recognised competitiveness reports is that of the World Economic Forum (WEF). The competitiveness studies are 'aimed at contributing to a better understanding of why some countries grow prosperous, while others are left behind' (WEF, 2007: xiii). Competitiveness is defined in various ways. For example, the 2001–2002 WEF report noted two distinct but complementary approaches: global competitiveness as 'the set of institutions and economic policies supportive of high rates of economic growth in the medium term' as 'the set of institutions, market structures, and economic policies supportive of high current levels of prosperity; referring mainly to an economy's effective utilization of its current stock of resources' (WEF, 2001: 16). Table 1.4 lists the WEF global competitiveness index rankings of the Nordic countries for 2000 to 2007/8. Several dates are provided so as to indicate the extent to which movement may occur in rankings which have been derived from the allocation of scores to a number of competitiveness factors that have been calculated from both publicly available data and an Executive Opinion Survey of, in 2007, 11,000 business leaders in 131 countries. The competitiveness factors are identified within three pillars: (1) basic requirements (institutions, infrastructure, macroeconomic stability, health and primary education); (2) efficiency enhancers (higher education and training; goods market efficiency, labour market efficiency, financial market sophistication, technological readiness, market size); and (3) innovation and sophistication factors (business sophistication, innovation).

In 2007 the WEF launched a travel and tourism competitiveness index (TTCI) that covered 124 countries around the world. According to the WEF (2007: xiii) the TTCI 'aims to measure the factors and policies that make it attractive to develop the [travel and tourism] sector in different countries.' The WEF TTCI is based on 13 'pillars' of travel and tourism competitiveness which, in turn, have been organised into three sub-indexes: (1) regulatory framework; (2) business environment and infrastructure; and (3) human, cultural and natural resources, that are comprised of 58 variables based on a mix of secondary and survey data. As with the Global Competitiveness Index the survey data was derived from the responses to the World Economic Forum's Executive Opinion Survey, and hard data were collected from various sources including UN reports, WTTC, IATA, Visa as well as consulting firms, such as Booz Allen. The rankings of the Nordic countries as well as the top 10 countries are recorded in Table 1.5.

Table 1.4 The Nordic countries in the World Economic Forum's global competitiveness index rankings 2000–2008

Country	Global competitiveness index rankings*							
	2000	2001	2002	2003	2004	2005	2006	2007–2008
Denmark	13	14	10	4	5	4	4	3
Finland	5	1	2	1	1	1	2	6
Iceland	23	16	12	8	10	7	14	23
Norway	15	6	9	9	8	9	12	16
Sweden	12	9	5	3	3	3	3	4
Out of		75	80	102	104	117	125	131

*Rankings are derived from each annual report rather than using any revisions of rankings that occur in the following or later years
Source: Hanell et al., 2002; World Economic Forum, Global Competitiveness Reports, various.

Table 1.5 Rankings of the Nordic countries with respect to the World Economic Forum tourism and travel index

Overall rank 2007	Country/ economy 2007	Regulatory framework rank 2007	Environment & infra-structure rank 2007	Human, cultural & natural resources rank 2007	Regulatory framework rank 2008	Business environment & infrastructure rank 2008	Human, cultural & natural resources rank 2008	Overall rank 2008
1	Switzerland	2	2	2	1	2	3	1
2	Austria	3	12	1	4	8	7	2
3	Germany	6	3	6	6	3	9	3
4	Iceland	5	8	5	3	9	36	11
5	United States	33	1	12	49	1	2	7
6	Hong Kong SAR	4	14	14	2	16	42	14
7	Canada	15	4	16	23	4	10	9
8	Singapore	1	11	42	7	13	37	16
9	Luxembourg	17	9	8	24	12	35	20
10	United Kingdom	21	6	10	26	6	5	6
11	Denmark	8	16	9	10	10	28	13
16	Finland	7	18	33	5	23	14	12
17	Sweden	19	13	27	9	15	8	8
23	Norway	9	21	40	8	19	20	17

Source: WEF (2007, 2008)

Figure 1.8 The Nordic countries
Cartography: D.K. Müller

However, it should be noted that the relative rankings of countries bear little relationship to the rankings of the overall number of international visitors a country actually receives. Although competitiveness is a significant policy goal there is still substantial confusion 'as to what the concept actually means and how it can be effectively operationalised ... policy acceptance of the existence of regional competitiveness and its measurement appears to have run ahead of a number of fundamental theoretical and empirical questions' (Bristow, 2005: 286). This is especially the case in tourism where there is already substantial evidence of the role of price competitiveness as a major determinant in tourism flows and

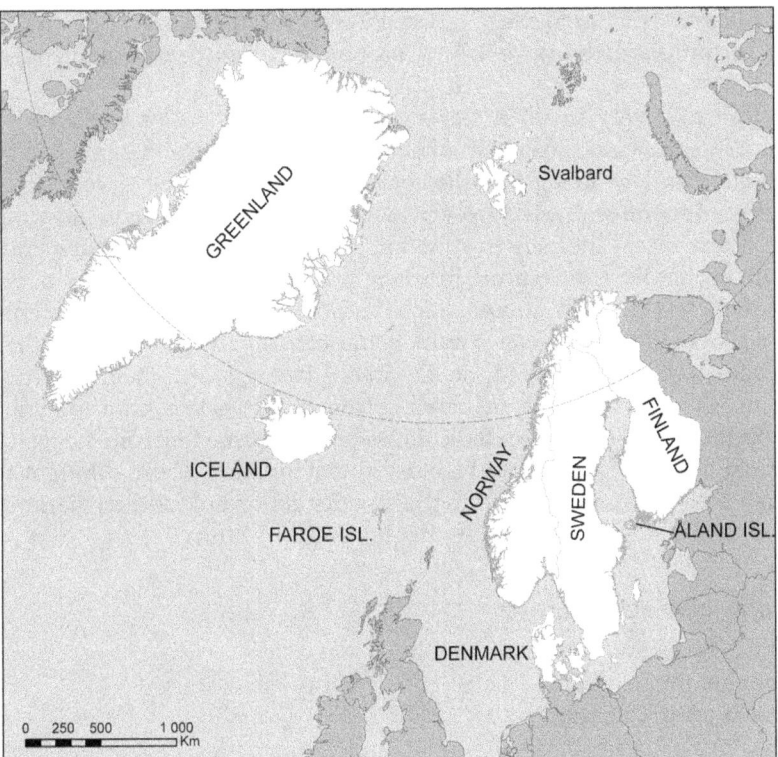

Figure 1.9 The Nordic countries and their autonomous territories
Cartography: D.K. Müller

where its parameters are clearly defined (Dwyer *et al.*, 2000a, 2000b). According to Hall (2007a) many of the factors that are used in the WEF index, although economically or socially significant in their own right, have no identified direct linkage to international tourism patterns in the tourism studies literature. While other factors that appear critical for international numbers, such as overall transport accessibility, exchange rates, border permeability and the contribution of domestic tourism to international tourism infrastructure as well as the characteristics of their markets arguably need to be given a greater weighting than factors such as how many World Heritage sites a country has. Nevertheless, despite such criticisms, it is likely that the WEF indices and rankings will likely draw significant attention from the media and from politicians for years to come.

World Economic Forum: http://www.weforum.org/en/index.htm.

Chapter Summary

This chapter has provided an introduction to some of the key concepts that will be used in this book. The first section discussed the Nordic idea and briefly outlined the historical development of the various political interrelationships that have made up the Nordic region. The section detailed some of the key concepts that are used in considering the nature of tourism, especially with respect to how it fits into broader patterns of human mobility, and how visitors and the tourism industry may be defined.

Although relatively small in terms of population, the Nordic countries are a significant international tourism market for a number of destinations as a result of the high standards of living that the countries share, with the exception of some extremely peripheral regions. However, inbound and domestic tourism is important for their economies, especially in rural and peripheral areas that have undergone significant economic restructuring in recent years. Moreover, tourism, particularly to the cottage and summer house, which have become increasingly winterised second homes in recent years, is an extremely important part of Nordic identity.

This book examines some of the key issues facing Nordic tourism. It looks at tourism in different environments ranging from the cities to coastal and wilderness areas. Specific chapters are also dedicated to second home tourism, the relationship between culture and tourism as well as winter tourism. These different tourism concerns are also understood within the broader context of tourism marketing and governance, and it is to these issues that the book will turn to first.

Self-review questions

(1) Why is tourism a sub-set of mobility?
(2) What are the geographical elements of a tourism system?
(3) How is a trip defined?

Essay questions

(1) Why is tourism more than just leisure travel?
(2) What are the implications of how the tourism industry is defined?

Key readings and websites

For an expansion of some of the issues raised with respect to approaches towards tourism studies and issues of definition see Hall (2005a, 2005b) and Coles et al. (2004, 2005, 2006) who examine a number of key dimensions of tourism as a form of human mobility. Coles and Hall (2008) provide a number of chapters that investigate international tourism from a business dimension as well as how tourism is a component of international trade in services, which provide yet other ways of assessing and categorising tourism activities and economic contribution. Smith (2004) provides a discussion of difficulties in defining tourism and the development of TSA approaches.

Table 1.6 provides a list of relevant Nordic and other websites from which up to date statistical, tourism and other data are available. Most of the statistical yearbooks of the Nordic states are available for downloading for free from their respective statistical agency website, while the annual Nordic Statistical Yearbook is available for downloading for free from the Nordic Council of Ministers site. Another useful source of statistical information is the Central Intelligence Agency (CIA) world factbook: https://www.cia.gov/library/publications/the-world-factbook/.

Table 1.6 Nordic government and other relevant websites

State	Site
Denmark	
Ministry of Foreign Affairs	http://www.denmark.dk/; http://www.um.dk/da
Parliament	http://www.folketinget.dk/
Prime Minister's Office	http://www.stm.dk/
Statistics Denmark	http://www.dst.dk/
Visit Denmark	http://www.visitdenmark.com/
Faroe Islands	
Parliament	http://www.logting.fo/
Prime Minister's Office	http://www.tinganes.fo/
Faroe Islands Enterprise (formerly Tourist Board)	http://www.tourist.fo/
Finland	
Finnish Tourist Board	http://www.mek.fi
	http://www.mek.fi/web/mekeng/index.nsf)
Government of Finland	http://valtioneuvosto.fi/etusivu/
Ministry for Foreign Affairs	http://formin.finland.fi/
Parliament	http://web.eduskunta.fi
President	http://www.president.fi/netcomm/
Statistics Finland	http://tilastokeskus.fi/index_en.html
Visit Finland	http://www.visitfinland.com/
Greenland	
Home Rule Government of Greenland (Grønlands Hjemmestyre)	http://www.nanoq.gl/
Statistics Greenland	http://www.statgreen.gl/
Visit Greenland/Tourism & Business Council of Greenland	http://www.greenland.com/
Iceland	
Iceland Naturally/Icelandic Tourist Board	http://www.icelandnaturally.com/; http://www.goiceland.org/
Ministry for Foreign Affairs	http://iceland.is/
Office of the Prime Minister	http://eng.forsaetisraduneyti.is/
Parliament	http://www.althingi.is/vefur/upplens.html
Statistics Iceland	http://www.statice.is/

(*Continued*)

Table 1.6 Continued

State	Site
Norway	
Government Norway	http://www.regjeringen.no/; government.no
Innovation Norway (replaced Norwegian Tourist Board)	http://www.innovasjonnorge.no/
Office of the Prime Minister	http://www.regjeringen.no/en/dep/smk.html?id=875
Parliament	http://www.stortinget.no/
Statistics Norway	http://www.ssb.no/
Visit Norway (Tourist site)	http://www.visitnorway.com/
Sweden	
Government Sweden	http://www.sweden.gov.se/; http://www.regeringen.se/
Official Gateway	http://www.sweden.se/
Prime Minister's Office	http://www.sweden.gov.se/sb/d/2058; http://www.regeringen.se/sb/d/1477
NUTEK (Swedish Agency for Economic and Regional Growth)	http://www.nutek.se/
Statistics Sweden	http://www.scb.se/
Swedish Travel & Tourism Council/Visit Sweden	http://www.visitsweden.com/
Åland	
Government	http://www.regeringen.ax/
Parliament	http://www.lagtinget.ax/
Statistics and Research Åland	http://www.asub.ax/
Visit Åland (Åland Official Tourist Gateway)	http://www.visitaland.com
Supranational institutions	
Arctic Council	http://www.arctic-council.org/
Council of the Baltic Sea States (CBSS)	http://www.cbss.st
European Travel Commission (ETC)	http://www.etc-corporate.org/
European Union (EU)	http://europa.eu/index_en.htm
IATA	http://www.iata.org/index.htm
Nordic Council & Nordic Council of Ministers	http://www.norden.org/
OECD	http://www.oecd.org/
United Nations World Tourism Organization (UNWTO)	http://www.unwto.org/

Chapter 2
Tourism Marketing

Learning Objectives

After reading this chapter, you should be able to:

- Understand the core elements of the marketing concept.
- Understand the nature of tourism services and products.
- Identify some of the key elements of marketing in the Nordic tourism system.

Introduction

Marketing is a key ingredient in tourism as it is one of the links between the desire and latent demand of the consumer and conversion into actual experience in the consumption of the tourism product (Duval, 2007). The current definition of marketing used by the American Marketing Association is that 'Marketing is an organizational function and a set of processes for creating, communicating and delivering value to customers and for managing customer relationships in ways that benefit the organization and its stakeholders' (American Marketing Association, 2008). To modify Kotler and Levy's (1969) definition of marketing in tourism management terms: marketing is that function of tourism management that can keep in touch with an organisation's stakeholders, read their needs and motivations, develop products that meet these needs, and build a communication programme which expresses the purpose and objectives of tourism management and tourism organisations. Whereas the customer is often the visitor, other stakeholders include the destination community, investors (private and public), sponsors, employees, interest groups and government. Tourism marketing is therefore much more than selling (Hall, 2007b).

Selling is a component of tourism marketing; but, selling follows rather than precedes tourism management's desire to create experiences (products) that satisfy its consumers. As Drucker (1973: 54) stated, in one of the classic texts on business management: 'Marketing is not only broader than selling, it is not a specialised activity at all. It encompasses the entire business. It is the whole business seen from the point of view of its final result, that is, from the customer's point of view.

Concern and responsibility for marketing must therefore permeate all areas of the enterprise.' Drucker's perspective is known as the marketing concept or marketing orientation and is a philosophy that champions the identification and satisfaction of customer needs, and the integration of marketing through the organisation. According to Cooper and Hall (2008), for tourism a marketing orientation implies that an organisation demonstrates four characteristics:

(1) a dominant marketing philosophy underpinned by research which demonstrates a focus on the consumer;
(2) encourages exchange and strengthens both networks and loyalty through recognising the importance of developing long-term customer relationships;
(3) a thought process accepting that strategic and tactical planning goes hand-in-hand and that encourages innovative thinking; and
(4) 'an integrated organisational structure geared to the organisation's goals of delivering value to the consumer through business-to-customer and business-to-business activities' (Cooper & Hall, 2008: 79–80).

This chapter examines a number of different dimensions of tourism marketing in the Nordic context. It first extends the notion of tourism as a service industry that was raised in Chapter 1 and highlights some of the implications of this for the tourism product and tourism marketing, especially with respect to the concept of co-creation. It then goes on to discuss issues of branding, particularly with respect to destination branding. The chapter then discusses marketing management and the marketing mix.

Tourism Services

Some of the commonly understood dimensions of services are that they are:

- *intangible* – they are experiences; although people can keep reminders of the experience such as souvenirs;
- *inseparable* – production and consumption of tourism services occur simultaneously;
- *variable* (also referred to as inconsistent or heterogeneous) – because tourism is geared towards the selling of experiences they vary substantially from one service experience to another; and
- *perishable* – the 'product' cannot usually be stored from one day to the next; if a hotel room has not been sold tonight, the opportunity for that sale is lost forever; experiences can be stored only in people's heads.

Pride and Ferrell (2003) also suggest that services imply client-based relationships and customer contact. Some of these dimensions and their characteristics with respect to consumption, production, product and markets are indicated in Table 2.1. Nevertheless, there are increasing concerns about the extent to which the above characteristics of services are truly generalisable between and even within different service sectors (Lovelock & Gummesson, 2004). These concerns are illustrated within the context of tourism in Table 2.2.

Table 2.1 Consumption, market, product and production characteristics of services

Service consumption	
Delivery of product	Consumption and production are coterminous in time and space, often requiring consumer or supplier to move to meet the other party.
Role of consumer	Services are consumer intensive.
Organisation of consumption	Often hard to separate production from consumption.
Service production	
Technology and plant	Low levels of capital equipment, heavy investment in buildings
Labour	Some services highly professional, often requiring high level of interpersonal skills; others relatively unskilled and often high degree of flexibility in terms of casual and part-time labour. Specialist knowledge may be important but rarely technological skills.
Organisation of labour process	High degree of variability with some workers often engaged in craft-like production while others have high degree of management control of details of work
Features of production	Production is often non-continuous and economies of scale are limited.
Organisation of industry	Some services state run-public services, large multinational firms operate in trade in international services with domestic-based firms often small scale with high preponderence of family businesses and self-employed.
Service product	
Nature of product	Immaterial, hard to store and transport with process and product hard to distinguish
Features of product	Often customised to consumer requirements
Intellectual property	Hard to protect and it is easy to copy many service innovations. Reputation is often crucial.
Service markets	
Organisation of markets	Some services delivered via public sector provision. Some costs are invisibly bundled with goods, for example, retail sector.
Regulation	Professional regulation and accreditation programmes in some services, government regulation of some service standards.
Marketing	Difficult to demonstrate product in advance.

Source: After Boden and Miles (2000), Hall (2005a, 2007b), Lovelock and Gummesson (2004), Pride and Ferrell (2003).

Table 2.2 The applicability of the characteristics of services to tourism

Characteristic	Service category involving			
	Physical acts with customers' bodies (e.g. lodging, passenger transport)	*Physical acts with owned objects (e.g. freight transport, cleaning and laundry, food)*	*Nonphysical acts with customers' minds (e.g. entertainment, interpretation)*	*Processing of information (e.g. internet booking, travel insurance, tourism research)*
Intangibility	Misleading, performance is ephemeral, but experience may be highly tangible	Misleading, performance is ephemeral but may physically transform object in tangible ways	Yes	Yes
Inseperability	Yes	No, customer usually absent during production	Only when performance is delivered live	Many exceptions, customers often absent during production
Variability	Yes, often hard to standardise because of direct labour and customer involvement	Numerous exceptions, can often be standardised	Numerous exceptions, can often be standardised	Numerous exceptions, can often be standardised
Perishability	Yes	Yes	Numerous exceptions, performance can often be stored in electronic or print form	Numerous exceptions, performance can often be stored in electronic or print form
Customer contact/client based	Yes	No, customer usually absent during production	Numerous exceptions	Many exceptions, customers may be absent during production

Source: After Hall (2005a, 2007b), Lovelock and Gummesson (2004).

Figure 2.1 The tourism experience at the intersection of consumption and production

Lovelock and Gummesson's (2004) concerns over the nature of services, reflect those of Hill (1999: 426) who stated that 'the distinction between goods and services has become erroneously and unnecessarily confused with quite a different one, namely that between tangible and intangible products'. Instead, Hill argued that the essential characteristics of services were that they cannot be produced without the agreement and possible active cooperation of the consuming individual or organisation, and that the outputs produced are not separate entities that exist independently of the producers or consumers. Hill's observations with respect to the nature of services (i.e. that services are separate entities that do not exist independently of the direct interaction between consumers and producers and they therefore cannot be stocked or have their ownership transferred), is significant for tourism marketing as it highlights the importance of the concept of 'co-creation' in which value is created in consumer–producer interaction (Hall, 2007b). An observation that not only has implications for the way in which the tourism experience is co-created where consumption and production meet (Figure 2.1), but also for our understanding of the different types of tourism products that are consumed.

The Tourism Product

As noted in Chapter 1 the concept of the tourism system implies that the tourist travels from their home location to a destination then returns home. Whereas the concept may seem deceptively simple, it actually has enormous implications for the way in which we understand tourism markets and products. For example, the movement over space and time means that the tourist is likely to consume a wide variety of services, thereby reflecting the complexity of service consumption and production identified in the previous section, as the tourist consumes different services at different stages of the trip. Furthermore, not only will the tourist consume different services, but they will have different experiences, which can be broadly categorised in a manner that corresponds with the physical stages of travel (Fridgen, 1984):

- travel decision making and anticipation;
- travel to a tourism destination or attraction;
- the on-site or at destination experience;
- return travel; and
- recollection of the experience and influence on future decision making.

Physical movement therefore goes hand in hand with changes in the psychology of tourism and there the motivations, expectations and experiences the tourist had (Hall, 2005a; Larsen & Mossberg, 2007). As noted above, such experiences are co-created between consumers and producers which highlights that products are continuously transformed in a process of producer and consumer interaction and that knowledge and reflexivity are used as means of differentiation between products (Prahalad & Ramaswamy, 2004). Such a situation also means that the value a 'product' has is in a continuous process of becoming. Products and their associated value are therefore socially constructed.

Indeed, it has been long recognised what actually constitutes a tourism resource depends on the motivations, desires and interests of the consumer and the cultural, social, economic and technological context within which those motivations occur (Clawson & Knetsch, 1966; Pigram & Jenkins, 1999). Tourism resources can therefore encompass a wide range of settings associated with space, topography, landscape, climates and expressions of culture. A tourism resource becomes a resource only if it is seen as having utility value, and different cultures and nationalities can have different perceptions of the tourism value of the same object or space. Moreover, the attractiveness or value of tourism resources can change over time, whether it be for a society, or event within the life of an individual (Hall, 2007b). This situation is well borne out in Case 2.1 on the North Cape.

Cases and issues 2.1: Prominent promontory: The social construction of North Cape

JENS KR. STEEN JACOBSEN

In August 1553 the British Willoughby expedition rounded Europe's dramatic northern extremity in search of the Northeast Passage to China. Some years later the name North Cape appeared on a map made by Richard Chancellor, a navigator of the Willoughby expedition.

This study deals with the desolate North Cape and its social construction as a tourism attraction. North Cape (*Nordkapp*) is a huge promontory rising abruptly over 1000 feet from the sea, on the Norwegian island of Magerøya in the Arctic Ocean. Compared with the relatively short history of modern tourism, international travel to North Cape has a long tradition. The headland is a principal attraction for a large part of the foreign visitors to Northern Scandinavia. Annually, more than 200,000 people come to see the Cape, predominantly during the never-ending Arctic summer day, the Midnight Sun period, from 11 May to 31 July.

It is commonly suggested that tourism attractions have inherent qualities. However, if one looks closer at most attractions, qualities that at first appear to be inherent seem mainly to be related to how places are comprehended (Leiper, 1990), indicating that they are socially constructed. MacCannell (1976: 44–45) has argued that sights (sites) are transformed into tourism attractions through five successive stages of what he has labelled 'sight sacralisation'. These phases are termed: naming; framing and elevation; enshrinement; mechanical reproduction; social reproduction. Drawing on the main tenets of this attraction theory, the principal features of the development of North Cape will be depicted (*cf.* Jacobsen, 1997).

Naming

The naming phase takes place when the sight, in this case also a site, is differentiated from similar objects as worthy of preservation. The phase includes reports acclaiming the values of the sight (MacCannell, 1976). North Cape has been a seamark for more than 1100 years. Until the middle of the 16th century, this headland was probably known as Knyskanes, which sounds much less internationally appealing. It may seem puzzling why North Cape was renamed. One possible explanation relates to the development of world maps in the 16th century (Bagrow, 1985), implying a new perspective for the understanding of the earth, easing navigation, and possibly also leading to new travel patterns. In earlier medieval maps the European world was constructed spatially around Jerusalem but with the new map systems, the centre of the world was changed – and apparently the peripheries as well. In this Eurocentric period, North Cape became a representation of the (northern) end of the world (i.e. the European continent) (Jacobsen, 1997). From the 18th century onwards, after the era of exploration, travel entailed coming from Europe rather than going to anywhere (Stratton, 1990), strengthening the travel interest in extreme peripheries like North Cape.

Written accounts of early visitors to the Cape expressed their passion for the experience of boundary and extremity. Such reports were found at least as far back as in 1664, at the priest Francesco Negri's appearance in the far north:

> And now I have arrived at North Cape, which is the outmost tip of Finnmark, and more than that I have arrived at the very end of the earth, as from here to the pole there is no other land inhabited by human beings. Having made it here, my curiosity is satisfied. (Negri, 1929: 357)

In the first part of the 19th century, the geologist Baltazar Keilhau depicted his arrival at the desolate promontory as an enjoyment of the vain idea of being at Europe's outermost tip, and so did several later visitors. In 1827 Robert Everest, of Himalayan fame, turned up at the Cape. However, the main growth of interest in both North Cape and northern Norway seems to coincide with the Swedish-Norwegian King Oscar II's coronal voyage in 1873, comprising a visit to the Cape (Jacobsen, 1994). Oscar II brought a large assemblage of fellow royalty, politicians, authors, and reporters along Norway's northern coastline, and this cruise had a sizeable impact: most newspaper and magazines in the world brought reports about this strange northern land of eternal day which the king had 'discovered' (Jones, 1957). Later, numerous prominent visitors such as painters, authors, explorers, politicians, princesses, kings and emperors have further enhanced the reputation of North Cape.

Framing and elevation

Framing is the placement of an official boundary around an attraction. Practically speaking, two types of framing may occur: protecting and enhancing. Elevation entails putting the object on display or opening it up for visitation (MacCannell, 1976). At the outset, opening the Cape for visitation was mainly related to improvements in the transport system. Regular steamship tours from Norway's northernmost town of Hammerfest were established in 1845, and a tourism steamship line from southern Norway was in operation in 1877 (Jacobsen, 1997). But there was no completely comfortable access to the summit until a new road was opened in 1956, replacing an 11-kilometre walk from the shores of Horn Bay. In 1927, the organisation Nordkaps Vel A/S was established, partly to protect the site but also to control a prospering tourism business. An official boundary was created when this organisation leased the promontory from the Norwegian State. Since 1929, North Cape has been under statutory protection as a nature reserve.

Enshrinement

When the framing material itself enters sacralisation, the enshrinement phase occurs (MacCannell, 1976). To a large extent, this phase can be interpreted as physical or morphological. A monumental column was erected in 1873 to the memory of King Oscar II's visit, and this was probably also the first on-site marker at the Cape. When the German Emperor William II made his first visit in 1891, a cairn was built as a memorial. A large globe, erected some decades ago, is probably the most important on-site marker today. The buildings at the site, however, seem not as yet to have become venerated.

Mechanical reproduction

The fourth stage in sight sacralisation is mechanical reproduction: the creation of prints, photographs, models, or effigies of the 'sacred object', which are themselves valued and displayed. It is mechanical reproduction that sets the tourists in motion on a journey to find the true object (MacCannell, 1976). Such reproduction of the Cape commenced around 1600, with copper engravings. Several artists have later rendered their impression of the cliff. Naturally, the portrayal of the Cape became more widespread when it became possible to print photographs. Today, pictorial representations of the Cape are frequently included in tourism marketing of both Norway and northern Scandinavia.

Social reproduction

The final stage of sacralisation, social reproduction, takes place when areas or institutions begin to name themselves after an attraction (MacCannell, 1976). The social reproduction had its breakthrough in 1950, when the surrounding municipality changed its name from Kjelvik to Nordkapp (North Cape) (Jacobsen, 1997). Social reproduction has continued up to the present. For instance, Nordkapp is an Italian sportswear brand.

Conclusion

This study has demonstrated that it is to a large extent possible to describe and explain the development of North Cape with the aid of the main tenets of MacCannell's attraction theory. However, it has been indicated that mechanical reproduction is actually the second stage in the advance of this site and it has also been established that different stages of sight sacralisation may occur at the same time (Jacobsen, 1997: 353). Most phases still continue, such as reports acclaiming the values of the site (the naming phase).

This analysis further indicates that attractions can be characterised by only very few, if any, inherent qualities. The fact that North Cape is actually not the northernmost point of Europe and Magerøya island indicates that mental maps might be more important to tourism than geographic maps. The distinctive form of the 1000 ft (307 m) high promontory has given North Cape advantages over the more northerly but low and thus less striking headland of Knivskjelodden. Karel Capek (1995: 157) said it like this: '... Europe has chosen North Cape as its northernmost point; it thinks that if this is to be the end, it should at least be respectable.' Borrowing a term from Lynch (1960), North Cape seems to have a high imageability, with a potential to evoking strong images in observers.

Tourists are often disappointed when an attraction does not measure up to their expectations but this occurs infrequently at the Cape, unless the promontory and its environs are shrouded in fog from the Arctic Ocean. North Cape seems to provide a stronger and more intense experience than most other places symbolising an extremity or the edge of the world, particularly when the Midnight Sun is observed from the north-facing plateau. An older symbol of the end of the European world, Gibraltar, apparently offers a similar experience to the one found at North Cape but at Gibraltar there are probably several other disturbing elements too close to the attraction nucleus. Most other European extremities are also closer to population centres and thus less exclusive destinations.

The tourist product should therefore be seen not just as the commercial package which the visitor purchases. Instead, the tourist product is an amalgam of factors, including physical resources, people, infrastructure, materials, goods and services, which, when brought together as part of the co-creation process, help create the tourist experience (Mossberg, 2007). This is often within specific destination areas, although it can be overall stages of the trip, even including in the tourist's home environment. Hall (2005a) argues that the tourist is simultaneously consuming at least four embedded products that are that range from individual service encounters/experiences, to firm level products, to destination products, and then to the total tourism trip product (Figure 2.2). It is even possible to divide the nature of the tourism product even further. For example, firms over multiple product packages each of which can be regarded as embedded within a business's overall product. Similarly, in light of what we know about the nature of services (see Tables 2.1 and 2.2) we can distinguish between physical acts with owned objects

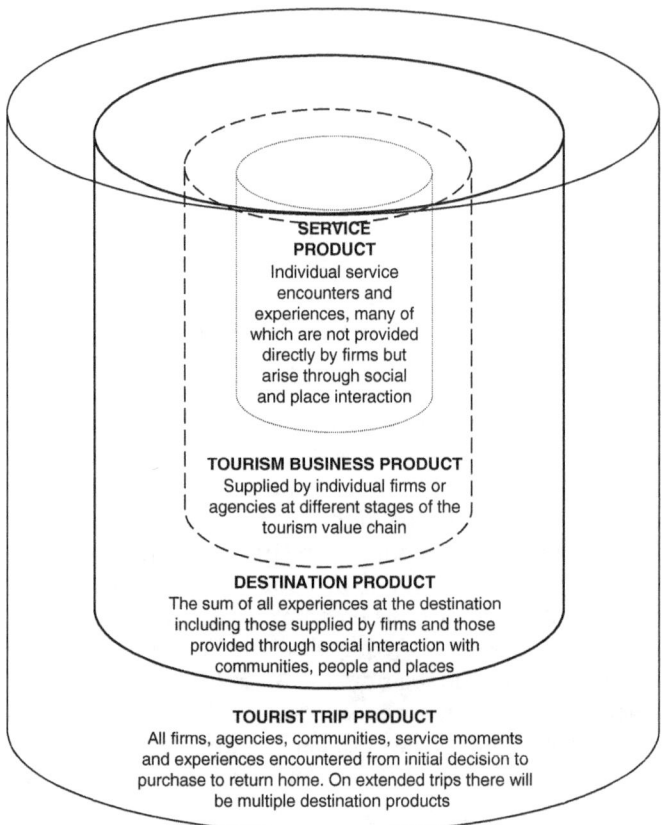

Figure 2.2 The embedded nature of the tourism product (after Hall, 2005a, 2007b)

(e.g. the purchase of food at a restaurant), physical acts with customers' bodies (e.g. the physical quality of the restaurant in terms of layout, chair, tables, table setting) and nonphysical acts with customers' minds (e.g. the entertainment, atmosphere and the information regarding the menu and wine list from the waiter) – all of which can occur as simultaneous acts of consumption and production and may be individual dimensions of the product. Furthermore, this can also all happen while the credit card payment is being made (processing of information) (Hall, 2007b).

Consideration of tourism products as being 'multilayered' raises a number of considerable issues for the tourism industry. First, it means that in order to 'package' products' successfully there has to be a high degree of internal and external coordination and communication within tourism firms and organisations as the trip product, and many destination products are networked products that are the result of the interactions between numerous producers and suppliers. Second, the networked nature of tourism creates substantial issues with respect to quality control over all the various elements and organisations, and the people within them, that help provide tourism products and experiences. Third, consumer satisfaction over a trip or destination product as a whole, and in some cases with individual businesses, is an amalgam of a series of co-created experiences that occur over time and that may be reflected upon and influence future travel plans and recommendations well after the tourist has returned to their home environment (Hall, 2005a).

The focus on experiences has also lead many tourism organisations to become interested in Pine and Gilmore's (1999) concept of the 'experience economy' in which experiences can be understood along two dimensions: passive to active; and absorption to immersion, with the ability of a good experience to 'transform' the visitor (Andersson, 2007). Cooper and Hall (2008) argue that within the framework of the experience economy, changing inter-generational values mean that consumers are seeking new meaningful and self-actualising experiences in their tourism consumption with experiences being created to be more authentic, personal, memorable and emotional as the tourist enters into a multi-faceted relationship with the destination and its human and cultural dimensions. However, another dimension of the social construction of tourist meaning and the focus of values as integral to co-creation and marketing is substantial focus of the role of brands in tourism, especially at the destination level, and it is to this that the next section will turn.

Brands and Branding

In marketing terms a brand represents a unique combination of product characteristics and functional and non-functional added values, which, taken together, have taken on a relevant meaning linked to that brand. According to Morgan and Pritchard (2002: 12) brand advantage: 'is secured through communication which highlights the specific benefits of a product, culminating in an overall impression of a superior brand. The image the product creates in the consumer's mind, how it is positioned, however, is of more importance to its ultimate success than its actual characteristics'.

The values that are associated with brands come from both 'deliberate' and 'accidental' sources. By this we mean, that although a tourist firm, particularly a destination marketing organisation (DMO), seeks to develop a particular brand for the destination it is promoting, much of the information a potential or actual visitor to the destination has with respect to understanding the brand will be derived from other sources, especially the media (especially film, television, newspapers, magazines and internet) and the comments of friends and relations (word-of-mouth). In some cases images created by media may be utilised by destinations for branding and promotion (Müller, 2006). For example, the song *Wonderful Copenhagen* from the 1952 film *Hans Christian Andersen* serves as the basis for the promotion of Copenhagen, along with related attractions such as the statue of the Little Mermaid (Plate 2.1) while the 200th anniversary of Hans Christian Andersen's birth in 2005 provided the basis for a year of promotion and special events in Copenhagen and throughout Denmark (see Case 2.2 on the coverage the bicentennial received). However, even though DMOs will try and

36 Nordic Tourism

Plate 2.1 The statue of the 'Little Mermaid' in Copenhagen

influence the media through public relations or even, in the case of films and television, product placement or seeking to have filming done at the location, the way in which consumers interpret the brand and the values of a destination or firm is social constructed by the viewer within the context of their own culture and experiences.

> ### Cases and issues 2.2: The Hans Christian Andersen bicentenary
>
> According to the Hans Christian Andersen Foundation the bicentennial event was covered by media in more than 60 countries. The media monitoring and evaluation agencies Durrants of London and Observer of Copenhagen monitored the media coverage in 35 selected countries on behalf of the Foundation providing 14,903 press clippings.
>
> The opening show of the bicentenary 'Once Upon a Time ...', which took place on 2 April 2005 at the Danish national football stadium in Copenhagen, generated worldwide awareness of the event. The show was televised live by Danish national broadcaster DR1 to a national audience of 1.6 million. Clips from the show were also featured by 60 international TV stations which are part of the EBU network. These clips, which were broadcasted in the days immediately following the event, were seen by 100 million viewers. Some of the TV stations that featured the news clips were the BBC; the German stations

ARD, and ZDF; AP TV; RAI (Italian TV); the Australian Broadcasting Corporation and Channel 9 (Australia); CNN; Euronews; HBO Latin (South American TV network). The British company 3DDTV has since sold the show to 29 countries across the globe. Along with Denmark these countries represent a minimum of 593 million potential viewers. The Foundation notes that in comparison, the 2001 Eurovision Song Contest was broadcasted to approximately 120 million potential viewers.

A valuation assessment of the media exposure generated by the bicentenary was undertaken in 35 countries and was estimated to generate a total value of DKK 3.12 billion (DKK 3,120,000,000) broken down into:

- Printed media (newspapers, magazines, etc): DKK 2.62 billion.
- Radio: DKK 12.01 million.
- TV: DKK 436.77 million.
- VNR (video clips produced and distributed internationally in relation to tourism promotion): DKK 31.88 million.

Derived from Hans Christian Andersen Foundation: http://www.hca2005.com/

In the case of the Nordic countries one of the most common images of the region in the wider-world is that of Vikings and their horned helmets. However, as many Nordic readers will be aware, horned helmets were extremely rare Viking headwear and instead are a product of the 19th and 20th century literary and other imaginations. In fact, it was Victorian Britain that really invented the Vikings as they are now often portrayed in the public mind. The term 'Viking' was virtually unknown until the beginning of the 19th century (the first Oxford English Dictionary reference dates from 1807) (Wawn, 2001) and it was the publication in English of Bishop Esaias Tegnér's *Frithiof's Saga* based on a 14th century Icelandic saga that captured the British public's imagination. The impact of the saga was profound as it was published in 16 English language versions during the 19th century, perhaps just as importantly many of the editions were published with accompanying pictures of Norsemen in horned helmets thereby creating an image of the Viking and of northern lands and values in the British imagination – an image, albeit socially constructed in modern terms, that has subsequently been reproduced in film, television and all sorts of media. Yet the Viking horned helmet well represents some of the issues associated with tourism branding and image, instantly recognisable, yet historically inaccurate but a potentially valuable form of commodified heritage (Halewood & Hannam, 2001). As Hannam (2003) recognised, the Anglo-American stereotypical representation of Viking heritage is of sea-faring, sexist and bloodthirsty men raping and pillaging. In contrast to this image, in the Nordic countries the dominant image of Vikings in popular culture finds fewer references to war and warriors. Instead, the Viking representation is very much concerned with the people who were regarded as pirates abroad, but lived in a well-ordered society at home. However, it is often the more bloodthirsty image that initially inspires Anglo-American tourists to visit Viking heritage sites or themed festivals.

In addition to issues of representation and authenticity of brands, and especially destination brands, another important branding issue is the potential for destination brand confusion. In the case of the Nordic countries brands exist at a number of scales ranging from the supranational (Scandinavia, Nordic) to the national, regional and municipal or attraction. Christmas time is a good example of brand confusion in the region with Greenland, Iceland and Finland all claiming to be the home of Santa Claus/Father Christmas. Lapland gives particular emphasis to being home of Santa Claus, and especially the city of Rovaniemi. However, the Santa Claus or equivalent brands are also promoted for tourism purposes by Enontekiö in northern Lapland, Gallivare in Swedish Lapland, Drøbak in Norway and Skutustadir in Iceland. Similarly, an examination of many national, regional and local brands in the Nordic countries highlights the similarities with respect to positioning in terms of using nature-based products.

Branding is only one promotion and marketing technique, but the successful utilisation of branding can have significant implications for the success of destinations in attracting their target visitor markets. Key attributes commonly identified as determining the significance of branding as a marketing tool include (Hall, 2007b):

- brands that can identify the product;
- brands that can offer the consumer confidence or security; and
- brands that add value while also offering an indication of price, quality and psychological benefits.

Ideally, brand promoters seek to occupy a market niche which is not occupied by any other brand so that their brand cannot be substituted by any other. However, in tourism terms this can be extremely difficult because of an incredibly congested market with respect to the promotion of destination images and attributes. Differentiation, the ultimate goal of branding, is absent from the many Nordic DMO marketing strategies if they are analysed at a Nordic level, because the key attributes of the brands are often very similar. Although many of the brands include some form of identification, many fail to symbolise price, quality or a summary of the destination offering. The branding strategies employed at national and regional levels in the Nordic tourism industry display many similarities in their tag lines, icons and themes, as well as in the components of their strategies. Furthermore, from a demand perspective, the lack of differentiation and brand proliferation can create feelings of confusion in the international marketplace in particular where the geographical attributes of many Nordic regional brands will not be so well known. Nevertheless, national and regional DMOs continue to seek to create a niche for their brands and promotions both domestically and internationally. Case 2.3 indicates some of the issues involved in the marketing of Iceland as a tourism destination.

Cases and issues 2.3: Marketing the 'natural' in Iceland

ANNA DÓRA SÆÞÓRSDÓTTIR and ANNA KARLSDÓTTIR

Neither the tourist industry in Iceland nor the Icelandic government has defined any special target group for Iceland as a tourist destination. The idea has been to keep as many options open as possible. The fact that Iceland is an island in the middle of the Atlantic, only accessible by plane or ship, known to be an expensive destination, does however automatically limit the target group.

Icelandic travel companies are mostly small, except for Icelandair, which is relatively small in international perspective in spite of being by far the biggest travel company in Iceland. Because of this the Icelandic travel industry has limited funds to spend on marketing Iceland as a tourist destination.

Marketing campaigns have, so far, mostly been in Iceland (for the domestic market), North America, the UK, Scandinavia and mainland Europe which reflects well current distribution of foreign visitors in Iceland. Branding has become increasingly popular by the tourist industry. In the year 2000 a project called 'Iceland Naturally' started in North America where clean and healthy products from Iceland and Iceland as an environmental friendly tourist destination were introduced. Annually US$ 1 million has been spent on the project by the Minister of Transport and some participating companies. This brand creation has been successful in North America and decisions have been made to continue the campaign there. In addition preparations have started for a similar project in Europe. The double meaning involved in 'Iceland Naturally', for example, how natural the Icelandic nature is and how logical it is to go there for a visit, works on many markets both of Latin and Germanic origin.

In order to make up for the seasonality problem increasing emphasis has been placed on marketing Reykjavík as an interesting destination, outside the three summer months, where the tourist can get everything a city can offer within reach of nature. The slogan 'Reykjavík Pure Energy' has been used for this purpose and some special happenings have been arranged off season such as the Airwaves music festival and Food & Fun, which is an event combining culinary skills, fresh natural ingredients and Icelandic outdoor adventure with the Reykjavik nightlife.

Advertisements for weekend trips to Reykjavík have been prominent not least due to their double meaning and sexual undertone in its marketing campaign. From 1980 and for a few years after Icelandair used a sexualised image of the myth about Icelandic beauty in an advertisement in Scandinavia with a picture of three beautiful women all wearing the same wool sweater and nothing else. In the period from 2001 until 2003 Icelandair had another campaign with a similar undertone. The campaign was for the British market and was published on their website, on posters in subway stations and leaflets. In the advertisements there were pictures showing the different trips they were selling as well as slogans describing the products or having historic appeal, which the target group should recognise. One of these advertisements showed a couple taking a bath in the Blue Lagoon covered with silicon on their face and the slogan used with it was 'Fancy a dirty weekend?' This slogan, as well as 'One night stand', 'Free dip every trip' and 'Two for one', was highly criticised by Icelandic women and other organisations claiming the advertisements made women into sexualised objects. According to Icelandair that was by no means their intention. They explained that the slogan in the Blue Lagoon advertisement was targeted at middle aged British people, supposedly referring to an old saying originating in Blackpool, meaning that the couple escaped from the children over the weekend.

In recent years product segmentation has generated increased numbers of agents active in marketing; consequently, the influence of Icelandair and the Tourism council has diminished. As a result the promotion of Iceland as a tourist destination has become more diversified and it has become more difficult to influence the marketing mix. Indirect advertisements such as inviting journalists for a visit and introducing to them the various products and experiences available in Iceland have been used quite extensively. The induced image the tourism industry wants to create gets influenced by the organic image the media generates as reporters who visit Iceland write about what they think is special or outstanding in some way.

The emphasis on the experience economy and authenticity as a function of tourism production means that consumer brand choice may be increasingly regarded as a statement about lifestyle and consumers' emotional relationship with a destination (or other tourism product) as much as it is about image. Therefore, some tourism marketers are focusing on product differentiation through brand loyalty and emotional appeal, rather than through discernible, tangible benefits. However, as Morgan and Pritchard (2002: 12) observed, 'mere emotion is not enough, the key is to develop a strong brand which holds some unique associations for the consumer'. Nevertheless, in the longer term, to be successful, brands and associated images must also be grounded in the realities of the destination product that the tourist actually consumes (Hall, 2007b). Therefore, according to Morgan and Pritchard (2002), to create an emotional attachment with consumers a destination brand must:

- be credible;
- be deliverable;
- be differentiating;
- convey powerful ideas;
- enthuse trade partners; and
- resonate with the consumer.

Yet achieving such brand strength is extremely difficult to do for three reasons (Hall, 2007a). First, destinations are seeking to promote themselves in an extremely congested and competitive marketplace in which there is often a high degree of substitutability and also much copycat marketing, and in which campaigns, brands and images perceived as successful in one destination can be adopted by other destinations (Anholt, 1998). Second, political pressures from different stakeholders within a destination might mean that some attributes of a destination are used or not used because stakeholders wish to have their own aspirations presented in a brand (and its associated campaigns), rather than marketers selecting attributes (Morgan & Pritchard, 2002). Third, and as noted above with respect to the Viking example, it can be very difficult for destinations to change their image and still remain credible in the mind of the consumer if the consumer already has a strong notion of what that image should be (Hall, 2007b) – even if the image is not grounded in the reality of the location or attraction. Case 2.4 on branding Denmark gives a very good outline of some of the difficulties involved in successfully branding a destination.

Cases and issues 2.4: Branding Denmark: Present strategies and future options

ANETTE THERKELSEN

At the break of the new millennium, the Danish Tourist Board launched their branding strategy for Denmark which were to secure Denmark a clear profile and with that a strong position on international markets in the years to come. Based on a wish to depart from the somewhat old-fashioned, fairytale image that Denmark had been associated with for decades, the Danish Tourist Board set about to create a more fashionable tourism profile. In that process inspiration was drawn from places such as Spain, Ireland and New Zealand which have been among the tourism destinations in the world that have poured most resources into branding initiatives. Under the headline 'Denmark. Enjoy!', three sets of values were identified based on supply and demand side analyses and discussions with central tourism actors (Table 2.3).

In a few words, the six elements were defined in the following manner: (1) cosiness signifies 'a particular Danish sense of warmth, well-being and

Table 2.3 Denmark: The brand

Cosiness	–	Unpretentious
Design	–	Talented
Oasis	–	Free

Source: Danmarks Turistråd (2000a).

togetherness'; (2) unpretentious means that 'Danes have an easy-going – often humorous – attitude towards life and authorities'; (3) design covers 'Danish furniture, architecture, art and literature ... a collective consciousness about form and function'; (4) talented signifies a people who are 'artistic, creative and well-educated'; (5) oasis entails that Denmark is 'a place where the harried (*sic*) traveller can relax'; and (6) free means 'space for the individual both in democratic and philosophical terms' (Danmarks Turistråd, 2000a).

The reasoning behind the three sets of values is that they should each contain a rational side (left hand side of Table 2.3) and an emotional side (right hand side of Table 2.3). The rational elements explain the way the tourism product is in terms of fact-based criteria, and the emotional elements describe Denmark's character in softer, more subjective terms (Danmarks Turistråd, 2000a). The values are eventually joined together in a vision that depicts the way Denmark should be experienced by others.

> Denmark should be seen by others as a cosy oasis in Europe. The visitor meets free and unpretentious people who possess a special talent in creating a society based on love of art, culture and social values. (Danmarks Turistråd, 2000a)

It is, however, worth contemplating the rationale behind these values and vision. Do cosiness, design and oasis constitute rational, objective facts about Denmark? Or do they rather represent Denmark from a specific, and with that subjective, angle that fits the purpose of the marketer? But under all circumstances by stating that the brand consists of an objective core, it gains and element of truthfulness for the actors that are to use these values in their marketing efforts.

In developing the branding profile, attention was also given to positioning Denmark in relation to competing destinations, particularly Sweden and Norway as these three destinations are often lumped together by foreign markets. Differentiation is ostensibly to be found in the combination of these three sets of values, not in the individual value, as other destinations can boast these in isolation. Surely, when we look around in today's branding jungle, numerous places promote themselves as either an oasis, a designer heaven or a place of warm and unpretentious people. A paradox, however, seems to be inherent in the fact that the branding manual encourages its users (i.e. national, regional, local organisations and private actors) to adopt all or just selected values which fit their given destination or product. Hence the uniqueness inherent the combination of values will be lost in those cases where selection takes place.

In terms of its approach to international markets, the branding strategy was envisaged to be global with localised aspects. Hence the values were to be used universally (i.e. on targeted markets) but mediated in different ways to meet local differences (Danmarks Turistråd, 2000a). In some target markets, the three sets of values would be more easily recognisable than on others where the level of explanation would differ. Take for instance the word 'cosiness' which research has established is an existing image in the north German market in relation to Denmark (Danmarks Turistråd, 2000b), whereas the British market hold no images of Denmark that resemble cosiness (Therkelsen, 2003). It would therefore take more of a concerted effort to establish this meaning in the British market than the German one.

After the initial implementation at the national level, the Danish Tourist Board worked on persuading the regional and local tourism organisations to follow suit, and major efforts were also put into convincing the private tourism trade to adopt the common branding strategy. Widespread scepticism was voiced particularly among local organisations and private companies when the brand was introduced (Andersen, 2001; Ooi, 2004). Tourism actors along the coastline and in rural areas in particular had difficulties in seeing any benefits in a modern design oriented branding profile. Nonetheless, five years later the Danish Tourist Board, which by now has turned into VisitDenmark, sees itself as successful in implementing 'Denmark. Enjoy!' at all levels of the Danish tourism trade, though there are still a few private companies with a strong brand profile of their own, as well as minor companies with virtually no marketing resources that stand outside the branding efforts (interview with VisitDenmark's brand manager, August 2005). Whether this positive assessment of a nationwide branding implementation can be confirmed by a thorough analysis of the marketing strategies of the various actors in the Danish tourism trade remains to be seen.

Another strategic choice inherent in the branding campaign is to start out as a communication project, establishing a clear message about Denmark and creating a branding tool box (consisting of pictures, design manual, logo, payoff and a Denmark film) to accompany it, which all actors in the Danish tourism trade are encouraged to make use of. The next step is then to expand the branding strategy into a product development project and hence to develop experiences in accordance with the promises that the values give (interview with VisitDenmark's brand manager, August 2005). An example of this is an educational programme developed for Danish inns where management and employees are to take a critical look at their products and gain inspiration for future development from the branding values and branding tools such as storytelling.

A future vision for the Danish branding strategy is to expand cooperation with other national actors that either promote Denmark abroad or use Denmark as a means to sell their products – actors such as the Danish Foreign Ministry, the Danish Export Council and InvestinDenmark. Synergy effects are envisaged to be the outcome of expanding the branding cooperation to other sectors and thereby applying the same core message about Denmark in multiple contexts abroad. Discussion are, however, still ongoing, as to how to cooperate, but according to VisitDenmark's brand manager 'Denmark. Enjoy!' has functioned as a source of inspiration for the other actors in their efforts of building their brands. The wish for increased cooperation in future is furthermore sustained by the managing director of VisitDenmark, who has publicly stated that Denmark has to think more strategically about its branding efforts abroad if it wants to make itself heard and seen in the global market place (Bülow, 2006).

The local level seems to offer some interesting possibilities and challenges for the national tourism brand in future. A city branding frenzy has swept Denmark over the past few years and hence numerous cities and provincial towns are in the process of positioning themselves as attractive places to live, work, invest and be tourists in, in both domestic and foreign markets (see website references at the end of the case). On the one hand, it appears to be a positive development for Danish tourism that more cities and towns have started to think in marketing terms about their places, and ask themselves questions such as: What are our resources? In what direction do we want our place to develop? Who are our present and future target groups? How do we position ourselves in relation to competitors? On the other hand, these local umbrella branding initiatives (Halkier & Therkelsen, 2004) pay little or no attention to the national tourism branding initiative, and hence VisitDenmark is facing a situation where a number of uncoordinated subbrands for Denmark is being created, which may cause considerable noise in the communication about Denmark. The synergy effects that are expected to materialise by the 'Denmark. Enjoy!' strategy may hence be endangered by these many local branding initiatives, and so Denmark may be returning to speaking with a multitude of voices towards potential target groups and hence remaining a diffuse entity particularly in international markets.

www.brandingdk.dk/web/branding
www.brandingesbjerg.dk
www.brandingaalborg.dk
www.randers.dk

Marketing Management

Overall marketing goals for tourism firms can range from increasing market share and raising product awareness and knowledge to increasing levels of profitability for particular products and services. In contrast, for non-profit tourism organisations marketing goals may include raising the profile of a site or increasing awareness of a particularly sensitive site for which visitation needs to be reduced, to bringing income into a park or local community. Ideally, goals and objectives should be formulated through the involvement of all levels of the tourism firms and DMOs after study of the external business environment, market analysis and segmentation, and, where appropriate, input from relevant stakeholder groups. In the case of DMOs this should include considerable consultation with the various communities and public within the destination as well as business stakeholders (Cooper & Hall, 2008).

The development of marketing strategies with clearly identified target markets and product mix is not the end of the marketing process. Tourism firms and organisations also have to ensure that marketing strategies can be implemented and the target market reached through the development of appropriate marketing and communication strategies and ensuring that human and financial resources are available for the development and promotion of the marketing product, while new relationships may have to formed with stakeholders (Cooper & Hall, 2008). For every major management action or responsibility that is required to give effect to the marketing strategy, plans of action should be developed which identify the communication strategy that is to be adopted. The plan of action will outline the required action, tasks, responsibilities, timeline for implementation, cost estimates and relative priority. Therefore, the plan of action becomes a valuable mechanism for ensuring not only the effectiveness of the marketing strategy but also ensuring that it is undertaken in as efficient a manner as possible (Hall, 2007a). This is an important task is the development of appropriate marketing mix strategies.

The design of an appropriate marketing mix strategy for tourism organisations and products consists of analysing market opportunities, identifying and targeting market segments and developing an appropriate market mix for each segment (Håkansson & Waluszewski, 2005). The mix includes the 'four Ps' of the marketing mix (Gummesson, 1994; Lindgreen *et al.*, 2004; Svensson, 2002) as well as other relevant Ps for tourism marketing (Hall, 2007b):

- *product/service characteristics*: the physical product, image, packaging and the service experience;
- *promotional decisions* concerning channels and messages: also referred to as the promotional mix or the marketing communication mix, this refers to such things advertising, sales promotion and public relations as well as the selection of media, including the web;
- *prices* to be charged for products/services, which may also be a surrogate for the value of the product or service;
- *places* and methods of distribution of products/services, including the availability of the product and services, including their image, location and accessibility. It also includes intermediaries, such as agents, wholesalers and retailers, who then sell to the consumer as well as associated tourism distribution channels.
- The *people* dimension refers primarily to the role of staff in providing experiences and therefore influencing the quality of the tourism product and consumer satisfaction. This includes not only issues of human resource management and emotional labour in the provision of tourism services but also the management and operation of family businesses (Andersson *et al.*, 2002).
- *Programming* refers to the variation of service or product in order to increase customer spending and/or satisfaction.
- *Partnership* refers to the cooperation that often develops between different tourism organisations and/or other organisations and stakeholders and the

mutual or shared benefits that may result. Sponsorship may be seen as a particular form of partnership in which the tourism organisation and the sponsor's seek a range of benefits from the sponsorship arrangement. As noted earlier in the chapter the networked nature of tourism production means that partnership is an extremely important aspect of tourism management and marketing especially at the destination level.
- *Packaging* is the combination of related and complementary services within a single price offering. Packages may be offered from within the products available in a single firm because of its vertical and horizontal integration but more often packages are developed through inter-firm and inter-organisational collaboration although often under the umbrella of a DMO. Roper *et al.* (2005) provide a discussion of how package tours are designed, distributed and produced in the context of the Norwegian package tour sector.

Cases and issues 2.5: Marketing Norway

ØYSTEIN JENSEN

Norway has traditionally been promoted on the international tourism markets through the fjords, the Vikings, heritage symbols such as the stave churches and the midnight sun. Based on a number of surveys of European tourists ordered by the national tourist organisations through the years the qualities linked to pure and unspoiled nature, solitude and space have generally been understood as central motives for travelling to Norway, for example, as promoted in the official brochure *Discover Norway* from The Norwegian Tourist Board/Innovation Norway. During the most recent years there has been an effort of increased focus on activities in the nature in the official promotion materials (e.g. *Hiking and Biking in Norway* from Innovation Norway).

Visiting structure and segments

In 2004 there were approximately 3.4 million overnight foreigners in Norway (Rideng & Dybedal, 2004) and about 75% of these (2.7 million) were on leisure/holiday and were consuming around 20 million overnight stays. This year represents a positive turn in the stagnation of the international tourism arrivals (purpose vacation) to Norway from 1998. The most important international tourism segments are Germany and neighbouring countries Sweden and Denmark (Rideng & Dybedal, 2004). As many as three out of 10 foreign tourist overnight stays were Germans, and after Sweden and Denmark the tourists from The Netherlands, Great Britain, United States, France, Italy and Finland had the biggest number of overnight stays.

Tourism to Norway is concentrated in the summer season (May–August) with 75% of the overnight stays by foreign tourists taking place in this period (Rideng & Dybedal, 2004). One of the bigger challenges for the Norwegian tourism industry has thus been to extend the tourist season beyond the summer, first of all with regard to non-Nordic Europeans. The majority of overnight stays by foreign tourists is made up of people coming by car or bus and a great deal of the continental tourists are individual tourists going on roundtrips. The relative number of foreign tourists travelling by tour buses (normally through tour operators) has demonstrated a decreasing tendency. The most frequent paid accommodation form among tourists is camping followed by hotels and cottages. Many foreign tourists also choose to go an on a cruise with The Norwegian Coastal Voyage along the coast of Norway between Bergen in the southwest and Kirkenes in the north (about half a million passengers all together in 2004, Statistics Norway, 2006b). Additionally, the ordinary international cruise traffic to the Norwegian fjords makes up an increasing market of its own with about 300,000 foreign passengers in 2004 (Rideng & Dybedal, 2004).

Destination marketing organisation, branding and industry involvement

The organisation of tourism marketing in Norway has been subject to several major reorganisations on a national level since the early 1980s. The general development can be described in terms of increased centralisation of the national marketing functions and decentralisation of product development and regional marketing. This has taken place with simultaneously reduced national coordination and increased regional autonomy on organisational and local governmental level.

The first major event was the establishment in 1984 of The Norwegian Travel Board (Norges reiselivsråd) which was the result of the restructuring of the former national tourism organisation (Landslaget for reiselivet i Norge) where international marketing as well as region marketing and product development had been coordinated by the same organisation. At the same time the international marketing function was separated and given to a new national organisation called Nortravel Marketing (NORTRA) and the internal national and regional tasks were left with the old organisation (Viken, 2002). This organisation was eliminated after a few years after a series of conflicts regarding functional roles and areas of competence and there were thus no overall organisation left for coordinating and facilitating internal national industry politics (Viken, 2002). This also represented the starting point of the development of autonomous regional tourism organisations many of which were converted into more commercialised, semi-public forms. In 1999 NORTRA changed its name to the Norwegian Tourist Board (Norges turistråd) and its main tasks were narrowed even more to international marketing through profile programmes and national branding. The marketing by the national tourism organisation was increasingly directed to the consumer level in selected countries and sub-segments although the task of facilitating relationships with foreign tour operators was kept. In 2004 the Norwegian Tourist Board was abolished and the task of international tourism marketing of Norway was integrated into a wider inter-sectoral organisation named Innovation Norway that was placed under the Ministry of Trade and Industry.

During the period before 2004 the tourism industry politics on central ministry level was characterised by reduced involvement and coordination efforts with an orientation towards a laissez-faire industry policy. The main task was to administer the annual grant to the national tourism marketing organisation and report to the national parliament. For 2006 this amount was 172.5 million Norwegian kroner which represents a considerable increase after a period of stagnation of grants up to 2005. Financial support of local product development has primarily been given by the Norwegian Industry Development Fund (SND) under Ministry of Local Government and Regional Development through its decentralised branches in the 19 counties of Norway (see Jacobsen et al., 1996). At the same time as there has been restructuring in Norwegian government involvement there has also been increased industry concentration leading to a situation where the agenda of the tourism industry and tourism strategies has been dominated by a few main tourism industry actors.

At the regional level several attempts have been made to find an effective organisation for the marketing and product development functions during the same period. With the increased decentralisation of the organisational structures and efforts towards more market orientation, various organisational solutions developed within different geographical parts of the country. One of the major structural developments took place through the establishment of the independent, regional, company-like tourism organisations ('landsdelsselskap') with the major task of marketing these areas. During the first part of the 1990s there where five such organisations. Their main roles relative to the broader local organisation structures varied between the different regions. For example, specialised 'Top of Norway' (northern Norway) gradually moved into marketing towards tour operators leaving open marketing towards free individual tourism to county organisations, such as Nordland Reiseliv, or other types of local organisations. In western Norway ('Fjord Norway') an opposite orientation took place. During a period of less than 10 years the importance of these regional organisations decreased dramatically and most of them disappeared leaving the ground to more local organisations, first of all to 'destination organisations', mainly located in the cities of Norway and in some of the most important tourism destinations. Today the only overall successful regional tourism organisation left is 'Fjord Norway' covering the four counties constituting the west of Norway. Surprisingly, no comprehensive analysis of 'what went wrong' with these wide regional tourism organisations has been undertaken, and the knowledge and the documented experiences of how to obtain an effective regional as well as a national structuring of the tourism organisation in Norway remains scarce and scattered.

As most hotels in Norway have tended to be more focused on filling up available capacity during the summer through selling rooms by quantity and low prices to the operators, the willingness to get involved in product development and in increasing the value of the local destination products has been rather lacking (Haukeland *et al.*, 1994; Jensen, 1998). Exceptions are the development of a few mass tourism attractions, such the North Cape Hall and the Artic Circle Centre as attractions for roundtrip tourism. Additionally, a number of historically-based, contrived attractions have been developed a number of which have been linked to previous Viking settlements. Despite of the importance of national parks within the Norwegian natural area administration, the use of national parks for tourism purposes have been quite limited (Haukeland & Lindberg, 2001), although the free access to the use of nature in this country represents a quality appreciated by visitors. Perhaps the most successful single Norwegian tourism product over the years has been the roundtrips (partial of full) with the Norwegian Coastal Voyage branded as 'The World's most beautiful voyage'.

The national brand promoted so far has initially been linked to six core values: natural, pluralism, fresh, friendly, free and surprise (Skjæveland & Tøsdal, 1999). These brand values were based on an analysis of a relatively limited sample of specific segments in European countries carried out by MarkUp Consulting. The values have been reflected in the logo and the brand value 'Norway – the natural choice' developed for international marketing and incorporated into the market communication material. On the regional level the most comprehensive efforts of destination branding have taken place in the western part of Norway, Fjord Norway, with focus on the integrating brand value 'explore' (see Hem & Iversen, 2004). One of the difficulties faced by the branding processes mentioned has still been the overall implementation of the brand values on local level (Hauge & Jensen, 2002).

Future developments

The direction of the future marketing strategies of Norway is for the moment under revision. One of the ambitions of Innovation Norway is to reconsider the brand profile based on a survey of the foreign consumer markets carried out by the Norwegian survey institute MMI. Some other contributions have previously been offered, such as the presentation of scenarios for 'tourism-Norway' in 2015 (Støre *et al.*, 2003) initiated by, among others, the Norwegian Tourist Board. In this work the wish among tourists to experience tranquillity (peace) or excitement are presented as two opposite trajectories for a market orientation of the Norwegian tourism industry. Though such types of contributions can be inspiring for a public discussion, the research-based knowledge about tourism underpinning this work is too scarce to use it as a base for deciding the future directions of tourism to Norway. In order to find a sustainable way of developing and marketing the Norwegian tourism product more thorough analysis, appreciation of already existing knowledge, evaluation of previous programmes as well as inter-disciplinary discussions are generally needed for coming to terms with the complexity of international and national tourism and stakeholders' interests.

Chapter Summary

This chapter has focused on the tourism marketing management and planning processes in order to highlight the challenges facing tourism organisations, and particularly DMOs, in reaching the market place. The chapter has also emphasised that the tourism experience is co-created and that a number of tools are used in tourism marketing and management, to continually try and add value to such experiences for the consumer. One of the implications of co-creation is that tourism resources and attractions are socially constructed and leads to a situation in which tourism organisations may try to influence, but cannot control, the values associated with their brands and images. The chapter has also highlighted that marketing is a communications strategy rather than selling. Nevertheless, marketing, especially at the destination level, is also highly contested, with different stakeholders having different positions as to how places should be marketed and the values and images that should be promoted. Furthermore, marketing and branding strategies, and even the very structure of organisations such as DMOs, shift and change over time (see Case 2.5 on Marketing Norway). At the destination scale, such changes are often associated with issues of governing tourism and it is to these issues that we turn to in the next chapter.

Discussion questions

(1) What are the key elements of tourism services?
(2) What are the implications of co-creation for understanding what makes a tourism resource or attraction?
(3) What are the eight Ps of marketing of the tourism marketing mix?

Essay questions

(1) Conduct a content analysis of the websites of (a) the Nordic countries, (b) Norwegian, Swedish and Finnish Lapland with respect to the images that are used. Discuss the implications for destination perception in the non-Nordic market.
(2) What are the similarities and differences of the branding strategies of the Nordic countries and what are the implications of these for tourism marketing?

Key readings and websites

Cooper and Hall (2008) provide a number of chapters that investigate tourism marketing from a services perspective at destination, firm and experience levels, while Håkansson and Waluszewski (2005) reinterprets the widely used 4Ps marketing concept. A special 2007 issue on tourism experience in the *Scandinavian Journal of Hospitality and Tourism*, 7 (1) provides an important Nordic marketing context and includes Andersson (2007) in relation to the experience economy and Mossberg (2007) with respect to a marketing approach to the tourist experience.

Gössling *et al.* (2008) is a very interesting paper that utilises the concept of co-creation to understand environmental issues in tourism.

Hans Christian Andersen bicentennial website: http://www.hca2005.com/
Innovation Norway: http://www.innovasjonnorge.no/
North Cape/Finnmark: www.nordkapp.no/www.visitnorthcape.com
Vikings in the British popular imagination: http://www.bbc.co.uk/history/ancient/vikings/revival_01.shtml
Wonderful Copenhagen/Visit Copenhagen: http://www.visitcopenhagen.dk/

Chapter 3
Nordic Tourism Governance and Planning Issues

Learning Objectives

After reading this chapter, you should be able to:

- Understand the multilayered nature of governance in Nordic countries.
- Understand the roles of government in tourism.
- Identify the tensions that exist between different elements of tourism policies.
- Understand the importance of different values, interests and institutional arrangements in tourism policy and planning.

Introduction

The role of the state is an important element of tourism management, planning and development, particularly so in the Nordic countries where there is a strong tradition of state involvement in regional economic and social development and in welfare provision. Indeed, the Nordic model of the welfare state is widely regarded as one of the political benchmarks of a Nordic identity. The Nordic welfare model, often highlighted by the large public involvement or what has been termed public service states (Rostgaard & Lehto, 2001), is regarded as a set of egalitarian institutions which not only give poorer elements of society access to a minimum standard of income and social services, but also bring those who would not otherwise have been poor closer to the general standard of their society, decrease the need of the well-to-do to develop exclusive services, and bring about some overall redistribution of income and other resources (Erikson *et al.*, 1987, in Kautto *et al.*, 1999). Although there have been differences in the policies of the Nordic countries, their common historical and cultural experiences have lead to some similar policy outcomes (Kautto *et al.*, 1999, 2001) and institutional arrangements, such as the twin features of localism and central–local integration

(Baldersheim & Ståhlberg, 2003). Nevertheless, the Nordic welfare state, along with other dimensions of government involvement in the social and economic sphere has been changing since the early 1990s along several fronts:

- Finland and Sweden joining the EU along with Denmark which has been a member since 1973. Norway which did not join and which cooperates, along with Iceland, under EEA (European Economic Area) rules has also been affected by integration but in a different fashion. The EEA is based on the same 'four freedoms' as the European Community with respect to the free movement of goods, persons, services and capital. Therefore, Iceland and Norway have also had to adopt part of the Law of the European Union and contribute financially to the European single market even though they have little influence on decision-making processes in Brussels. Both Iceland and Norway have also fully implemented the Schengen agreement with respect to the abolition of physical borders among European countries and the subsequent freedoms of human movement.
- There have been changes to the domestic political power structures, and hence policies, with respect to changes in power between conservative and social democratic parties which although not ending, at least has challenged the position of social democrats in government.
- The political change in the Nordic countries has mirrored change in other Western countries with respect to greater emphasis on market solutions and less reliance on government to solve economic and social problems (Lindblom, 2001; Lindvall & Rothstein, 2006). In some cases this change has lead to the privatisation of public infrastructure or, where not privatised, the running of public owned organisations in a more corporate manner (Schneider *et al.*, 2005). Such shifts in philosophy have also gone hand-in-hand with economic globalisation and restructuring that has meant significant changes in the industrial base in both urban and peripheral regions as greater international economic competition has led to the development of new industries and the loss of old ones unable to compete with cheap exports from emerging economies.

The broader changes in the role of the state in Nordic countries have both direct and indirect effects on tourism, including changes to the institutional arrangements of tourism policy and destination; the distribution and objectives of funding for tourism related development projects; policy goals, for example, the relative balance between public and private good objectives; the relationship between the public and the private sectors; the provision and management of tourism related resources; and the extent to which leisure and tourism is regarded as a social welfare good. There has long been a tradition of active labour market measures by the governments of the Nordic countries but, particularly in rural and peripheral areas, tourism has come to be regarded as a much more important component of the economic development policy mix than it was in the 1980s. This has resulted in tourism increasingly becoming a component of regional development programmes and as an instrument of economic policy although the means by which this is achieved has experienced change over time.

The chapter is divided into several section. First, it outlines some of the traditional roles of government with respect to tourism. However, it then goes on to note that there has been a shift in the notion of government – which is especially important in the Nordic context – so that the term governance is increasingly applied with respect to understand public sector involvement in tourism at various scales. Nevertheless, regardless of the level of governance some common planning themes still emerge with respect to tourism. These are then discussed with respect to the, sometimes conflicting, goals of planning as well as their implementation and evaluation. The final section examines some of the issues surrounding the politics of tourism and stress that tourism planning and policy should be understood as being part of broader political processes.

The Roles of Government in Tourism

The particular role that government plays in tourism is dependent on the jurisdiction in which it occurs, the economic significance of tourism in the economy of the jurisdiction, political culture and history, regulatory and legislative powers and inter-governmental relations. This means that different jurisdictions, that is, national or municipal governments, will have different powers and goals with respect to tourism. That noted, there are a number of common themes that can be identified in discussion of actual or potential government roles. The forerunner to the UNWTO, the International Union of Tourist Organisations (IUOTO) (1974) identified five areas of public sector involvement in tourism: coordination, planning, legislation and regulation, entrepreneur and stimulation. Two other roles can also be recognised, a social tourism role, and a broader role of public interest protection (Hall, 2008). These are detailed below.

Coordination

Coordination is needed in order to avoid duplication of resources both within and between government tourism agencies as well as between government and the private sector. Because tourism is a broad field that connects with the interests of a wide number of different government bodies and many private sector actors, coordination has come to be regarded as one of the most important roles that government has in tourism, not least because they also have financial resources to encourage coordination. In the Nordic context coordination, especially in central–local relations, relies upon a range of 'regulatory regimes' which vary greatly, from legal scrutiny of individual local decisions to joint normative and cultural influences that guide employees and decision-makers. Other methods of coordination may include incentives, contracts, grants, competition, citizen rights and growth agreements between central and regional governments (Baldersheim & Ståhlberg, 2003). Interestingly, they argue that in Denmark, Finland, Norway and Sweden there is a move to reduce rule-oriented control and instead shift towards 'softer' forms of control and supervision such as information dissemination, and joint analysis and evaluation, factors that will be returned to in further detail below. Some of the features

Table 3.1 Regulatory regimes with respect to coordination of central–local relations in Denmark, Finland, Norway and Sweden

Method	Denmark	Finland	Norway	Sweden
Constitutional protection of local self-government	Yes	Yes	No	Yes
Tax levels set by	Local freedom	Local freedom	Ceiling by government	Local freedom; voluntary moratorium in the 1990s
Volume of transfers fixed through	Consultations central–local	Consultations central–local	Consultations central–local	Consultations central–local
Initiative for review of legality	Committee initiative at the county level and complaints from concerned parties	Complaints from concerned parties	Prefect initiative and complaints from concerned parties	Complaints from concerned parties
Decision-making with respect to review of legality	Collegial	Administrative court	Bureaucratic	Administrative court
Budgetary review	No	No	No	Yes, automatic, merit review
Land use plan approval	Coordinated	Consultations	Fragmented, objections/veto	Consultations, objections
Economic development initiatives	Mostly up to regions and municipalities	Move toward regional agreements	Rudimentary central-regional coordination	Planning/semi-contractual agreements

Source: Derived from Baldersheim and Ståhlberg (2003).

of regulatory regimes relating to coordination and control of central–local relations are outlined in Table 3.1 with respect to the mainland Nordic countries.

Planning and strategy

Public planning for tourism occurs over different scales from the international to destinations or that of specific environments and sites as well as different policy focus (e.g. economic, social, environmental, regional, urban, rural, land use, marketing, labour force) and is an important aspect of government activity with respect to tourism development. However, the growth of stakeholders in tourism has meant that increasingly planning is being regarded as a form of direction or strategy setting rather than a centralised form of command and control.

Regulation

Because of its sovereign role with respect to legislation government has enormous regulatory powers, whether actual or potential, with respect to tourism. Importantly, many of the regulatory influences on tourism are not tourism specific. For example, general common regulatory measures including visa entry, industry and labour regulation, environmental protection and taxation policy will significantly influence tourism growth and development although they will usually have been designed without tourism being regarded as a major policy concern.

Entrepreneurial state

In most European countries government has long had an entrepreneurial role in tourism. Although entrepreneurship is often primarily seen as a private sector activity governments have historically acted to develop business opportunities where the private sector has not been willing to undertake investment risks. This means that national and local governments have often acted to own and operate businesses such as airlines, travel companies and accommodation along with what is widely regarded as more traditional ventures such as national parks, art galleries and museums as well as transport and visitor infrastructure. In the Nordic countries government ownership in transport infrastructure, such as railways and airlines, has long been significant for tourism although several of these ventures have now been corporatised or privatised. Nevertheless, the 'entrepreneurial state' is still strong at the regional level in terms of municipal support for tourism related ventures, including ownership of tourism infrastructure and attractions, such as museums, art galleries, sports stadia and event centres.

Stimulation

The stimulation role of government is similar to the entrepreneurial function but has a service focus. Three major sources of stimulation are identified: (1) the provision of financial incentives such direct subsidies to private firms; (2) government funding of research that is made available at nil or low cost to the private sector; and (3) financial support for, or the direct undertaking of, destination or attraction marketing and promotion. In addition to visitor promotion many governments also aim at encouraging investment in tourism attractions and facilities in their region. For example, the web site of the Tampere Region in Finland (see http://www.pirkanmaa.fi/english) is fairly typical of many of the Nordic regional councils. At the time of writing the website not only provided a presentation on 'Tampere – A City of Great Congresses' (produced by the Tampere

Convention Bureau) but also the business pages highlighted that 'The diversity of enterprise structure and the sheer number of innovative companies make the region an attractive prospect for the location of new enterprises'. The opening page clearly indicated in its 'welcome to the Tampere region' that Tampere was the place for business, lifestyle and pleasure: 'The region provides its residents with opportunities of work and prosperity, good traffic connections and attractive living conditions in an environment that combines the centrally located towns with clean nature and rural landscapes close at hand.' However, although the promotion of investment opportunities is an important stimulation role that government plays, tourism promotion with respect to encouraging visitation is usually recognised as a separate function.

Promotion

Destination marketing and promotion has long been one of the main activities of government in tourism whether at the national or local level. Many countries and regions directly cover the costs of destination or attraction marketing either in full or in part even though private tourism businesses also clearly benefit from such promotion. One of the primary justifications for this is that tourism promotion represents a 'public good' given the high degree of fragmentation that exists between the various elements of the tourism industry and the notion that the various economic and social benefits of tourism are accrued by society at large (Cooper & Hall, 2008). In some cases such destination promotion may be regarded as appropriate by the tourism industry because of the tax they pay as well as the general contribution that the tourism industry can make to the overall imaging or branding of a destination (see Chapter 2) thereby potentially leading to increased international investment, national pride, or benefits for other export sectors in terms of reinforcement of national and/or regional brands.

Social tourism

Social tourism refers to the provision of tourism opportunities for those that are economically or otherwise disadvantaged. Principles of social tourism has historically had significant influence within the Nordic welfare state systems which have had goals of universal accessibility to welfare benefits (Sandell & Sörlin, 2000). For example, Hjalager (2005) highlighted that in the Danish case welfare state interventions have had, and still have, substantial implications for tourism in terms of product and service innovations, citing the example of the Danish Labour Market Holiday Foundation which, since it was established, has set new standards in the Danish tourism sector. Perhaps of wider influence has been support for second home construction in the Nordic countries as a means of increasing household access to tourism and leisure participation (Müller, 2004a) (see Chapter 8 for further discussion on second home development in the Nordic countries).

The public interest

The final role of government in tourism policy is that with respect to the public interest. Notions of the state in Western countries have developed in such a way that their actions are meant to be undertaken with respect to notions of a broader public interest rather than to a narrow sectoral or other interest. This has been an especially important ideal in Nordic countries where concepts of universalism and the welfare state are often entwined with that of public interest (Kildal & Kuhnle, 2005). However, the concept of the public interest has undergone significant change in recent years as the role of the state at various levels has come to be debated along with the policy instruments by which policy goals are achieved (Hall, 2008). Central to this discussion has been the shift from concepts of government to that of governance. It is to this concept that we will now turn.

From Government to Governance

Governance has become an extremely significant concept in contemporary considerations of tourism planning and policy (Hall, 2008). The concept of 'governance' as opposed to 'government' reflects that rather than the implementation of government policies being undertaken solely by a government agency, there is now a greater emphasis on partnership by government departments and ministries with:

- other departments and ministries;
- state-owned or part-owned organisations that operate on a commercial basis (para-governmental organisations);
- the private sector; and
- non-governmental organisations.

Such partnerships have developed as a result of decentralisation of government, the growth of outsourcing, institutional restructuring, EU membership and the arrangements of EU-support programmes requiring partnerships, shift in the philosophies of the role of government including what may be described as 'a crisis of government' and recognition that some complex problems – such as environmental change – may require the involvement of as many stakeholders as possible into the development and implementation of policy solutions. In such situations the role of government agencies and departments is to steer organisational networks and partnerships in a required direction rather than adopt a more directive command approach (Hall, 2005a).

However, the concept of partnerships is not undisputed. It is argued that it sets the traditional democratic organisation of decision-making aside leaving greater influence to powerful stakeholders (Hall, 2008; Westholm, 1999). Also, available funding schemes from regional, national and supranational institutions govern the agenda of local planning committees because of the lack of other independent funding. In addition, Westholm (1999: 153) points out following disadvantages:

- partnerships are often exclusive and lack responsible representatives for the public;

- partnerships are conservative and inadequate in handling conflicts due to their aim of creating consensus;
- partnerships hide *de facto* privatisation; and
- partnerships are unclear regarding responsibilities and power.

Moreover, partnerships entail the risk that government in fact is withdrawing its involvement in tourism and transfers main responsibilities to private stakeholders. Müller (2006) demonstrates this for the case of literary tourism in Sunne in Sweden. Here the main touristic attractions, the literary sights related to the work of Nobel prize winner Selma Lagerlöf, are poorly developed because of the objections of nature-based tourism stakeholders dominating the local public privately-owned tourism promotion company.

Rhodes (1996, 1997) identified a number of characteristics of governance:

- interdependence between organisations;
- the concept of governance is broader than that of government and includes a role for non-state actors (i.e. private sector, non-government organisations and interest groups, including organisations in the voluntary sector);
- the boundaries between the public and other sectors is now far more opaque;
- the need to exchange resources and negotiate shared purposes between organisation members of networks leads to ongoing interaction between members of networks;
- the rules of interaction within a network is set by its members with trust being critical for the maintenance of network relationships;
- many networks have a significant degree of autonomy from government and are self-organising;
- although the state does not necessarily occupy a privileged position it can imperfectly and indirectly steer networks.

The characteristics of governance and the growing significance of partnership relations for state agencies has meant that both the development as well as the implementation of government policies has become extremely complex (Cooper & Hall, 2008; Hall, 2008). Such a situation is clearly of relevant to tourism not only because of the complexity of tourism related policies, that is, with respect to economic development or environmental change, but also because destination promotion and planning by its very nature necessitates the development of networks and partnerships (see Chapter 2). Furthermore, it is argued that because state authority, power and legitimacy have ceased to be bounded on a strict territorial basis which is the basis for sovereign governance, the governance of key cultural, economic and financial issues

> will be handled more and more by the transfer of goal-specific authority from states to regional or multilateral organizations and to local or subnational polities. Hence, the governance of key issue areas will be maintained not by territorial state-bounded authorities, as in the past, but rather by a network of flows of information, power and resources from the local to the regional and multilateral levels and the other way around. (Morales-Moreno, 2004: 108)

Such multilateral governance means that the traditional functions of the nation state are now being undertaken at other scales including the

- sub-national, that is, local and regional government, such as municipalities;
- supranational and international, that is, organisations with regulatory capacities that lie in the international sphere. In the Nordic context there are several good examples including the European Union, the Nordic Council of Ministers and the Barents Euro-Arctic Council, while at the international level a United Nations body such as the World Tourism Organization is also a significant influence in some policy areas.
- Trans-territorial, that is, state related organisations that have territorial boundaries that cross national borders. Such developments are relatively common in Europe with trans-territorial bodies having received considerable financial support from the EU. In the Nordic region the Øresund region of Denmark and Sweden is one of the best known along with tourism promotion and development at the Finnish-Swedish border (Ioannides *et al.*, 2006; Prokkola, 2007).

Interestingly, commentators such as Baldersheim and Ståhlberg (2003) have argued that the Nordic model of government and policy making and implementation is actually incomprehensible unless understood in terms of multi-level governance. This is because in the Nordic countries, the provision of public services, including many tourism resources, is usually channelled through local authorities with local governments accounting for approximately two-thirds of all public expenditure and employing around two-thirds of all public employees. The role of local authorities as well as the importance of multilayered governance in the Nordic region is further enhanced with consideration of the role of supranational bodies such as the EU, also including institutions that have their origins more within the activities of the Nordic countries, such as the Euro-Arctic Council and the Nordic Council. Collectively these bodies provide an overlapping web of institutional arrangements for the Nordic countries and their regions (see Table 3.2).

The activities of the various supranational bodies are extremely important for tourism because they influence tourism directly with respect to specific tourism related projects as well as indirectly via work in policy areas such as the environment, sustainable development, human mobility and trade in services. Again, much of their work also has a strong regional focus. For example, the Nordic Council of Ministers Tourism initiated an Ad Hoc Working Group (2003) that developed a *Road Map For Sustainable Tourism in the Nordic Countries* although its influence was rather limited. Nevertheless, the Nordic Council's activities with respect to environmental cooperation, knowledge sharing and encouragement of intra-Nordic travel and trade likely have a much larger influence on tourism.

The Arctic Council has noted the importance of tourism within the region in relation to both economic development and its relationship with environmental change including the development of a sustainable model for Arctic regional tourism and a programme to improve knowledge of the Arctic marine environment as well as cruise ship tourism. Similarly, tourism influenced by the trade and environment

Table 3.2 Supranational institutions of the Nordic region

Body	Established	Role	Denmark	Finland	Iceland	Norway	Sweden	Other
Arctic Council	1996	A high level intergovernmental forum to provide a means for promoting cooperation, coordination and interaction among the Arctic states, with the involvement of the Arctic Indigenous communities and other Arctic inhabitants on common Arctic issues, in particular issues of sustainable development and environmental protection in the Arctic.	Full (with the autonomous regions of Greenland and the Faroe Islands)	Full	Full	Full	Full	Canada, Russian Federation, United States of America; permanent participants are Aleut International Association (AIA), Arctic Athabaskan Council, Gwich'in Council International, ICC, Raipon and The Saami Council
Barents Euro-Arctic Council	1993	A forum for intergovernmental cooperation in the Barents Region that was established in order to 'provide impetus to existing cooperation and consider new initiatives and proposals' (Kirkenes declaration, 11 January 1993).	Full	Full	Full	Full	Full	Russian Federation

(Continued)

Table 3.2 Continued

Body	Established	Role	Denmark	Finland	Iceland	Norway	Sweden	Other
Council of the Baltic Sea States	1992	Is an overall political forum for regional intergovernmental cooperation. The members of the Council are the 11 states of the Baltic Sea region as well as the European Commission.	Full	Full	Full	Full	Full	Estonia, Germany, Latvia, Lithuania, Poland, Russian Federation and the European Commission
Nordic Council/ Nordic Council of Ministers	1953 (Nordic Council of Ministers established 1971)	The prime ministers assume overarching responsibility for Nordic cooperation within the framework of the Nordic Council of Ministers. The Council of Ministers is based on mutual understanding between peoples; it applies the consensus principle and is not a supra-state body. The Nordic Council consists of a number of councils on different policy areas.	Full (and autonomous regions of Greenland and the Faroe Islands)	Full (joined Nordic Council 1956) (and Åland)	Full	Full	Full	

activities of both the Barent Euro-Arctic Council and the Baltic Council. However, although important at a regional scale, the investment of the Nordic supranational institutions is tiny in comparison with the EU. For example, compared to the Nordic Council of Ministers' budget for transnational and cross-border cooperation activities, described as Interreg programmes, the annual economic contribution from the EU is 23 times greater. The EU Interreg III programme is made up of three strands:

- cross-border cooperation between adjacent regions;
- transnational cooperation between national, regional and local authorities;
- interregional cooperation.

Non-EU members Iceland, Norway, the Faroe Islands and Greenland are able to participate as close neighbours provided they fund their own participation. Norway has historically participated as much as possible, the Faroe Islands and Greenland participated for the first time in the Northern Periphery programme while Iceland joined Interreg programmes for the first time in 2002 in the case of the Northern Periphery programme (Hanell *et al.*, 2002: 70). Table 3.3 outlines some of the tourism related components of the Nordic Interreg IIIA and B programmes for 2000–2006. The EU territorial cooperation policies for the 2007–2013 period will also likely be of significance to tourism. As of 2006 the indicative financial allocation for the period from Nordic states was Denmark 103 million Euro, Finland 120 million Euro and Sweden 265 million Euro (Neubauer *et al.*, 2007).

However, one of the key lessons to be learned from examining the activities of national and supranational institutions is that there is a need to differentiate between tourism policy, that is, policies that have been developed specifically for the purpose of managing, regulating or promoting tourism, and policies that affect tourism, that is, public policies that either through their scope or because of their specific intent affect or influence tourism such as policies for the protection and maintenance of biodiversity which represents the key resource for ecotourism (Hall, 2006a). Indeed, it should be noted that within many significant policy areas such as regional development or environment conservation, policy is often comprehensive in design so that it deliberately combines a number of single industry or issue concerns within an integrated approach. Therefore, in many ways, outside of regional promotion and marketing the bulk of public policy that affects tourism is not usually categorised under tourism policy. For example, Hilding-Rydevik *et al.* (2005) noted that with respect to sustainable development at the regional level much work is related to tourism: 'Peripheral regions are believed to be afraid of being over-exposed and at the same time have a decrease in population. Sustainability is believed to be a strategic matter relating to tourism and settlement. Quality-tourism is a key-word and the nature and environment of a region must be attractive in order to tempt visitors and settlers, so the argument goes' (Hilding-Ryderik *et al.*, 2005: 51–52).

Given the significance of multi-level governance that is often implemented at a regional and local level it should be no surprise that much tourism-related public policy is related to regional policy and to the periphery in particular. As Hanell *et al.* (2002: 16) note, 'Due to a special situation in countries with vast territories and small populations, a special policy sector called "periphery policy" has arisen in Norden. Periphery policy is partly recognised as something different from the

Table 3.3 Nordic Interreg IIIA and B programmes 2000–2006

Programme region	Countries	Managing & paying authority	Total financing Euros	EU (ERDF) proportion	Tourism	Heritage	Image & marketing	Sustainable development/environment	Transport	Business connections/entrepreneurship
Kvarken–MittSkandia	Finland, Norway, Sweden	Västerbotten County Council, Sweden	56.7 m	23.9 m (5.7 m Nor.)						
Karelia	Finland, Russia (Republic of Karelia)	Regional Councils of Pohjois-Karjala, Kainuu and Pohjois-Pohjanmaa (rotated annually)	56.4 m	28.2 m					x	
South-East Finland	Finland, Russia	Regional Councils of Etelä-Savo, Etelä-Karjala and Kymenlaakso (rotated)	82.9 m	21.8 m				x	x	x
Southern Finland Coastal Zone	Finland, Estonia	Regional Council of Varsinais-Suomi, Finland	31.4 m	14.1 m	x			x	x	
Nord (North Calotte, Sami, Kolarctic)	Finland, Norway, Sweden, Russia	Regional Council of Lappi, Finland	118.5 m	47.2 m (28 m Nor.)	x				x	x

Nordic Tourism Governance and Planning Issues 65

Table 3.3 Continued

Programme region	Countries	Managing & paying authority	Total financing Euros	EU (ERDF) proportion	Tourism	Heritage	Image & marketing	Sustainable development/ environment	Transport	Business connections/ entrepreneurship
Øresund Region	Denmark, Sweden	Hovedstatens Udviklingsråd, Denmark (managing authority), NUTEK Sweden (paying authority)	61.6 m	30.8 m			x			
Sweden–Norway	Sweden, Norway	Jämtland County Administration Board, Sweden	111.2 m	32 m (48.8 Nor.)			x	x		x
Sweden/ Finland Islands	Sweden, Finland	Landskapsstyrelsen (Åland's government & administration board)	18.4 m	8.6			x	x		
Fyn-K.E.R.N. (Kiel, Plön, Schleswig-Holstein)	Denmark, Germany	Fyns Amt (Denmark)	21.3 m	9.9 m	x			x		x
Sønderjylland/ Schleswig	Denmark, Germany	Sønderjyllands Amt (Denmark)	28.3 m	13.8 m	x			x		x
Storstrøms Amt – Kreis Ostholstein/ Hansestadt Lübeck	Denmark, Germany	Storstrøms Amt (Denmark)	19.1 m	9.6 m	x					

(Continued)

66 Nordic Tourism

Table 3.3 Continued

Programme region	Countries	Managing & paying authority	Total financing Euros	EU (ERDF) proportion	Tourism	Heritage	Image & marketing	Sustainable development/ environment	Transport	Business connections/ entrepreneurship
Baltic Sea Region	Denmark, Finland, Sweden, Germany, Estonia, Latvia, Lithuania, Norway, Poland, Russia	Rostock, Germany, and Karlskrona, Sweden (Common Secretariat). Investitionsbank Schleswig-Holstein, Rostock, Germany (Joint Financial Body)	194.5 m	97.1 m (12 m Nor.)		x		x		
Northern Periphery	Finland, Sweden, Scotland, and Faroe Islands, Norway, Iceland, Russia	Vasterbotten County Administration (Sweden) (managing and paying authority); Faroese Representation in Denmark (host of Joint Programme Secretariat	47.2 m	21.3 m				x	x	x
North Sea	Denmark, Belgium, Germany, Netherlands, Sweden, UK, Norway	Danodh National Agency for Enterprise and Housing (managing authority). County of Viborg (paying authority and host of secretariat)	264.5 m	122.8 m (19 m Nor.)		x		x		

other aspects of regional policy... much of the regional policy attention is focussed on periphery policy problems, and it is here that Norden except Denmark deviates from EU countries.'

Strengthening national cohesion between regions is a significant policy concern in all regions; however, increasingly it is regarded more with respect to economic development and investment growth across all regions rather than welfare and redistribution in favour of the weaker regions. Nevertheless, the means by which this can be done is limited by EU or EEA competition rules and other regulations. According to these rules, direct public support to individual businesses is in principle prohibited. However, there are exceptions to the rules for regions with low levels of economic development. In Nordic regions that are considered weak from an EU or EEA perspective, or have a very low population density, a maximum of 30–35% net investment support is allowed for small and medium-sized businesses. For regions that are considered weak from a national perspective a maximum of 20–25% net investment support is allowed (the aid ceilings are 5–10% points lower for companies with more than 250 employees). Outside the support areas, up to 10% investment aid is allowed for SMEs independent of location. The population in Nordic support areas that received investment aid in 2000–2006 is substantial: Denmark (17.6%); Finland (41%); Iceland (38.1%); Norway (25.3%); and Sweden (15.9%) (Hanell et al., 2002: 31–32) much of which was directed into tourism related businesses.

Interestingly, differences do appear between industrial and regional policy with the former having a narrower realm of policy concern which also tends to be not as integrated in scope with broader social and environmental concerns. For example, with respect to Iceland which has a substantial focus on tourism, 'The central objective[s] of Icelandic regional policy are to ensure settlement that will utilise the natural resources of the country, to strengthen settlement in areas that offer potential for a varied and profitable economy and can offer modern services, and to slow the migration stream from the outlying areas to the capital region' (Hanell et al., 2002: 39).

The 2002–2005 Icelandic Strategic regional development policy and plan had five main objectives:

- to alleviate the differences in living standards and economic opportunities between inhabitants of different regions and to generate for people living in the peripheral regions as gainful living conditions as possible;
- to assist communities in the peripheral regions to adjust to rapid socio-economic changed by strengthening the local authorities and supporting economic development initiatives, education, services and infrastructure;
- to improve living conditions in peripheral regions by strengthening the largest communities that are the most attractive for people to live in;
- to facilitate the development of local cultures;
- to support diversified industrial and economic structures, to alleviate differences in business conditions between regions and to assist companies in the peripheral regions to make the best use of their business opportunities

with due regard to the objectives of sustainable development and responsible environmental policies (in Hanell *et al.*, 2002: 39).

This can be contrasted with the tourism industry policy of the Faroe Islands Ministry of Trade and Industry (2001) which is the policy directive that guides the relevant daily work of the Faroese Government and the Faroe Islands Tourist Board.

> The overall industrial policy of the Faroe Islands envisions a self-supporting economy, built upon many different profitable sectors of commercial endeavour, which gives rise to progress and well-being within a framework of free international competition, and which endorses both vigorous development and the protection of the environment as the country's highest priorities. (Ministry of Trade and Industry, 2001: 5)

> The role of government vis-à-vis the tourism industry, as stipulated in the overarching industrial policy, is to create a stable framework within which the industry can operate. (Ministry of Trade and Industry, 2001: 6)

Similarly, the Finnish Ministry of Trade and Industry commented that their draft national tourism strategy launched in May 2005 had

> ... the primary goal of improving appreciation of tourism in Finland and enhancing the sector's competitiveness and profitability, based on strategic choices and measures. The strategy aims to provide Finnish tourism with a shared direction in which the sector's business will be developed. (Ministry of Trade and Industry, 2006: 24)

The tensions between regional and industry perspectives in tourism policy are reflective of broader policy and interest concerns over the role of tourism development, particularly with respect to goals of sustainability, which is a tourism policy issue of considerable significance in the Nordic context that cuts across a range of tourism activities (e.g. Clement, 2004; Eligh *et al.*, 2002; Halme, 2001; Kaltenborn *et al.*, 2001; Lee, 2001; Müller & Jansson, 2007a, 2007b; Saarinen, 2003, 2005; Saarinen & Hall, 2004). Although some of these tensions occur at the national level between different national government agencies usually responsible for environment and trade and industry, the development of new regional development governance structures as a result of some of the trends noted at the beginning of the chapter, means that there are greater central–regional partnerships in policy-making, with the local and regional levels increasingly taking over responsibility for implementation (Baldersheim & Ståhlberg, 2003). In such a context there is therefore a strong need to not only understand the governance of tourism policy and planning but what the process should look like.

Tourism Planning

When we discuss concepts of tourism planning in this book we are usually referring to planning exercises that are conducted by public bodies rather than by individual tourist firms. Such firms ideally do also undertake planning exercises but their goals are usually geared towards profit and/or return on investment for the firm whereas public planning is looking for a return, economic or otherwise, as a public good. Planning for tourism also occurs in a number of different:

- *forms* (e.g. development, infrastructure, land and resource use, organisation, human resource, promotion and marketing);
- *structures* (e.g. different government, quasi-government and non-government organisations);
- *scales of governance* (international, transnational, national, regional, local, site, sectoral and personal);
- *spatial scales* (international, supranational, national, regional, local and site);
- *temporal (time) scales* (for measuring change, development, implementation, evaluation and satisfactory fulfilment of planning objectives) (Hall, 2008: 14).

Furthermore, planning within public agencies is rarely exclusively devoted to tourism per se, for example, although extremely important for tourism, national park agencies have a range of other responsibilities with respect to biodiversity and environmental conservation as well as focusing on attracting and managing visitors. Instead, planning for tourism tends to be an amalgam of economic, social, political and environmental considerations which reflect the diversity of the factors which influence tourism development (Gunn with Var, 2002; Hall, 2008). Perhaps because it is such a combination of considerations, as illustrated by the potential tensions between economic and sustainable development noted above, there are a range of broad approaches or traditions in public tourism planning that each have their own focus (Hall, 2008):

(1) 'Boosterism', the notion that tourism is inherently good.
(2) An economic, industry-oriented approach that emphasises the economic dimensions of tourism development and which in recent years has focused on urban and regional economic development and competitiveness.
(3) A physical/spatial approach that emphasises the environment as a tourism resource and the need for its conservation as well as the ecological basis for development. The approach also has strong connections with resource management and integrated regional and spatial planning traditions.
(4) A community-oriented approach that emphasises the role the members of the destination play in the tourism experience and which is increasingly focusing on the role the community plays in capacity building as well as the social dimensions of tourism.
(5) A sustainable tourism approach that seeks to develop an integrated development strategy that balances economic, social and environmental considerations along with broader ethical, quality of life and welfare concerns.

Within the Nordic countries all of these traditions exist; however, the multilayered governance architecture of the region and the political culture that places substantial emphasis on partnership and public participation means that boosterism is not a planning approach that is consistently expressed by local agencies over time. Instead, broader concerns over economic, environmental and social dimensions and trade-offs tend to moderate the planning process. Yet regardless of the focus of tourism planning a common element in contemporary public planning is attention to strategy including the identification and articulation of different planning and policy choices and their effects. Peter Hall describes the emphasis of this 'new planning' in respect of

> tracing the possible consequences of alternative policies, only then evaluating them against the objectives in order to choose a preferred course of action; and, it should be emphasized, this process would continually be repeated as the monitoring process threw up divergences between the planner's intentions and the actual state of the system. (P. Hall, 1992: 229)

To a great extent such a rational approach dominates the logic behind much tourism planning in the Nordic countries and follows the steps in the planning process identified by Anderson (1995):

(1) Identify issues and options.
(2) State goals, objectives, priorities.
(3) Collect and interpret data.
(4) Prepare plans.
(5) Draft programmes for implementing the plan.
(6) Evaluate potential impacts of plans and implementing programmes.
(7) Review and adopt plans.
(8) Review and adopt plan-implementing programmes.
(9) Administer implementing programmes, monitor their impacts.

In such a process there is as much emphasis on evaluation as there is in the formulation of goals and the involvement of stakeholders, with the results of such evaluations often being problematic in terms of the value of the process or project. For example, the report of the Nordic Council of Ministers Tourism Ad Hoc Working Group on sustainable tourism acknowledged that there would be a substantial gap between policy-making and implementation:

> Several problems were acknowledged throughout the process of Action Plan development. These were related to: (a) securing the representation of the relevant stakeholders in the workshop; (b) securing the participation of representatives from all Nordic countries; (c) securing the dissemination of the proposed strategy and action plan to tourism SMEs in the region. (Nordic Council of Ministers Tourism Ad Hoc Working Group, 2003: 12)

Interestingly, even despite these problems the report was still produced. A more thorough evaluation is illustrated in Case 3.1 that examines an EU project in Denmark.

Cases and issues 3.1: Evaluating an EU-project on improving sustainable competences in micro, small and medium sized Danish tourism enterprises

PETER KVISTGAARD

From the mid-1980s to the present, the tourism sector in Denmark – as well as in other parts of the world (Vernon *et al.*, 2005) – has seen an increasing number of short and long-term development projects in the form of public–private partnerships. In Northern Jutland, the northernmost region of Denmark, 143 tourism development projects have received funding from the European Regional Development Funds as well as national funding in partnership with private investors up until the writing of this case. Only a few of these have been evaluated, be it formally or informally. An exception to the rule seems to be a regional tourism action programme that was active from 1994 to 1997 with funding from the County of Northern Jutland and evaluated in 1997 (Dansk Turismeudvikling I/S & Strange, 1997). However, the evaluation was rather limited, only stating how many projects had received funding, for what purpose and with what estimated effect. The evaluation did not:

- specify the point of departure;
- state the overall purpose of the evaluation;
- state evaluation criteria and grounds for choosing the criteria;
- specify how the evaluation was constructed and how the data should be interpreted and by whom; and
- state the logic behind the evaluation.

Furthermore, it was unclear whether or not the evaluation was carried out as an internal or external evaluation as evaluator was both the secretary and also administrator of the action programme.

The main body of the tourism development projects in Northern Jutland especially and in Denmark generally has never been evaluated formally by external evaluators; nor, so it seems, have they been evaluated internally by the organisations responsible for implementing the projects. Based on this astonishing lack of evaluation, it is in general very difficult to conclude with any kind of certainty whether or not the projects have actually had the effects that were promised and wished for. So far, there exists relatively little hard and/or soft 'evidence' of the positive or negative consequences of using public money for tourism development projects in Denmark. This, to many students of public policy, constitutes a serious problem. The aim of this contribution is to pass on thoughts and lessons from a formative impact evaluation of a specific tourism development project in the region of Northern Jutland, Denmark, one exception to the apparent rule.

The case study: Improving sustainable competences in Northern Jutland

The scene. Northern Jutland is the smallest region in Denmark with a population of 495,000 people, covering 6173 km². Sixty percent of all firms have less than five employees, and approximately 8000 people are employed full time in tourism (2003). There is one big city, Aalborg, with 163,228 inhabitants. In 2003, Northern Jutland had the highest number of overnight stays in Denmark amounting to close to 7 million. The main type of accommodation is self-catering accommodation such as summer-houses, camping, holiday centres, and nature and beaches attract tourists mainly from Denmark and Germany.

Actors. The regional tourism development agency – Midt Nord Turisme (MNT), 25 micro, small and medium sized tourism enterprises, and an external university based evaluator.

Purpose. A halfway impact evaluation in order to shed light on two issues: (1) A status on the project concerning the participating enterprises' work with sustainable competences – has the project so far had impacts on the participating enterprises' environmental attitudes, knowledge and behaviour? (2) Input for the next steps in the project – how can the project be improved, and can the resources be used better in order to meet the objective of the project?

Resources. Enough to carry out the necessary work to shed light on the above issues.

Time frames. The project ran from 2004 to 2007. This halfway evaluation ran from September to December 2005.

Objective. The objective of the project is to develop sustainable competences in as many enterprises as possible in order to have at least 20 enterprises live up to the criteria of the European Union classification programme for sustainable tourism enterprises, the EU Flower. This EU-certification programme is made to ensure that certain minimum standards are met concerning the use of energy, water, handling of waste and the use of cleaning materials.

The evaluation process. Initially, the evaluator tried to find out about the overall and specific purpose of the evaluation: what type of evaluation was wanted and why; the significance of, and to, third parties; and time frame. It was decided to use an actor model as the evaluator quickly realised that focus should be on the participating enterprises and their experience with, and results from, the process so far. An actor model focuses particularly on user evaluation and establishes a forum for the participating enterprises to freely express pros and cons of the project. Table 3.4 shows what specific considerations preceded undertaking the evaluation.

Point of departure. A qualitative method was used in order to establish the forum. Three focus group interviews with 17 persons were conducted and tape-recorded with interviewees guaranteed anonymity. This makes up a 68% representation of the total number of people from the enterprises involved in the project. In order to make sure that the interviews revolved around the

Table 3.4 Evaluation model for evaluating sustainable competences focusing on process and users

Dimension	Specific considerations
Point of departure Participatory evaluation model	The aim of the evaluation is to develop new knowledge that can lead to problem solving actions
	The role of the evaluation is to empower the actors to solve their own problems
	The analytical focus is on the actors' perceptions of problems and solutions
	The method is self evaluation, dialogue, confrontation and negotiation
	The role of the evaluator is to act as a process consultant
Actor model User evaluation is the centre of attention	Problem: are the users satisfied and what can be learned from the process so far in order to make improvements?
	Criteria: the users formulate the criteria
Evaluation logic Logic based on dialogue	The evaluation is carried out as a tool for mutual understanding of the benefits of the project
	The evaluation is constructed as an arena for dialogue
	The evaluation is an opportunity for discussion about meaning, values and challenges

Source: Inspired by Foss Hansen (2003).

same issues, an interview guide was developed. This guide focused on connections between actors' attitudes, motivation, knowledge, habits and actual behaviour concerning sustainability. The evaluator looked for patterns of both positive and negative meaning, patterns of meaning and divergent meaning.

Results of the evaluation. The interviews showed a remarkable consensus concerning a number of issues.

(1) All interviewees stated that their improved competences concerning sustainability to a very high degree are the result of the sustainability coach's work and commitment. This coach had been brought in with the special task of aiding the enterprises. The enterprises state that they do not have the necessary resources alone to take on the comprehensive task of changing procedures, filling in papers and following up on day-to-day business without the help of the coach. In other words, each enterprise has so few resources that it is paramount for them to have someone to help and guide them. This structural situation was understood not only by the coach but more importantly also by Midt-Nord Turisme (MNT)

who initially allocated the necessary funds for the coach. This makes the project stand out.

(2) Almost all the enterprises stated that they have experienced an increase in their motivation for working with sustainability, mainly due to the cooperation with the highly committed coach. This increase in motivation has affected the behaviour of the enterprises in a positive way reducing expenses for waste disposal, electricity, and water, in some cases dramatically.

(3) All the enterprises mentioned that the biggest weakness of the project so far is their own willingness to put in the necessary man-hours, especially regarding the 'overwhelming piles of paper' that each enterprise must fill out in the EU-Flower certification process. In this connection it must be mentioned that an established chat room on the internet for experience exchange has not been used satisfactorily, in the view of both the enterprises themselves and the MNT.

(4) The enterprises state that a central project improvement would be making customers aware that these enterprises invest time and money in sustainability. In other words, the enterprises would appreciate a stronger focus on the marketing aspects of working with sustainability.

(5) The enterprises state that there is a direct correlation between their participation in the project and the environmental improvements that they have carried out in the first half of the project period.

(6) The enterprises state, with one clear exception, that they cannot establish a clear correlation between their environmental improvements and improved sales. In short: at present customers do not choose these enterprises because of their work with sustainability. This is perhaps the biggest problem for the project as such, as it was one of the major arguments in the original application for EU and regional funding that the project would increase the number of overnight stays in hotels and trailer parks and the number of visitors in museums and restaurants.

(7) By the end of 2006, 10 tourism enterprises had received the EU-Flower certification.

The evaluation was all in all very positive, and the project has been expanded so that more enterprises can join the project, primarily because of the project itself, but also because of the positive evaluation (Kvistgaard, 2005).

Discussion. An evaluation like the above example can serve a number of purposes of which only a few are mentioned here (see Hall & Jenkins, 1995).

(1) It can serve practical and functional purposes. In this case it clearly serves the purpose that the responsible organisation learns about the effects of the project through the voiced input from the participating enterprises. Thereby the responsible organisation can make the necessary changes, perhaps emphasise other aspects than those foreseen.

(2) It can serve the purpose of making the participating enterprises more aware of the value of participating in the project. Maybe the consequences become clearer, maybe it becomes clearer that there are a number of common issues between the participating enterprises that they can work together to solve.
(3) It can also serve political purposes as an evaluation can make the sponsors of the project (in this case the EU and the region of Northern Jutland) aware of the positive impacts of the project.

Often, so it would seem, there is a wish by the principal for a sort of 'cause and effect change over time' evaluation. However, this wish is often formulated too late in the process. In the case study it was too late to measure concrete effects concerning the environmentally friendly behaviour of the participating enterprises, as there was no starting point. Measuring any kind of difference must necessarily involve a starting point. If x was 0 in 2004, how high/low is x in 2005 (hard evidence)? Or have the participating enterprises expressed any changes in terms of attitudes and behaviour in the course of the project (soft evidence). In this study it was impossible to do this.

One might speculate as to why so few tourism policy projects in Denmark generally and in Northern Jutland especially are evaluated formally and externally. One reason could be that many projects are too small to include an evaluation process. There is simply not enough money in the project. Another reason could be that there is no tradition for evaluating tourism projects. In other sectors such as health and education, evaluations are institutionalised activities included in project budgets from day one. A third reason could of course be that in many projects, sponsors (for instance, the EU) do not require that an evaluation be carried out. Consequently, one can only speculate about the effects of the projects as the responsible organisations, understandably, tend to present favourable results in order for the organisation to be able to apply for funding for other projects later on.

All in all, both from an academic and a societal standpoint it is objectionable that tourism development projects apparently do not get evaluated that often. Without some sort of evaluation it is very hard to say whether or not each project results in consequences that tourism as such and society in general would deem satisfactory and worthwhile.

Guldbrandsen's (1997) evaluation of tourism projects in Greenland also highlighted a number of the issues noted by Kvistgaard in Case 3.1 including the value of undertaking such research as well as a conclusion that what was needed was 'specific and integrated regional strategies that are community and marketing oriented rather than community based' (Guldbrandsen, 1997: 130). Nevertheless, it is interesting that the study concluded with two questions that apply equally to many other tourism development projects: 'What scale of tourism development should a region choose to follow? And what scale provides a positive impact on the local community?' (Guldbrandsen, 1997: 130).

The issue with such questions is that they highlight the political nature of the tourism planning and policy process. It is to this that the final section of the chapter will briefly turn.

The Limits of Rational Planning: The Importance of Politics in Tourism Policy

One of the great struggles that tourism graduates have is to marry idealised models of management and planning in texts and the realities of planning in 'the real world'. The problem really is one of representing planning processes and how 'messy' they can be and, clearly, the larger the problem to be solved, and the area in which it is located, the more stakeholders and therefore, solutions, there may be to fix it. By their very nature many tourism planning issues at the regional level are quite complex as a balance between economic and environmental and social benefits and impacts are sought, as well as the public–private mix of benefits and costs of any development. One of the difficulties of such situations is that planning and policy are inherently political, not necessarily in the sense of elections and political parties as significant as they can be, but political in the sense that policy and planning choices are also choices about values, with different stakeholders and interests holding different values with respect to policies, plans, decision-making and implementation.

At all stages of the planning and policy process interest groups (e.g. tourism industry associations and networks, chambers of commerce, conservation groups, community groups), significant individuals (e.g. local government representatives, business and community leaders), members of the bureaucracy (e.g. employees within tourism organisations or economic development agencies) and others (e.g. consultants, academics, affected and interested members of the general public) all influence and perceive tourism policies and their potential affects in significant and often substantially different ways. Tourism policy and planning are therefore consequences of the political environment, values and ideologies, the distribution of power, institutional frameworks and decision-making processes (Hall, 2008). Indeed, Arnesen (2007) discusses the extremely important question of who has a reasonable claim to be involved in public policy processes leading to a decision on how a particular resource should be used? On what claims do rights to be involved actually rest? As he notes with respect to national parks in Norway:

> ... 'involvement' it is today something of a buzzword in natural resource management. Rhetorical disputes between various conflicting groups in outfield land use on 'who owns nature' are regularly popping up. It is seen in disputes between urban based conservation movements (e.g. in disputes on conservation of large carnivores) and rural based farmer groups. So, to take a rhetorical

stance: One the one hand we have ecological knowledge that says that there really does not exist patches of land or species that are self-contained. All connects to all, so we all should have a say in land use. Ecology, we could say, have de facto defined nature in toto as a common. One the other hand, someone does own patches of land, and are de jure in their right to decide more than those not owning? In outfield land issues, what should be given priority? Who should be involved, how and to what extent? (Arnesen, 2007: 126)

Some of these issues, particularly with respect to encouraging collaboration between different interests are indicated in Case 3.2 on area protection in Norway.

Cases and issues 3.2: Protection and equity?: Local and indigenous encounters with the grand scheme of area protection in Norway

TOR ARNESEN & JAN ÅGE RISETH

The grand scheme of nature protection area enlargement

The plan for a significant increase of national parks and large nature protection areas was accepted by the Norwegian Parliament (Stortinget) in 1993. Most land is crown land and commons – land with an intricate system and time-honoured traditions of split property rights – some 'forged' into formal agreements and juridical arrangements others 'floating' as locally accepted and transferred 'ways of doing things' and practices of tolerated use. Together they reflect local and indigenous communities' life forms.

The grand scheme assumes a transfer of jurisdiction from being under local domain control to central domain of formalised control. This transfer of jurisdiction is not efficient in translating 'floating and informal' systems of rights and use, resulting in local and indigenous powerlessness within their traditional territories (Riseth, 2005). Increased attention is required for better understanding the social processes involved. It is not uncommon, we claim, for the nature protection bureaucracy to underestimate democratic aspects of domain control transfers.

Involvement processes are foregoing decisions on domain transfer in nature protection areas. In order to reduce conflicts, stakeholders were invited to join in on a consensus within a given framing of power and authority: shifting management regime powers from local to central authorities. However, this has partly resulted in organised resistance (Table 3.5). Why do invitations to join consensus feed conflict rather than dampen it? Are insistences on more local community domain control a reactionary threat to nature conservation? The answers are respectively: 'Because it's unfair', and 'no'.

Figure 3.1 illustrates a remarkable situation between nature types threatened and thus potentially in need of protection, and areas actually protected. The gist of the illustration is a lapse of reason in answering why so little focus

Table 3.5 National park processes in mid- and north Norway

National park	Year established	Process, outcome and management
Forrolhogna (FH)	2001	Local Management Trial. No change in perspective and praxis.
Junkerdalen (JD)	2004	Partnership process before decision promoted compromise. Continued urban recreation pressure. Indigenous impotency.
Blåfjella-Skjækerfjella (BS)	2004	Strong local opposition and distrust during process. Local Management Trial. Management Plan: Reinforced interest contradiction between indigenous and locals. Indigenous boycott of management board.
Tysfjord-Hellemo (TH)	Break in process	Strong Indigenous opposition stopped process. Negotiation of a new basis required.
Øvre Dividal-Extension (ØD)	Proposal ready	Strong local distrust. Locals supported by local and regional political bodies. Bureaucracy overruled locals using biodiversity arguments.

on protecting coastal and off-coast areas with high biodiversity value and high development pressure? Why – as the present grand scheme aims at – protect more mountain areas with generally significantly less biodiversity and low development pressure? To answer this one has to acknowledge that the rationale for national parks – often called 'the jewels' of area protection – is not primarily based on ecology or biology, but on symbolism: nature per se in National Parks is of course neither 'national' nor 'parks'. The name denotes a perception by an elite and a norm attributed to the area in a political 'ceremony' celebrating the nation states' most emblematic sceneries. Thus, national park issues needs to be seen in a national state perspective, for example, establishing border delimitation between Norway and Sweden in 1751 was crucial in the building of the nation state and the enclosure of all in a nation state ideology, including the formal colonisation of Sapmi (Sámiland). National Park policy falls in line with this long nation building historical tradition, now repeated on the arena of emblematic nature scenes.

Though there are biological and socioeconomic arguments in the 'ceremony' – notably allowing for some area-extensive species and controlling development of physical infrastructure – it is a very shaky argument indeed to hold that these alone call for a domain transfer to central control. This is of course seen by rural societies experiencing domain transfer resulting in

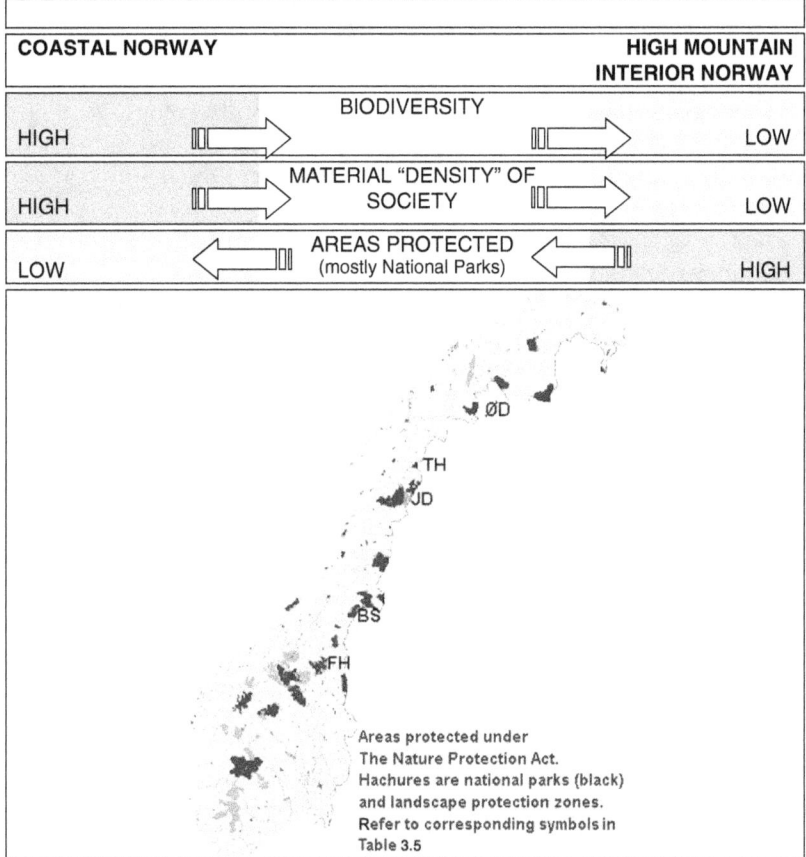

Figure 3.1 The Norwegian paradox of protection need and policy

impotence were traditionally empowered and it is seen as an injustice the present consensus-invitation cannot undo. Add to this the following question: If locals demand more hands-on control than present legislation allows for in running protected areas, is this a threat to a sound ecosystem management? Does a sound management inevitably call for the strong central authority? Not at all.

Ecosystem management vs. the old 'red tape games'?

Ecosystem management is concerned with how discretion should be designed in order to come to grips with ecological challenges. A shift towards ecosystem management acknowledges that management regimes have to adapt in iterative cycles to situations and demands (Grumbine, 1994). This requires shifts in area protection management styles found operative in Norway today, from a dominant hierarchical ideology to a collaborative approach.

Danter *et al.* (2000) focus on different categories of changes required in ecosystem management approaches:

- Professional emphasis: Inclusion of biology and ecology to deal with non-linear changes.
- Interdisciplinary cooperation: Change from the 'this-is-my-turf-expert' approach to 'reasoning-in-networks'.
- The role of decision-making should change from risk-aversion to adaptive management. Adaptive management ranges from provisional decision-making, management experiments, local solutions and subsidiarity ideals, to risk taking and organisational learning. Changes of this category require a move from present a top-down control to field-level empowerment, collaboration, reasoning-in-networks and learning organisations.
- The impact of ecosystem management theory on present centralised land conservation management regimes is of the outmost importance and not sufficiently appreciated in current biocratic and centralistic bent power structures. The extent, the heuristic and political function of local involvement called for in ecosystem management theory is not adequately recognised. From an ecosystem management perspective locals should share – even play a major role – authority in nature conservation management regimes.

Conclusion
We argue for a redesign of local involvement in collaborative regimes, where running control is kept locally, and defining discretion and monitoring control involves central authorities. Local and indigenous people should administer a substantial authority within an ecological management approach to aim for a sustainable use of areas protected at some defined level and for some defined purpose. Today national parks in Norway are mainly a 'playground' for 'the old red tape games' in central government. It should be transformed to a sustainability concern for those living in and off the land. What this requires is amending the Nature Conservation Law, and a reform of management modus operandi to incorporate ecosystem management insights.
See http://www.utmark.org (in Norwegian).

Given such a situation one of the most significant tasks in tourism planning is trying to find a way to identify and, where possible, find common ground between the various value positions. Such a situation is not easy and inevitably some stakeholders may not be happy with outcomes. However, possibly one of the most significant dimensions of such processes within the Nordic countries are that decision-making processes are relatively transparent and there is a very low level of corruption. Furthermore, the institutional arrangements have considerable checks and balances with respect to government actions. One implication may be the relative inaction, or slowness, of the planning process in certain circumstances while attempts at consensus construction occur. Nevertheless, the trade-off between quality of process and quality of outcome is an important consideration in tourism decision-making especially in situations where it may be impossible or very difficult to 'undo' decisions or their impacts once they have been made.

Summary

This chapter has emphasised that the Nordic countries have a particular institutional context in which policy making occurs that also affects tourism planning and policy. Key themes include notions of universal access, welfare state, close partnership between central and regional governments, and the much stronger role of regional and local government in implementation than in many other jurisdictions. In addition, a key theme is the idea of multilayered governance, a concept which has grown in importance since the early 1990s because of the increased role of the EU in Nordic public affairs although, as the chapter notes, there are a range of more Nordic oriented supranational bodies such as the Nordic Council, Arctic Council and the Barents Euro-Arctic Council. All of these bodies, together with national and regional institutions are influencing tourism planning and development in the Nordic region, and also constitute a significant long-term area of research.

Within this context a number of roles of government are identified as are different traditions of tourism planning. A key point that not all agencies have the same perspective on tourism development and that tensions often exist between economic perspectives and those with broader social and environmental concerns, particularly with respect to seeking to enhance more sustainable forms of tourism.

Self-review questions

(1) What are the roles of government in tourism?
(2) What is multilevel governance?
(3) What are the main traditions of tourism planning?

Essay questions

(1) What is the value of undertaking evaluations of tourism plans and policy?
(2) Who should be involved, how and to what extent in tourism planning decisions in relation to a national park?
(3) What roles do values play in tourism planning?

Key readings

On tourism planning and policy in general see Hall (2008) or Gunn with Var (2002); an older though well-cited work is Hall and Jenkins (1995). Rossi *et al.* (1999) provide a useful guide to evaluation. For tourism policies and plans in the Nordic region the best source is government and institutional websites, at the regional and local level. Although not tourism specific, Neubauer *et al.* (2007) provide one of the best overviews of regional development strategies in general in the Nordic region while the Nordregio website is a rich source of information on spatial planning and regional development in the Nordic context, including changes in local and regional government jurisdictions.

In addition to the websites identified in Chapter 1 see:
Arctic Council: http://www.arctic-council.org/
Barents Euro-Arctic Council: http://www.beac.st/
Council of the Baltic Sea States: http://www.cbss.st/
Nordregio: http://www.nordregio.se/

Chapter 4
Urban Tourism

Learning Objectives

After reading this chapter, you should be able to:

- Understand the concept of the urban tourism in a Nordic context.
- Understand key concepts of place marketing and urban imaging.
- Identify the key functional elements in the tourist city.

Introduction

Although urban centres have long attracted visitors as centres of commerce, trade and transport, it is only since the late 1970s that cities have consciously sought to develop, image and promote themselves in order to increase the influx of tourists. Before that time tourism was generally regarded as something that tended to occur in rural or wilderness areas. However, following the de-industrialisation of many industrial and waterfront areas in the 1970s and 1980s as a result of technological innovation (especially communication, container shipping and mechanisation of heavy industry), and free trade, tourism has been perceived as a mechanism to regenerate urban areas through the creation of urban leisure and tourism space. As a result of these changes cities may be described not only as places of industrial production but also 'increasingly as centres of control, interaction, creativity and enjoyment' (Burtenshaw *et al.*, 1991: 70). This process has been almost universal in the industrialised nations (e.g. Garcia, 2004; Page & Hall, 2003), and is particularly apparent in many of the larger Nordic cities, and even many of the smaller ones. For example, consider the extent to which the former industrial waterfronts of cities such as Copenhagen, Göteborg, Helsingborg, Helsinki, Oslo, Oulu and Stockholm have been turned into sites of tourist attractions, retail and leisure (Plates 4.1–4.5). Indeed, such a situation even led Harvey (1988, cited in Urry, 1990: 128) to ask: 'How many museums, cultural centres, convention and exhibition halls, hotels, marinas, shopping malls, waterfront developments can we stand?'

84 Nordic Tourism

Plate 4.1 Waterfront development in Helsingborg, Sweden

Plate 4.2 Waterfront development in Copenhagen, Denmark

This chapter examines urban tourism in the Nordic context. Although the Nordic image tends to be one grounded in nature the reality is that the Nordic countries are actually highly urbanised, with the majority of people having their main place of residence in cities and towns rather than in the forests or fells. Images grounded in lakes, that is, Finland; fjords, Norway; snow, Lapland, are powerful marketing

Plate 4.3 Waterfront development in Helsinki, Finland

Plate 4.4 Historic waterfront in Oulu, Finland

images but they often fail to convey how significant urban centres are for tourism both as attractions and activity spaces in their own right, as well as gateways for those that are seeking more natural experiences. The chapter is divided into three main sections. The first section briefly defines the concept of urban tourism and some of its constituent elements. The second section looks at the ways in which urban centres are loci of place competition and place reimaging. The final section

Plate 4.5 Historic waterfront in Stockholm, Sweden

utilises a modification of Burtenshaw *et al.*'s (1991) identification of a number of overlapping functional areas in the tourist city to discuss the way in which tourism embedded in urban form and function.

Urban Tourism

At its simplest urban tourism can be defined as tourism in urban areas. However, such a self-evident statement fails to convey the complexity of urban tourism as a phenomenon and the issues that arise in its analysis. As Page and Hall (2003) noted in their introduction to cities as places for tourism, urban tourism can be construed in a number of different yet related ways:

- in relation to processes of urbanisation in which tourism is both affected by and contributes to, including urban regeneration and place competition;
- as tourism within an urban environment in which the built character of the urban environment leads to tourism activities, forms and processes that are qualitatively and quantitatively different from non-urban areas, including service encounters and experiences at the interface of particular forms of consumption and production;
- as destinations in their own right;
- as gateways for tourist entry to a country or region.

In addition, there are some very specialised types of urban tourism relating to capital city tourism (Hall, 2002), particularly for national capitals, and the tourist historic city (Ashworth & Tunbridge, 2000), for example, Old Rauma, an old town with medieval street pattern and wooden houses, which still today forms a major part of the centre of the city of Rauma on the west coast of Finland, was selected on the World Heritage List in 1991 as a representative of a 'Nordic wooden town' (Vahtikari, 2004).

The issue of defining urban tourism perhaps reflects Wöber's (1997: 30) observation that 'an urban product is what the market accepts as such', or, to paraphrase Ashworth and Voogd (1990), every urban product is an assemblage of selected resources which are bound together through interpretation, for example, the presentation of the product to consumers through a range of communications media and approaches, such as place marketing. However, the spatial borders of the particular tourism product which the consumer is interested in often does not neatly correspond to the administrative and political boundaries of a municipality or region. Indeed, the core urban tourism product may be confined to a central business district (CBD) or a waterfront development with the remainder of the district not being of interest to the majority of visitors (Page & Hall, 2003). Such situations often lead to the development of cooperative marketing and governance measures between municipalities and government agencies in order to try and overcome administrative boundary issues that the visitor does not recognise in their understanding of a destination.

As well as definitional issues in a conceptual sense there are also issues with respect to urban tourism statistics. In her review of the definition and compilation of European city tourism statistics Ostertag (2007) notes that the comparability of statistical measurement of city tourism has many shortcomings and, as a result, even relatively basic tourism data such as nights, arrivals, number of beds, number of accommodation establishments, occupancy ratios and length of stay may vary significantly between cities. Furthermore, there is considerable variance over city size as well as sampling area, that is, historic city, central city, official city limits or area larger than the official city limits but which are perceived as being part of the conurbation ('greater city'). Nevertheless, the clearly substantial issues in assessing urban tourism in a comparative and definitional sense do not negate the importance of urban tourism as a field of study. The next section will examine the significance of the field with respect to macro-tourism policy and marketing issues of place competition and reimaging.

Place Competition and Reimaging

Place competition is regarded as a key concept in seeking to understand the way in which cities and regions market and develop themselves in the contemporary economic and political environment. One of the most well cited works in the field is that of Kotler *et al.* (1993; also see Kotler *et al.*, 1999; Kotler & Gertner, 2002), according to whom we are living in a time of so-called 'place wars' in which places are competing for their economic survival with other places and regions not only in their own country but throughout the world:

> All places are in trouble now, or will be in the near future. The globalization of the world's economy and the accelerating pace of technological changes are

two forces that require all places to learn how to compete. Places must learn how to think more like businesses, developing products, markets, and customers. (Kotler *et al.*, 1993: 346)

The notion of place competition has had considerable impact on the way governments at the national, regional and municipal level consider their capacities to attract and retain capital, people and firms. The idea of competition emphasises the notion that there are other locations that are also seeking to do the same with place marketing being the tool by which successful competitive strategies are achieved.

> European communities are in active competition with each other, and also leading places, due to a lack of skilled marketing abilities, can lose their vitality to survive in the place competition, when the practices of place marketing are not mastered. There is the necessity of 'place excellence' among places (Kotler *et al.*, 1999) ... Place marketing is used for multiple goals, such as to build a positive image for the place and attract enterprises, tourists, institutions, events etc. Today, places need to attract tourists, factories, companies and talented people, as well as find markets for their exports, and this requires that places adopt strategic marketing management tools and conscious branding (Kotler & Gertner, 2002: 253) ... Past place promotion strategies no longer work in the rapidly changing markets, and in the new place competition situation ... In order to compete effectively, places must develop a real marketing approach. The place competition is global, and all places whether located in Europe, Asia, Latin America or the USA, need to develop new capabilities to survive in the competition. Consequently, places must produce services that current and potential citizens, companies, investors and visitors need. (Rainisto, 2003: 11–12)

Kotler *et al.* (1999) argued that in Europe alone there are over 500 regions and over 100,000 single communities competing for the same scarce resources. Similarly, Jansson and Power (2006: 10) comment 'The realisation of the competitive realities they face has driven many, if not most, Nordic cities of various sizes and locations into more or less extensive image and branding work.' Undoubtedly, such a discourse is reasonably persuasive given the extent to which place competitiveness has become a significant policy goal for many locations (see Case 1.3 with respect to evaluations of Nordic competitiveness at the state level). For example, in an OECD report on the competitiveness of Stockholm as a site for business, attention was paid, among other things such infrastructure, to the importance of developing the city's brand and reputation as a business centre.

> To play an even bigger role in Europe and globally, the Stockholm region will need branding to increase its international visibility. Compared to many other European cities, Stockholm is far less known. Stockholm may have to rely on regional branding to attract foreign resources (businesses and workers). Regional branding could serve to overcome the disadvantages of being located at the edge of European and global markets as well as increase its currently low international visibility. (OECD, 2006: 80)

The OECD report utilised a study of familiarity with cities as business locations, in which Stockholm was ranked 23rd, to reinforce the importance of improved branding for Stockholm. (In the same study of European cities Copenhagen ranked 22nd, Oslo 29th and Helsinki 30th) (OECD, 2006). In the same way that national competitiveness ranking has become increasingly common and is gaining domestic and international media attention (see Case 1.3), Jansson and Power (2006: 11) argue that 'It is therefore important that cities and regions pay attention to such rankings and attempt to move up in them.' Indeed, the Financial Times' *fDi* magazine, annually picks European, Scandinavian and Northern Cities of the Future based on 28 individual criteria from nominated European cities and regions (see Case 4.1). However, place promotion is also important for smaller places.

Cases and issues 4.1: Place competition: 'Cities of the Future'

The Financial Times Group *fDi* magazine conducts an annual competition for nominated cities and regions to join their lists of European cities and regions of the future. The competition is held in three stages. In part one, regions and cities put forward bids to be selected as the best location in their country or region (e.g. Scandinavia). The winners of the first round then have the chance to compete in broader geographical categories which for the Nordic countries is Northern Europe. Finally, the best locations in Europe compete for *fDi* magazine's top awards: European City and European Region of the Future. Awards are also given for the best locations in each individual category such as Most Cost-Effective, Best FDI Promotion Strategy, and Best Economic Potential (see Table 4.1). Locations are ranked by 28 individual criteria (Table 4.1) in seven main categories (economic potential, cost effectiveness, human resources, IT and telecommunications, transport, quality of life and FDI (foreign direct investment) promotion). They receive three points for coming first, two points for second place and one point for third. Where answers are not based on numerical data that could be ranked automatically, a panel of judges vote for the top three locations, also awarding three points for first, two for second and one for third (*fDi*, 2006).

The results of such exercises are significant in part because of benchmarking but perhaps more so as opportunities for place promotion. In 2006/7 Copenhagen was the 'winner' for the Northern Europe Scandinavia section with Oslo the 'runner-up'. This, and previous, success was given considerable coverage by Copenhagen economic development and promotion agencies such as Copenhagen Capacity (CopCap).

fDi website: http://www.fdimagazine.com/

In a survey of 220 Swedish municipalities Niedomysl (2004) found that 70% of municipalities had conducted place marketing activities with the attraction of new inhabitants being a high priority including, for a small percentage, the attraction

Table 4.1 Categories and criteria for 'cities of the future'

Category	Judgement based	Numerically based
Economic potential	• Economic potential • FDI deals	• GDP • GDP growth • Level of inward investment
Cost effectiveness		• Out-of-town office rent • Centrally located office rent • Industrial rent • Secretarial salaries • Middle management salaries • Manual labour rates
Human resources	• Universities	• Number of [year] graduates • Percentage of population with a degree
Transport	• Transport	
IT and telecommunications		• Phone lines per 1000 people • Mobile phone ownership (% of population) • Broadband connectivity (% of population) • Maximum broadband speed available • Telecommunication charges
Quality of life for expatriates	• Housing • Hospitals • International schools • Natural and cultural heritage	
Best FDI promotion strategy	• Promotion strategy • Three biggest attractions for FDI • Incentives • Infrastructure and urban planning projects	

Source: Derived from *fDi* (2006).

of second home residents. Increased amounts were being spent on place marketing by municipalities, who are becoming more advanced in their use of branding techniques in addition to redevelopment. Tourism and leisure factors were also significant as part of the factors used to attract in-migrants in the form of quality of life, living environment, outdoor leisure opportunities and natural beauty suggesting that municipalities were also very aware that place marketing exercises have to be supported by on the ground realities, thereby reinforcing the relationships between marketing and urban management and planning.

Nevertheless, while all this is happening there is still some uncertainty as to what the concept of competitiveness actually means at the place level, whether national, regional or urban, 'and how it can be effectively operationalised ... policy acceptance of the existence of regional competitiveness and its measurement appears to have run ahead of a number of fundamental theoretical and empirical questions' (Bristow, 2005: 286). It is, as Markusen (1999: 870) would say, a relatively 'fuzzy concept': 'characterizations lacking conceptual clarity and difficult to operationalize. In some cases, no attempt is made to offer evidence at all. Elsewhere, evidence marshalled is highly selective. Methodology is little discussed'. Indeed, Niedomysl (2004) found in the Swedish case that there was little real evidence that such campaigns have been effective, at least by the time his study was completed. Similarly, Jansson and Power (2006: 15) suggest:

> We should be careful then about what sort of commercial methods and logics we try to adapt to complex urban areas. This, of course, does not mean that we should neglect the importance of image for cities or regions. Neither does this means that we should ignore the fact that everyone else seems to be busy improving their images. It simply means that be should we acutely aware of the differences between selling a product and improving a cities image or brand.

Jensen (2005) also emphasises that there are at least four differences between branding a product and a city:

(1) There is a difference in the number of stakeholders and their related interests; branding a city or a place includes a complex web of preferences.
(2) It is a hard task to negotiate a legitimate local value base with local participation. This is almost never an issue when branding a product.
(3) Branding a city or a place usually has to follow the paths of existing notions or historical identities of a place. Most products do not have the same depth of history or associations to consider.
(4) The consumers of an urban brand are often more diverse than the consumers of a normal product since urban branding has to serve diverse groups of potential investors, residents and tourists.

Nevertheless, more often than not, the need for attention to competitiveness is presently taken as a given by policy-makers concerned as they are with the deceptively simple question as to why some places are doing better than others (Hall, 2007c). Tourism is embedded into the place competition discourse in a number of ways (Hall, 2007c). First, tourism can be seen as part of a form of competition to attract visitors. Second, tourism is often an intrinsic element of place marketing and imaging. Third, tourism is an enabling component of place development with respect to infrastructure.

The implications of the different, but often interrelated tourism competitiveness elements, along with place competition overall having significant implications for cities and their management and governance. For example, although the traditional commercial and retail functions of central business districts are still important, leisure and tourist functions are becoming increasingly significant (Garcia, 2004). The ramifications of such an approach are far-reaching, particularly with respect to the way in which cities and places are perceived by some stakeholders as products to be 'packaged' and 'sold' (Hall, 2005a) in order to be able to 'compete' (Löfgren, 2000).

One strong theme that has been recognised in the urban tourism literature since the early 1990s with respect to place competition is that of urban imaging or reimaging (given that some places may actually be trying to change their existing images that consumers and stakeholders may have, including potential visitors and investors in tourism). The principal competitive aims of urban imaging strategies are to attract tourism expenditure; generate employment in the tourist industry; promote local goods and services for export; foster positive images for potential investors in the region, often by 'reimaging' previous negative perceptions. This can include the promotion of 'new' places such as cross-border regions such as the Øresund region of Denmark and Sweden. The final competitive aim is provide an urban environment which will attract and retain the interest of professionals and white-collar workers, particularly in 'clean' service industries such as finance, education and communications, while also ensuring a supply of service workers for tourism and other support service industries.

Urban imaging processes are characterised by some or all of the following:

- develop a critical mass of visitor attractions and facilities, including new buildings/prestige centres, for example, the Arabianranta Art and Design Quarter in Helsinki, iconic or signature buildings, such as the 'Turning Torso' in Malmö and the newly built opera house in Copenhagen;
- host hallmark events, for example, European City of Culture, such as Stavangar in Norway in 2008;
- develop urban tourism strategies and policies often associated with new or renewed organisation and development of city marketing, for example, Wonderful Copenhagen and CopCap; and
- develop leisure and cultural services and projects to support the marketing and tourism effort. This usually includes the creation and renewal of museums, design and art galleries and the hosting of art festivals, often as part of a comprehensive cultural strategy for a region or city, aimed at promoting tourism as well as the notion of creativity. For example, Bayliss (2007) notes potential changes to the planning system designed to facilitate Copenhagen's transformation into a 'creative city'.

Some of these urban imaging processes can be illustrated with reference to place marketing in the Copenhagen region over a period from 1992 with the founding of 'Wonderful Copenhagen' through to 2003 (Table 4.2). What the table illustrates is that there are a number of different place brands that aim to connect with different markets as well as the positioning of Copenhagen within a broader regional context, particularly with respect to the Øresund region for which governance structures and business networks have developed in parallel fashion to place marketing and branding strategies. In business and industrial terms Øresund has been marketed as a European 'hub of excellence' with the region also being branded a 'human capital' so as to reinforce the positive social, economic and natural environment of the region, including a well-educated workforce and the geographic concentration of certain industries (Cooper & Hall, 2008). Jensen (2005) also notes that this was an attempt to add a 'softer' dimension to a project that often seemed

Table 4.2 Copenhagen and place marketing

Year	Place marketing practice	Significant events
1992	No systematic place marketing activity in place before 1992.	The city decides to invest DKK 20 million annually in city marketing. 'Wonderful Copenhagen' founded.
1993	Copenhagen Capacity (CopCap) focuses on information technology, biotechnology and the environment. Strategy was built as the inward investment agency of Copenhagen.	
1994	Copenhagen Capacity is officially established by the city of Copenhagen, the municipality of Frederiksburg and the counties of Copenhagen, Frederiksborg and Roskilde.	'Green City Denmark' established. The Øresund Region – model for cooperation across borders Medicon Valley established (a member-based regional organisation).
1997		Medicon Valley Academy (MWA) was set up to be a network organisation in Medicon Valley. Wonderful Copenhagen and CopCap initiate a 'Copenhagen Hotel Development Network' project.
2000	Medicon Valley – brand gets established as 'the umbrella' of the Copenhagen-Malmö region.	The Greater Copenhagen Authority, HUR (Hovestadens Utvicklingsråd) is created. HUR acts as the funding body of CopCap. Copenhagen Eventures Wonderful Copenhagen's event department, is created as a result of an agreement between Copenhagen Council and Wonderful Copenhagen.
2001	A new long-term strategy for CopCap puts Øresund at the centre to attract new investment to Copenhagen.	
2002	Øresund 'The Human Capital' brand.	The twin-cities of Copenhagen and Malmö invest a total of DKK 100 billion in an interaction model. Cooperation with Gothenburg under development.
2003	'Business is Easy' taken as a new leading slogan. New brand 'Øresund IT – The human tech-region' markets IT in the Øresund region, besides 'Medicon Valley'.	

(Continued)

Table 4.2 *Continued*

Year	Place marketing practice	Significant events
2004	'The greater Copenhagen area is to be one of Europe's most attractive and dynamic city regions for both leisure and business travellers' – Wonderful Copenhagen vision.	
2007–9	'Copenhagen/Öresund – one destination, two countries'.	New Wonderful Copenhagen Strategy – strong focus on experiences, culture and creativity – Wonderful Copenhagen at the centre of multiple networks

Source: Rainisto (2003); Wonderful Copenhagen (various).

to be more about infrastructure and business, particularly given its initial brand of 'Medicon Valley'. In addition, other programmes, some of which were funded by the EU, such as 'Cultural Bridge 2000' to parallel the opening of the Øresund Bridge were also created to reinforce regional identity. Along with the 'human capital' brands, such measures were designed to contribute to identity construction in the region – a 'work in progress' (Jensen, 2005: 28) that was supposed to identify a particular 'Øresund Way of Life' (Øresund Network AB, 2002):

> Balance and harmony are keywords for life in the Øresund Region. The people of the Øresund live in one of Northern Europe's fastest growing business areas, while the good life awaits them right on their doorstep ... According to an American survey based on 160 countries, Denmark is the best country in the world to live in. Sweden is not far behind in fourth place. The social systems are well developed and take good care of the sick and the old. Crime is limited and unemployment is low. Interest in taking care of the environment is expressed in the unspoilt countryside and in urban development... Accessibility characterizes the Øresund Region. The tailor-made road and rail network blurs the borders and puts the entire region within reach ... The quality of working life an leisure activities makes the Øresund Region peerless. (Øresund Network AB, 2002: 9 in Jensen, 2005: 28)

Nevertheless, the success of such imaging and identity creation has not been without its critics. For example, Berg and Löfgren (2000: 12) commented, 'The only thing that actually unites many of the actors it that they agree upon working under the label "the Øresund Region" and that they often share a specific optimistic rhetoric we call *Ørespeak*.' Similarly, according to Ek:

> Through the use of maps and regional statistics, the Øresund region was represented as a fact rather than a vision or dream, and the 'Øresund citizen' was characterized as a well-off, mobile, postmodern cosmopolitan with a high degree of cultural capital and a larger curiousity about the world inherited

from the Vikings and their interest in trade and travel (the common image of the Vikings as ravagers and warriors is quite misleading). (Ek, 2005: 79)

Even though the notion of 'a human capital' is used in both internal and external promotion, Ek (2005) argues that this brand/identity has never successfully been adopted by the region's citizens. Indeed, Maskell and Törnqvist (1999) provide a cautionary note with respect to the long-term nature of place and identity building and how it intersects with some of the short-term needs of place competition:

> It took generations to build national identities and establish complex national innovation systems favouring growth and prosperity, including at a regional level. It will take years of hard work to amalgamate two countries' distinctive innovation systems into one, even when most formal barriers have been eroded. It will take even longer for a common cross-border regional identity to form. Only then will the full potential of cross-border synergy begin to materialise. (Maskell & Törnqvist, 1999: 11)

In fact, the above comments on identity construction only serve to reinforce the that the prevailing competitiveness discourse is often output related and, arguably, pays insufficient 'attention to the broader non-tradeable modalities of competitive behaviour which may characterize regions' (Bristow, 2005: 295), such as culture and identity, even though issues relating to the cultural and social dimensions of place have long been recognized as critical to innovate capacity (Hall & Williams, 2008). Or, to modify Hewison (1991: 175), you cannot get a whole way of life into an Ikea shopping bag.

Of course, the fact that not everyone can be a winner does not mean that competition is without value. Rather it suggests that there are both benefits and problems inherent in place competition. However, it is significant that within regional development, tourism is often seen as part of an imitative 'low road' policy in contrast to 'high-road' knowledge based policies. With Malecki (2004: 1103) noting that 'The disadvantages of competition mainly concern the perils that low-road strategies build so that no strengths can prevail over the long term, which presents particular difficulties for regions trying to catch up in the context of territorial competition based on knowledge.' Low-road strategies are regarded as being focused on 'traditional' location factors such as land, labour, capital, infrastructure and locational advantage in relation to markets or core elements of production as well as direct central or local state subsidies to retain firms, with more intangible factors, such as intellectual capital and institutional capacity being secondary. Such low-road strategies of regional competitiveness are general regarded as being tied into property-oriented growth strategies that focus on the packaging of the place product, reimaging strategies and the gaining of media attention (Hall, 2007c; Hall & Williams, 2008).

In contrast to the low-road approach that repackages other place marketing approaches, such as the 'creative city', Malecki (2004) argues that a high-road approach of genuine entrepreneurship and innovation is possible although it also more difficult to travel. Instead, the regional infrastructure of the high road is regarded as requiring both hard (communications, transport, finance) and soft (knowledge, intellectual capital, trustful labour relations, mentoring, worker-welfare orientation) infrastructures in order to encourage innovation (Malecki, 2004). It is this emphasis on the human dimensions of innovation and regional development that makes it hard to achieve as the elements are not so visible as in the case of the development of a new stadium or an opera house. Yet in both paths tourism is regarded as important. In the low road tourism is arguably a focal point of the development strategy in which increasing numbers of visitors is regarded as a sign of competitive success in its own right. In the high road approach tourism is not a main focus, instead tourism is regarded as an enabler, that is, in terms of accessibility, connectivity, business and convention travel, and quality of life, to achieve a broader set of urban and regional development strategies (Hall, 2007c; Hall & Williams, 2008). Nevertheless, regardless of the approach that is used with respect to tourism and regional development, urban form and process will be affected and it is to these issues that we will now turn.

The Tourist City

According to Page and Hall (2003) 'the urban landscape changes in light of the intersections of visitor demands and public and private supply in a manner which can be both identified in space and in terms of the actions and behaviours of visitors'. Nowhere is this perhaps better evidenced than in Burtenshaw *et al.*'s (1991) concept of there being overlapping functional areas in the tourist city. Four areas were identified:

- 'the historic city' (historic monuments, museums, art galleries, theatres, concert halls);
- 'the culture city' (museums, art galleries, theatres, concert halls);
- 'the night life city' (theatres, concert halls, nightclubs, entertainment, red light districts, cafes, restaurants); and
- 'the shopping city' (cafes, restaurants, shops, offices).

To which Page and Hall (2003) added two other functional districts:

- the 'business city' which is evidenced by convention and meeting halls and trade centres and associated hotel accommodation, developed as a result of the demands of business travellers and conference and exhibition delegates; and
- the 'sports city' characterised by sports and event facilities such as stadia.

In addition they noted the special case of capital cities which also have specific functional requirements. These different types of functional areas will be briefly discussed below.

The historic city

Older city cores and historic districts, often including cultural institutions such as museums, are an important tourism resource. All the Nordic capitals, for instance, have extremely significant historic districts. For example, Gamla Stan (Old Town) in Stockholm retains the medieval street layout and is also the home to a number of palaces and churches of interest to visitors. The tourist dimension of the district is also reinforced by the number of tourist oriented shops and restaurants that have located there. Similarly, Helsinki also retains an historically significant waterfront as well as a significant tourist attraction in the form of the World Heritage listed former naval fortress and base of Suomenlinna. Although World Heritage sites are not as significant for historic city promotion as in Germany or southern Europe, there are a number of urban World Heritage sites of which Bergen in Norway and Rauma in Finland are probably the most well recognised. Table 4.3 records the urban World Heritage sites in Nordic countries.

Table 4.3 Urban world heritage in Nordic countries

Country/ urban area	Site and description	Date	Member OWHC
Denmark			
Helsingør	Kronborg Castle. Kronborg Castle is an outstanding example of the Renaissance castle that played a highly significant role in the history of northern Europe. It is also world-renowned as Elsinore, the setting of Shakespeare's Hamlet.	2000	
Roskilde	Roskilde Cathedral. Built in the 12th and 13th centuries, this was Scandinavia's first Gothic cathedral to be built of brick and it encouraged the spread of this style in northern Europe. Porches and side chapels were added up to the end of the 19th century, providing a clear overview of the development of European religious architecture.	1995	
Finland			
Helsinki	Fortress of Suomenlinna. Built in the second half of the 18th century by Sweden on a group of islands located at the entrance to Helsinki's harbour, this fortress is an interesting example of European military architecture of the time.	1991	
Rauma	Old Rauma. Situated on the Gulf of Bothnia, Rauma is one of the oldest harbours in Finland. Built around a Franciscan monastery, where the mid-15th-century Holy Cross Church still stands, it is an outstanding example of an old Nordic city constructed in wood.	1991	x
Norway			
Bergen	Bryggen. International port ('Exterior office' of the Hanseatic League). On the west coast of Norway, the Bryggen harbour quarter occupies the north eastern bank of Vagen harbour.	1979	x
Røros	Mining town. Røros, also known as 'Bergstaden Roros,' is located 100 km south of Trondheim in the midst of an extensive mountainous terrain. It was developed on both banks of the Hyttelva River, where copper was extracted.	1980	x

(Continued)

Table 4.3 *Continued*

Country/ urban area	Site and description	Date	Member OWHC
Sweden			
Falun	The mining excavation known as the Great Pit at Falun is the most striking feature of a landscape that illustrates the activity of copper production in this region since the 9th century. The 17th-century planned town of Falun with its many fine historic buildings, together with other industrial and domestic settlements of the Dalarna region, present what was one of the world's most important mining areas.	2001	
Karlskrona	Karlskrona is an outstanding example of a late-17th-century European naval city. The original plan and many of the buildings have survived intact, along with installations that illustrate its subsequent development up to the present day.	1998	
Luleå	Gammelstad, at the head of the Gulf of Bothnia, is the best-preserved example of a 'church village', a unique kind of village formerly found throughout northern Scandinavia. The 424 wooden houses, huddled round the early 15th-century stone church, were used only on Sundays and at religious festivals to house worshippers from the surrounding countryside who could not return home the same day because of the distance and difficult travelling conditions.	1996	
Stockholm	The Royal Domain of Drottningholm stands on an island in Lake Mälaren in a suburb of Stockholm. With its castle, perfectly preserved theatre (1766), Chinese pavilion and gardens, it is the finest example of an 18th-century north European royal residence inspired by the Palace of Versailles.	1991	
	Skogskyrkogården. This Stockholm cemetery was created between 1917 and 1920 by two architects, Gunnar Asplund and Sigurd Lewerentz, on the site of former gravel pits overgrown with pine trees. The design blends vegetation and architectural elements, taking advantage of irregularities in the site to create a landscape that is finely adapted to its function. It has had a profound influence on funerary design in many countries.	1994	
Visby	Hanseatic Town of Visby. Visby is located on the northwest coast of Gotland, the largest island in the Baltic Sea. About 100 km from the coast of Sweden, the site is terraced and its port, although today silted up, is ice-free.	1995	×

OWHC = Organization of World Heritage Cities.
Source: Derived from UNESCO World Heritage List http://whc.unesco.org/en/list.

The culture city

The historic city and the cultural city clearly have substantial overlap as concepts and as functional areas of the city. Although culture can be broadly defined as a way of life or more technically a social field of meaning production, the term is used more narrowly with respect to the notion of the 'cultural city'. In this context the cultural city usually refers to a cluster or designated space for institutions of culture, such as museums, galleries and theatres, or to cultural precincts that may combine institutions with other public and private sector culturally oriented organisations, such as design and media businesses, or to a specific ethnic enclave, the most well known type being a Chinatown. In the Nordic context the cultural city concept applies much more to the former institutional connection than any enclave. The city of Kouvola in Finland, however, has been seeking to expand its Chinese trade connections as a result of rail traffic between Kouvola and Asia and has supported the development of a China Centre (Huuskonen, 2006). More usual is the cluster of cultural institutions seen in many of the Nordic cities where art galleries, museums and theatres are co-located, for example, Göteberg with the cluster of the Museum of Art, the City Theatre, the City Library and the Concert Hall (also home of the city's Symphony Orchestra) surrounding the Götaplatsen at the southern end of the main street of the city, Kungsportsavenyn.

The night life city

The function areas of a city can also be distinguished by their temporal as well as spatial occurrence. Many readers will be familiar with the notion of a nightlife or entertainment district that is characterised by clubs, restaurants, theatres and cafés that function in the evening and night time. These areas are also sometimes referred to as 'urban playscapes' characterised by 'new bar, pub, hybrid bar/clubs and café concepts and themes and stylised environments aimed at an assumed set of consumer demands, lifestyles and tastes' (Chatterton & Hollands, 2002: 101). From a place competition perspective, 'the attractiveness of branding as a strategy stems from its ability to offer a wider "lifestyle" experience, to increase uniformity and hence reduce costs and overheads, to increase a feeling of consumer choice safety, convenience and reliability' (Chatterton *et al.*, 2002: 19). However, as in all aspects of place marketing, it is questionable whether there will be long term economic benefits for the local economy if the uniqueness of a city's nightlife, formerly characterised by 'local experiences', becomes eroded and a lack of distinctiveness in comparison with other city's nightlife districts occurs.

The shopping city

Shopping plays an important part in the urban tourist experience with retail usually being one of the main areas of visitor expenditure to a location after accommodation and, another associated part of the shopping city, eating out. Many city downtown's have areas that are clearly characterised as tourist shopping areas by the number of souvenir shops, money exchange stores and signs in non-local languages. However, more general retail is often an attractive component of domestic and short-break tourism, with only some of the larger centres with substantial international connections being a focal point of retail for international visitors, for example, Copenhagen, Stockholm and Helsinki. Cross-border tourism is not as significant as it used to be in the Nordic region as a result of gradual levelling of many taxes between the countries although, for example, there is still some travel by Swedes on the ferry from Helsingborg to Elsinore (Helsingør) in Denmark to take advantage of lower alcohol prices (see also Chapter 7).

Perhaps one of the most interesting cross-border shopping developments in the Nordic region was the opening of the world's northernmost Ikea store in Haparanda, Sweden, on the border with Tornio, Finland, in 2006. The store has bilingual signage and has developed a retail attraction area that not only includes northern Finland and Sweden but also Norway and Russia. The store has become a positive symbol for the development of the region and has generated significant employment, especially as a result of other firms co-locating in the city (Enberg, 2005). Duval-Smith (2007) reported that the regional boom had cut Haparanda's unemployment from 10% to 3.8%. According to Haparanda Mayor Sven-Erik Bucht:

> The whole of northern Sweden was begging for a store. But we came up with the Barents region concept and there was just no gainsaying our arguments... What's happening is what we call the ketchup effect: you bash and bash the bottle, and nothing comes. Then suddenly, splat, there's red stuff everywhere. (quoted in Duval-Smith, 2007)

The business city

Non-commuting travel for reasons of work is termed business tourism. The focus of business tourism is the development of B2B (business to business) relationships as well as attendance at meetings, conferences, conventions and exhibitions. The business city in a tourism sense closely overlaps the CBD as well as other business locations and centres. Most major cities in the Nordic region have purpose-built conference and exhibition centres usually developed through municipal funding with the development of such centres related to the overall business significance of a location and their accessibility by transport (Page & Hall, 2003). The International Conference & Convention Association (ICCA) ranked Copenhagen in its top 10 meeting destinations for 2006–2007. (The ICCA rankings cover meetings organised by international associations which take place on a regular basis and which rotate between a minimum of three countries.)

The Union of International Associations (UIA) also produce annual statistics on meetings organised or sponsored by international organisations. Although rankings are incomplete they again suggest that Copenhagen is a leading international meeting city along with Helsinki and Stockholm. Key to the number of meetings are also the countries chairing of the EU presidency as well as the hosting of European Union and international bodies (Table 4.4).

Table 4.4 Ranking of Nordic capitals for hosting of international meetings

Year	2001	2002	2003	2004	2005	2006
Copenhagen	9	6	8	6	10	
Helsinki	18	21	13			6
Oslo			24			
Stockholm	13	12	12			

Note: from 2004 on UIA press releases on international meeting statistics only gave top 10 cities.
Source: Fischer (2007); Union of International Associations (various).

Business tourism is attractive to cities for a number of reasons (Swarbrooke & Horner, 2001):

- business travellers usually have a higher level of per head of spending than other types of tourism;
- the business traveller is a core market for airlines and hotel chains; and
- the specialised nature of business travel has led to the development of a range of ancillary services and infrastructure which generate employment and income.

Therefore business, convention and hotel and accommodation infrastructure tends to be co-located, usually in a city core area (and often close to the nightlife city), although there has been some business tourism clusters developed closer to airports or other transport infrastructure. For example, Helsinki-Vantaa Airport is positioning itself as a European gateway to Asia (and vice versa) and has developed new hotels, meeting and convention facilities and a world trade centre. This is part of a broader development referred to as Aviapolis, a 42 km^2 area with the airport as the hub but with high levels of accessibility by train, public transport and car. The Aviapolis development is also interesting as it might potentially spread the employment opportunities of the hospitality sector which tend to be highly concentrated. For example, in the case of Copenhagen, itself a strong focus of urban tourism as discussed above, employment change in the hotel and restaurant industry in the outer city between 1994–2002 (5.3%) was far behind that of the inner city (34.5%). This compares with an overall change in employment of 12.7% and 11.7% respectively (Hansen & Winther, 2007), thereby emphasising the significance of the spatial concentration of tourism related businesses such as hotels as well as their connection to broader tourism and place marketing strategies (see Case 4.2).

Cases and issues 4.2: Development of hotels in Nordic urban centres: Expansion, diversification and yield maximising strategies

SZILVIA GYIMÓTHY

In the wake of the general economic upswing, the Scandinavian hotel market has experienced record growth since 2005. Both accommodation volume and

Table 4.5 Key indicators of the Swedish hotel market in 2006

	Sweden total	Göteborg	Malmö	Stockholm
Hotels	1.917	58	30	118
Available rooms/day	93.000	6.838	2.996	13.423
Turnover, mSEK	13.800 (10%)	1.391	553	3.427
Sold hotel rooms (million)	16.7 (5%)	1.639 (6%)	712 (6%)	3.404 (7%)
Occupancy rate %	49.2%	69%	65%	66%
Average room rate (ARR), SEK	827 (8%)	849 (7%)	777 (2%)	1.007 (3%)
Revpar (SEK)	407 (4%)	557 (11%)	506 (7%)	699 (8%)
Number of days over 70% occupancy	23	183	173	203

Source: Adapted from Alexandersson and Hirsch (2007: 10, 11).

prices increased at a rapid pace, particularly in the urban centres of Stockholm, Göteborg and Malmö (Table 4.5). The accommodation capacity in Stockholm grew by 30% in the past six years, which is almost three times higher than the average national growth rate. International franchisors are entering the Nordic hotel market and sign an unprecedented number of contracts with local operators, for example, Scandinavian Hotel Group. Within six years, Marriott is planning to open 15 new hotels under different sub-brands in the Nordic countries. At the same time, Choice has launched an aggressive expansion campaign introducing its sub-brand Clarion, which has convinced eight hotels in smaller provincial centres (affiliated formerly with Park Inn and Radisson SAS) to leave the Rezidor Group (Hirsch, 2007). Hoteliers mention better environmental policies and a better focus on Nordic nearmarkets among reasons for shifting chain affiliation to Choice, which is surprising, knowing that Scandic (Scandinavia's long-established and only hotel brand) stand for similar values. Nevertheless, by the end of 2007, the number of hotels run under Choice-brand in this region will outnumber those affiliated with Scandic. Looking at the operating profiles of new or planned establishments, the trend is moving towards a diversified hotel market, introducing new concepts, such as extended hotel stays, executive apartments, specialised conference centres and lifestyle hotels to Scandinavian customers.

The benevolent economic cycle has also resulted in higher room prices. In Sweden, average room rates (ARR) rose by 12% since 2000 (827 SEK), which is three percentage points more than the average price increase in Sweden (Alexandersson & Hirsch, 2007: 10). Comparing the behaviour of congress hotels in Stockholm, Göteborg and Malmö, it seems that the three cities pursue different strategies to maintain growth. Stockholm still intends to construct more five-star hotel accommodations and exclusive meeting centres in order to increase its share of upmarket products. Profiting from the capital location, Stockholm hotels follow a skimming strategy, and are still very successful in attracting long-haul travellers and megaconferences. When booking through international booking systems, hotels are able to claim non-discounted rates. Some of them raised their prices by 15–40% within two years, giving priorities to full price paying segments, rather than to volume boosters, such as bus groups. On the other hand, hotel bookings in Göteborg and Malmö are based on seasonal business, smaller congresses or a result of negotiated prices with corporate customers, like Volvo or Tetra Pak. These are based on long-term contracts, where even a 3–4% price level raise is difficult to achieve. Therefore, hotels and destination organisations in these urban centres work hard to host prestigious, volume-enhancing events and have also succeeded in attracting the Athletic World Cup and America's Cup yacht racing. Media attention and free publicity during these events has already given yield and the induced effects are expected to continue in the future.

The tactics described above give different results on the bottom line. Stockholm's discriminating price segmentation results in higher RevPAR (Revenue per Available Room) levels, however, occupancy rates in the Swedish capital fell by five percentage points compared to those of 2000. In Göteborg and Malmö, the average room rate is 20% lower than in Stockholm, however, they can boast higher occupancy levels. Several analysts are concerned about the escalation of room prices, discriminating pricing strategies combined with rapid capacity expansions in the Swedish capital. Pandox' CEO, Anders Nissen predicts that skyrocketing prices will upset corporate travel organisers, while Stockholmsmässan's director, Tom Beyer reports complaints from conference bookers, noting that improvements in service standards and hospitality lag behind the price increase (Swartz, 2007). If professional congress organisers (PCOs) turn their back on Stockholm while the expected global economic depression reaches the Nordic region, overcapacity and overpriced hotel rooms may have severe consequences. A diversification strategy combined with cost rather than revenue management may – in the long term – prove to be a safer approach than charging higher prices during an economic upswing.

The sport city

Sports are also an important element of the attractiveness of urban areas (Chalip & McGuirty, 2004; Harrison-Hill & Chalip, 2005). They achieve this through their capacity to gain media coverage as well as contribute to leisure participation and spectating. Sports are also economic contributors to a location through three means:

(1) Attraction of spectators and participants from outside of region.
(2) Construction of sports facilities and venues.

(3) Maintenance of sports facilities and venues.
(4) Attraction of sponsorship and funding from outside of region.

As hallmark tourist events, sports events have assumed an important role in international, national, and regional tourism marketing strategies, providing host communities with opportunities to secure prominence in the tourism marketplace for short, well-defined, periods of time. Usually the larger the city the more prominent the sports facilities although because of their locational needs some of the winter sports facilities are exceptions to this (see Case 10.1 for a study of the development of Ruka resort in Finland). Because of their characteristics sports events, like other hallmark events, are different in their appeal from the attractions normally promoted by the tourist industry, because they are not continuous phenomena although their required infrastructure usually is. Indeed, in many cases, the hosting of events are strategic responses to the problems that seasonal variations in demand pose for the tourist industry. However, the ability of an event to achieve this objective depends on the uniqueness of the event, the status of the event and the extent to which it is successfully marketed within tourism generating regions. Yet, in the long term, the full benefits of the development of clusters of sports facilities will be a function of the capacity to generate use, for example, many hockey stadia are also used for exhibitions, concerts and meetings, as well as other events. This is an issue raised in more detail in Case 4.3 on the long term benefits of the Lillehammer Winter Olympic Games held in 1994.

Cases and issues 4.3: The status of an Olympic town: A dozen years thereafter

THOR FLOGNFELDT JR

Lillehammer has been well known for organising winter sport events for more than a century. For example, the Birkebeiner Ski Race across the mountain from Rena to Lillehammer has been organised since 1932. Covering some 54 km with a back pack of 3.5 kg it is an event limited to 14,000 participants on the third Sunday of March every year, which is sometimes also held in conjunction with the Cross-Country World Cup.

In the early 1980s the interior of Norway ('Innlandet') was suffering from being located in an 'oil shadow' and not receiving the benefits of the oil industries that the western parts of the country had. In response the idea was formed that a big project had to be launched to emerge from the oil shadow. Initially, Lillehammer sought to bid for the 1992 Winter Games that were eventually held in Albertville, France, but in September 1988 Lillehammer was awarded the games in an Olympic meeting in Seoul, Korea. This meant a very strict planning regime was required (Lesjø, 2003) until February 2004 in order to be able to successfully host the games. However, with a guarantee from the government and recognition that the cultural component should be equal to the sports dimension by Premier Brundtland, spending rose from an initial estimate of 1.8 NOK billon up to 7.7 NOK billion plus additional road and construction money of more than 2.5 NOK billion. All of this was justified by the 'after-use' of both venues and other installations, with the planning strategy stressing that the after-use was the main goal of the exercise, not the Olympics itself.

The '16 days of glory' were a great success and is often regarded as one of the 'best winter games ever'. The following is a summary of the status of the impacts of the Games written some 12 years on.

Who were the 'winners'?:

- The Lillehammer University College and an 'entrepreneurship house' Fakkelgården (The Torch Estate) are after users of the Radio/TV-centre and the Press Centre. These were the central units for 7000 TV-workers and another 7000 press journalists during the Games and were equipped with as up-to-date media systems that were then available. The buildings were rebuilt for new uses just after the Games and, of course, a national symbolic structure used for 16 days cost more than permanent university college buildings. Altogether the benefits were 2500 additional students, a national movie school and 300–400 new high skilled jobs.
- The larger farms in the area got money for upgrading their dwelling and service rooms to accommodate VIP groups, mostly from sponsor firms and international companies. Many of these have continued to host meetings and social events (weddings, anniversaries) and the area around Lillehammer is taking an important part of this market both from the capital Oslo and abroad.
- The cultural and art providers received substantial new investments: a new block for the Art Museum; a concert hall and a new block plus other investment at the Maihaugen museum (from 2006 the Norwegian Olympic museum – a Hall of Fame institution is also within the care of Maihaugen); restoring the old Bank Building from 1894 as a local cultural event centre plus restoring other cultural buildings.
- The transport infrastructure benefited through improved road systems both to and within the area and improvement of the railroad system from Oslo, including better services to the new Gardermoen Oslo Airport which reopened as a commercial airport in 1998, 80 km closer to Lillehammer and with complete railroad access.
- The water and sewage systems were improved to the highest quality including connections to potential second homes and alpine resort areas. This meant new development possibilities totally financed by the national government.
- All those firms that could win in the development process without having to over-invest in structures that had no demand in the after games period.

Another set of 'later winners' would be:

- The sport organisers who invested in new mass competitions, including the Birkebeiner Mountain Bike Race held in August; 92 km on roads and paths across the mountain range with more than 12,000 bikers taking part. IngaLåmi the 15 or 30 km female only cross country race is both a ski and a very social event with up to 3000 participants. The Birkebeinerløpet, the cross country foot race, is held in September. The course is 21 km with the start at Sjusjøen and finish in Lillehammer.
- Event companies that have been active in selling 'soft Olympic competitions' in the Olympic venues. Their markets are often national and international companies who want to measure different national bodies or divisions against each other in a friendly but competitive way.
- High school and college sport students – Lillehammer has been turned to a very popular education place for winter sport athletes who want to combine education and training – including international students. The Norwegian Sports Federation and Olympic Committee assist these projects.
- Design firms – the Olympic design programme managed something new, to sell Olympic designware years after the games – but not at reduced prices. Some items are still selling 12 years after the games.

Then there are many 'after-use losers':

- Those investing in hotels without any good market connection. Journalists sought hotel standard accommodation during the Olympics and in the long preparation period. However, the primary demand of the holiday tourism markets in Scandinavia is for self catering units, not hotel rooms. This means that beds had to be converted from hotel quality to self-catering ones. Some had been constructed to do this in an easy way (the Hafjell area), others had not. After the Olympics, hotels have been closed down and new self catering accommodation units have been built.
- Restaurants and bars were losers from the management point of view. In a long reconstruction period, however, students and other frequent users have been able to press prices down due to an over-capacity.
- Those organising professional sport events need strong sponsorship. This is not easy to acquire after the games. The free-style skiing venues were regarded as some of the best in the world, and the competitors and sport federation wanted it used for World Cup Events. The lack of sponsorship, however, meant that no new competitions have been held there. The venue has now changed to other activities. Competitions in ski hill jumping, Nordic combined and alpine skiing have been frequently organised; luge and bobsleigh occasionally; ice hockey once (World Cup finals) while other sport venues have only been used locally. This means that the capacity is too high for the actual audience.

Summing up

Most pre-Olympic research (from potential organisers) showed that the demand would be high for after-use of venues and other investment. This was at best a 'half-truth'. The design of venues should have been more focused on local or national needs instead of giving all the potential Olympic spectators the opportunity to be present during the event. The big costs after the Olympics are:

- sponsoring international events;
- managing the yearly maintenance of the venues, and a refreshing of some;
- keeping accommodation and food and beverage services in full demand;
- seeing that many potential customers are losing the memories of best games ever.

But it was a wonderful occasion for a small town to have, if the country could afford it.

Summary

This chapter has examined some of the issues surrounding urban tourism in the Nordic countries. It has highlighted that urban tourism is embedded within broader processes of place competition and place marketing. It was noted that place branding has become an important part of such urban and regional place competition but it was also noted that the relative success of such strategies was debatable, especially when placed within a broader context of what tourism as a tool of regional development was actually trying to achieve. The chapter then went on to examine the notion of functional areas of the city as a way of describing some of the specialised areas of a city that perform important tourism roles. Several of the issues regarding the response of places to restructuring and economic change will also be discussed in the next chapter, although in the context of the rural environment.

Discussion questions

(1) What are the different functional areas of the tourist city?
(2) What are the characteristics of urban imaging or reimaging strategies?
(3) What are the differences between branding an product and a city?

Essay questions

(1) To what extent are Nordic towns and cities participating in place wars in relation to tourism?
(2) Does place marketing work?

Key readings

Page and Hall (2003) provide an overview of urban tourism. Kotler *et al.* (1993, 1999) remain essential reading on place competition. Hall and Williams (2008) discuss issues of tourism and regional innovation in the context of place competition, while Hall (2007c) also critiques the concept of place competition. Löfgren (2000) provides an interesting analysis of the issues associated with place competitiveness with respect to smaller cities in the Nordic periphery with a discussion of Trondheim's strategies. Bayliss (2007) notes the implications of the arrival of creativity on the urban agenda of Copenhagen. Several papers in O'Dell (2005) discuss urban imaging and, as O'Dell described them, 'urban experiencescapes' in relation to tourism and to the Øresund region in particular.

Aviapolis: http://www.aviapolis.fi/en.php
Copenhagen Capacity: http://www.copcap.com/
European Cities marketing: http://www.europeancitiestourism.com
Organization of World Heritage Cities: http://www.ovpm.org/en/
Øresund region: www.oresund.com
UNESCO World Heritage Site: http://whc.unesco.org/en/1
Wonderful Copenhagen corporate site (Danish only):
http://www.wonderfulcopenhagen.dk/

Chapter 5
Rural Tourism: Tourism as the Last Resort?

Learning Objectives

After reading this chapter, you should be able to:

- Understand the concept of rurality.
- Understand some of the reasons behind change in the Nordic countryside and tourism's role in that change.
- Identify some of the relationships between rural tourism and sustainable tourism.

Introduction

The economic transition of rural areas has turned increasingly towards tourism production. This chapter introduces the basic elements and conceptual dimensions of rural tourism and aims to discuss the issues and challenges of rural tourism and the economic transition process of rural areas towards tourism. In addition the connection of growing tourism and sustainability is addressed. The chapter is divided into four main sections. The first section discusses the changing role of rural areas. The second section defines the concept of rural tourism and related terms. The third section places rural tourism in the context of sustainability in rural development and the final section introduces a case study with concluding remarks. Second homes are also an important aspect of rural tourism. Such is their significance that they are discussed separately in Chapter 8.

The Changing Nordic Countryside

During the past decades rural areas in Nordic countries have faced the same pressures for economic transition as the rest of the world. These pressures include processes such as continued out-migration, and especially a loss of the younger and educated population, technological development of agricultural methods and changes in agricultural policies and subsidy systems. Both politically and economically, rural areas have become more open to global processes and forces, and new industries such as tourism. Traditionally strong primary industries such as agriculture, forestry and fishing have decreased dramatically in the course of globalisation and future employment potential has become limited (Aronson, 1993; Butler & Hall, 1998; Cloke & Little, 1997).

As a result of the economic transition of rural areas the primary economies have gained a negative reputation concerning their future employment capacity. The economic transition and related social changes have been more rapid and dramatic in periphery areas. In Finnish Lapland, for example, the number of persons employed in forestry alone decreased by 70% between 1985 and 1994 (Ministry of Agriculture and Forestry, 1996). With the exception of Iceland this trend has been similar in all the Nordic countries (see Table 5.1).

It is perceived that the market for rural tourism is growing across Nordic countries and in many places tourism, recreation and related services have been viewed as replacement industries for traditional rural livelihoods (Hall, 2007d; Müller & Jansson, 2007a, 2007b). Nowadays, tourism is widely regarded as an effective source of income and employment for rural communities (see Case 5.1 on tourism in the Åland Islands). Positive future prospects of rural tourism development are often based on the global economic significance of the whole tourism industry and the estimations of the present growth of alternative forms of tourism in particular (see WTO, 1996). Indeed, several authors have forecasted further growth in rural tourism (see Komppula, 2004; Nilsson, 2002), and the European Community and politicians within member states have placed a crucial role for tourism in rural development initiatives (Roberts & Hall, 2001) (see also the role of tourism and related areas in Interreg programmes in Chapter 3) (European Commission, 2000). This has produced relatively high expectations but it has also potentially created some misinterpretations and over-estimations of the contributions which tourism can make to rural economies.

Table 5.1 Employment in agriculture, forestry and fishery (persons in thousands) in 1990–2005

Country	1990	1995	2000	2005
Denmark	137	114	95	85
Finland	214	162	138	122
Iceland	—	13	13	11
Norway	128	107	96	78
Sweden	150	132	117	94

Note: Finland includes Åland.
Source: Nordic Statistical Yearbook (2007).

Cases and issues 5.1: Tourism in the Åland Islands: An overview

EVA HOLMBERG-ANTTILA

Åland (Ahvenanmaa) is an autonomous, demilitarised region of Finland comprising more than 6500 islands. In 1905 the population of Åland was about 22,000 people living on 150 islands. Today the population is about 26,200 and only 65 islands have inhabitants throughout the entire year. The main island of Åland makes up 70% of the total land area and it is the home for 90% of the population. Over 40% live in the only town, Mariehamn (Government of Åland [GoÅ], 2006).

Åland's attractiveness in the tourism market: A historical overview

Cruise ships started operations in the Baltic Sea in the late 1950s and, due to this, Åland became easily accessible for Swedes and Finns. The number of tourist arrivals exceeded 1 million in 1972 and 1.5 million in 1990. In the middle of the 1990s Åland's tourism sector faced a severe crisis as the cruiser liner Estonia sunk in 1994. People became aware that not even huge passenger ferries are completely safe. At the same time both Finland and Sweden were in a deep economic recession. Moreover, the weak value of the Swedish Crown made the tourist services of Åland expensive for the Swedes. The number of arrivals dropped to around 1.1 million in 1994–1995. Some recovery took place in the late 1990s; in 1998 the number of arrivals exceeded 1.6 million and the magic 2 million figure was almost achieved in 2004 (ÅSUB, 2005).

Tourism and the economy

In comparison to many other peripheral islands, the economy of Åland is not dominated by primary production nor tourism, but by the shipping sector that contributes to about 40% of the regional GDP. Shipping has made Åland one of the most prosperous areas in northern Europe (Kauppalehti, 2005). Since the economic depression in the early and mid 1990s the economy has been doing extremely well. The unemployment rate has been under 2.5% in the years 2000–2004 and in the years 1999–2002 the growth of the regional GDP was on average 1.8% per year (ÅSUB, 2005).

Nevertheless, the economy of Åland is highly dependent on tourism activities, some of them produced outside Åland. The shipping sector sells tourism services all over the Baltic Sea. As Figure 5.1 indicates, the direct income the land-based tourism sector received from tourism in 2003 amounted to 3.1% of the whole GDP whereas the corresponding numbers were 34.8% when the whole direct tourism income was considered. Totally (directly and indirectly) tourists' spending on land-based tourism services contributed to 4% of the GDP. The total economic impact of tourism (both direct and indirect income effects) on the Åland economy corresponded to 41% of the whole GDP. The employment effects of tourism are significant as well (Figure 5.1). Directly, land-based tourism activity created 5.5% of the employment in Åland. Together

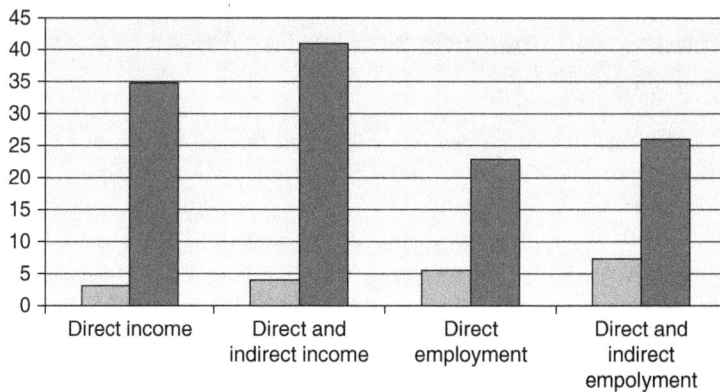

Figure 5.1 The impact of tourism on the economy of Åland in 2003 (%)
Note: Tall columns illustrate impacts of tourism and shipping combined, short columns tourism alone.
Source: ÅSUB (2004b: 2).

with the indirect employment effects of tourists visiting Åland tourism created 7.3% of the total employment of Åland in 2003. If the shipping sector is included the direct employment effects of tourism were 22.9% and the total employment effects 26% (ÅSUB, 2004b).

Even if the contribution of the land-based tourism sector to the economy as a whole seems small its importance should not be underestimated. A main concern is though that the land-based tourism sector has been struggling with profitability problems for several years. The number of registered accommodation nights decreased by 9% from 2003 to 2004 (Turun Sanomat, 2005) even though the number of arrivals grew by 11.5%. The future is not expected to bring any relief for the land-based tourism sector since as a tourism destination Åland is facing several challenges.

A tourism product that is out of date?

As in many other peripheral tourism destinations the tourism product of Åland relies on a beautiful landscape and clean, safe environment. More than 100 companies offer accommodation services (Kauppalehti, 2005) but the level of service quality and the overall standard of the facilities in Åland are often not satisfactory to visitors (GoÅ, 2004). Many entrepreneurs offer tourism services as a sideline business and lack commitment in developing their businesses. Shopping is expensive and the offer of, for instance, apparels is limited. Compared, for instance, to the west coast of Sweden the nightlife of Åland is also quiet during the main tourist season in July.

Another problem is that Åland does not possess enough interesting services attracting tourists outside the high season, resulting in poor profitability and lack of diversity in the tourism industry (Ålands Turistförbund [ÅTF], 2004).

Many facilities are closed during wintertime. The occupancy rates are generally very low, for example, the average occupancy rate of hotels in 2003 was only 30.6% (ÅSUB, 2004a). An extension of the season has been a main concern for the regional Tourist Board as well as the Government of Åland. As a remedy, the Government of Åland had, after almost 10 years of discussion, finally voted for financing of new conference facilities in Mariehamn in 2004 (GoÅ, 2004).

Dependence on an efficient transport system

In Åland the land-based tourism services are strongly dependent on an efficient and, for the tourists, relatively affordable transport system. When Finland was negotiating its EU membership, Åland chose to join even though this was not matter of course. One of the main issues for Åland was a continued possibility to allow the selling of tax-free products on all the ferries visiting Åland at a time when the EU was discussing the abolition of the whole tax-free system. In Åland it was perceived that the cruiser liners' right to sell tax-free goods could guarantee efficient transport to and from the islands. In recent years the taxes on alcohol have been decreased both in Finland and Sweden and the possibilities to purchase tax-free goods will not necessary be an incentive to visit Åland in the future. Moreover, so far it has been relatively cheap to travel to Åland, since the passengers have spent a lot on board (GoÅ, 2004). If this spending decreases, the cruisership companies will face a situation where it is necessary to raise the fare prices.

Dependence on excursionists

Another problem facing Åland is that many tourists today are excursionists (visiting Åland for less than 24 hours). These tourists have no real opportunities to spend money on tourism services produced in Åland (ÅSUB, 2004b) due to the time schedules of the cruise liners. In 2004 it can be estimated that at least 50% of the registered tourist arrivals in fact were tourists hardly seeing anything else of Åland than the ferry terminal. From the perspective of the Åland economy, it has been favourable that companies registered in Åland so far have managed the traffic to and from Åland. However, these companies have been very reluctant to give up their own interests in favour of the mainly small companies producing tourism services on land.

The future of Åland as a tourism destination

In 2003 a tourism strategy aimed at developing Åland as a tourism destination was prepared. The main aims of the strategy were to promote a profitable and environmentally sustainable tourism, to lengthen the season and to achieve a better regional balance (GoÅ, 2004). The Government of Åland has taken a major role by committing itself to finance several new investments. It is alarming that even if there is no lack of private capital in Åland the private sector has lately showed limited interest in new investments (Nya Åland, 2005). Many companies also have problems with intergenerational change due to their weak profitability. It seems in many ways that Baum and Twining-Ward (1997) were right when they stated that Åland as a tourism destination seems to have matured. It is not likely that Åland in the future can manage to attract more tourists than today, but this might not even be desirable. The main competitive advantages of Åland are its idyllic atmosphere and clean environment. Compared to many other peripheral regions in Europe Åland can be placed in a positive situation given its economic success is not dependent only on the further growth of tourism, but on the decisions made by the shipping companies that are the most important employers and greatest contributors to regional GDP.

'Rurality' and Rural Tourism

Many academic and policy texts tend to portray the rural areas as homogenous and static spaces, ones which seem to exist in a timeless zone outside the flows, activities and contradictions of modern and urban life (Murdoch & Pratt, 1997). However, rural areas and 'rurality' are as complex, multidimensional and virtuous as any other social and economic spaces in the contemporary world in which we live (e.g. Cloke, 2003). According to Cloke (1992, 2003) many ongoing societal changes have influenced the nature of rurality. For example, the increased mobility of people, goods and information have changed the autonomy of communities (see Marsden, 1999). Obviously, peripheral local communities have never really, been economically, socially or politically independent, but their ability to control the processes affecting everyday life have dramatically changed due the globalisation and the intensity of interdependency and mobility (see Hall & Page, 2006; Müller & Jansson, 2007a). As a result Nordic rural areas are increasingly utilised by non-rural users and actors. Changing rural tourism and the growth of second homes in rural areas, for example, are both reasons and results of these kind of processes (see Hall & Muller, 2004, and Chapter 8).

There is no universal agreement for the definition of rural areas or rural tourism. Different countries define rural areas based on various national criteria concerning population clusters. In Denmark and Norway population agglomerates of less than 200 people are considered rural, while in Finland the limit is less than 500 people, for example (see Palttila & Niemi, 2003). In general, rural areas have been publicly associated with certain functions such as agriculture, low number of population, dispersed settlement patterns and peripherality (Cloke, 1992). OECD (1993: 11) states that rural areas are comprised of the people, land and other resources in the open country and small settlements outside the strong economic influence of major urban areas. In this respect rural represents something opposite to urban and city and is characterised by relatively small settlement patterns and low population density.

However, rural areas are not always restricted to any particular land use, economy or degree of economic wealth. Rather than defining rurality through certain objective measures, such as maximum population limits, it can be seen more as a socially constructed idea that characterises it and also differentiates it from urban in specific, but culturally changing, contexts. As an idea rurality refers to different landscapes and a way of living in different cultures and economic spaces; what is considered rural in central Europe may reflect urban or semi-urban in northern Sweden or Iceland.

In tourism, rural areas have become places to be consumed and where production is based on establishing new commodities for (urban) visitors or reimagining and rediscovering places for tourism (Cloke, 1992; Hall & Page, 2006). As a concept rural tourism is challenging to define. Narrowly, rural tourism can be limited to certain tourism activities located to specific places such as farms. More generally rural tourism can be defined according to its relationship with its environment (Sharpley & Sharpley, 1997). According to Lane (1994) rural tourism should ideally be located in rural areas and be functionally rural and rural in scale that is, based on small-scale enterprises, traditional social structures and ways of living, agrarian economies and relatively natural settings. The pace of development and change should be traditional in character, growing slowly, organically and be connected with and controlled by local family units. Rural tourism should also represent the complex pattern of the rural environment, location, history and economy (see Hall & Page, 2006).

In reality, many forms of rural tourism (or tourism in rural areas in general) do not correspond with these high principles reflecting more the ideal version of rural tourism (Sharpley & Sharpley, 1997). In practice there are various kinds of rural tourism activities and attractions which can differ by their location, style of management, degree of integration with the surrounding social structure, intensity of use and holiday type. The Council of Europe (1988), for example, has listed a wide range of leisure activities related to tourism in rural areas which do not necessarily meet the ideal situation presented by Lane (1994) (Table 5.2).

Despite the specific character of activities and functions rural tourism should be closely integrated in wider rural development goals. According to Hall and Jenkins (1998) rural tourism should:

(1) sustain and create local incomes, employment and growth (stimulate local business);
(2) contribute to the costs of providing economic and social infrastructure;
(3) encourage the development of other industrial sectors; for example, through local purchasing links (rural tourism should serve as a recognised factor in regional economic development);
(4) contribute to local resident amenities and services; for example, sports and recreation facilities, outdoor recreation opportunities, and arts and culture; and
(5) contribute to the conservation of environmental and cultural resources.

Many of these development goals reflect similar principles to sustainable tourism or alternative tourism (see Fennell, 1999). Although, as noted by Sharpley and Sharpley (1997), it should not be surprising that rural tourism is often confused or equated with the idea of sustainable tourism.

Table 5.2 The range of tourist leisure activities related in rural areas listed by the Council of Europe

Touring		Water-related activities	
• hiking • horseback riding • touring in caravans	• cycling • donkey riding • cross-country skiing	• fishing • swimming • river tourism (e.g. houseboats, barges) • sailing	• canoeing and rafting • windsurfing • speedboat racing • facilities on the 'aqualand' type, etc.
Sporting activities		Activities on dry land	
• pot-holing	• rock climbing	• tennis	• golf
Discovery-type activities		Aerial activities	
• local industrial, agricultural enterprises	• craft enterprises	• light aircraft • hot air balloons	• hand-gliding
Cultural activities		Health-related activities	
• archaeology • restoration sites • artistic workshops	• folk groups • course in crafts • cultural routes	• fitness training	• health resorts

Source: Council of Europe (1988: 25–26).

Sustainable Dimension of Rural Tourism

In addition to the ongoing socioeconomic changes in rural contexts the idea of sustainable development has strengthened the transition process of rural areas towards tourism and related modes of production (Marsden, 1999). Compared to forestry and many other primary economies, such as mining and fishing, tourism is also often regarded as a 'softer' and therefore more 'sustainable' option for the environment (see Saarinen, 2003). However, as noted by Butler (1999), tourism may not always be the most favourable or wisest use of natural or cultural amenities and resources in specific locations in the long term, and 'sustainable tourism' may in practice be an unsustainable and unequal process for the original rural communities (see Bianchi, 2004).

For a relatively long time tourism has been understood as a road to rural and community development (see Lane, 1994; see also Chapter 3) but since the early 1990s it started to represent a tool to put sustainable development into practice on a local scale (Berry & Ladkin, 1997; Hall, 2008). In tourism sustainability refers generally to development that meets the needs of present tourists and host regions while it protects and enhances opportunities for the future (Hall & Lew, 1998). While there is no consensus on the exact definition of the concept or its relation to the idea of sustainable development (see Saarinen, 2006; Sharpley, 2000) the following three elements are usually identified and linked to sustainable tourism: (1) needs of the host population; (2) satisfaction of tourists; and (3) safeguarding the natural environment (Cater, 1993; see also Butler, 1999). In rural contexts the conceptual link between tourism and sustainability leads to a tourism industry that sustains local rural economies without damaging the environment on which it depends (Countryside Commission, 1995; see also Lane, 1994).

However, while tourism may represent a potentially sustainable form of economic development it may also create new problems and environmental conflicts. For example, in Norway conflicts have emerged between traditional rural economic activities (commercial farming and forestry) and the interests of increasing recreational visitors (Larsen, 2001) (see also Case 3.2 by Arnesen and Riseth). Different users with their specific interests literally compete for the utilisation of the same areas for often quite mutually exclusive values and practices. It is expected that these kinds of conflicts between 'hard' and 'soft' or 'old' and 'new' uses of the environment will become increasingly common in rural areas in future as it is already apparent in all Nordic countries (see Hall & Müller, 2004a, 2004b; Puijk, 2000; Saarinen, 2005; Vail & Heldt, 2004).

Farm Tourism

One of the oldest forms of tourism in rural areas, especially in Nordic countries, is so-called *farm tourism*. According to Nilsson (2002) farm tourism has its ideological roots in the 19th century romanticism of nature and early 20th century social travel. Social travel, or its modern equivalent term social tourism, is a diverse concept that underlines the supply of holiday opportunities at an affordable price that make vacations possible for all social classes (see Chapter 3). Contemporary commercialised farm tourism is a small scale activity in which the farmer is the main industrial actor (Plate 5.1a and b). Farm tourism has a strong emphasis on the host–guest relationship where interaction between the visitors and locals is regarded as an intense mixture of 'the host's private life and the guest's experiences' (Nilsson, 2002: 10). Farm tourism or farm-based tourism is often equated with agritourism which is used to describe those tourism products which are strictly connected with the agrarian environment, products or stays (Sharpley & Sharpley, 1997: 9). However, agritourism may also cover broader elements such as festivals, museums and craft workshops, and food related tourism, for example, integrated into an agrarian context. Case 5.2 examines a culinary tourism project in Iceland.

Cases and issues 5.2: Culinary tourism project in Northern Iceland

GUÐRÚN GUNNARSDÓTTIR and LAUFEY HARALDSDÓTTIR

Attracting tourists to rural areas is a complex task in a competitive environment where major destinations are firmly established in the consumer's mind.

Plate 5.1 Rural tourism is traditionally a small scale and family-owned business in Nordic countries. Farm tourism businesses in Norway: (a) showing a property where horse riding is available; (b) showing a farm that offers artistic space and events in summer

Figure 5.2 The location of the study area, Skagafjörður, northern Iceland

Regional imagery is increasingly used to brand rural destinations, especially with the promotion of quality local food products (Williams, 2001). The Rural Tourism Department at Holar University College in cooperation with University of Guelph, Canada, launched a culinary tourism project in 2004, which aims to promote a regional culinary identity for the Skagafjörður region in Northern Iceland (Figure 5.2).

Tourism in Iceland has grown substantially over the last decades and is now a major source of foreign revenue. From 1990 to 2000, international tourism in Iceland increased by 117%; a yearly increase of 8% (Þjóðhagsstofnun, 2000). However, many rural regions lag behind in terms of tourism development and have not been successful in translating their resources into viable tourism products. This is the case for Skagafjörður where surveys among tourists demonstrate lack of a distinctive destination image, and is an area that tourists travel through to reach the north of Iceland (Guðmundsson, 2005; Gunnarsdóttir, 2005). On the other hand, a survey of tourists in the region showed a high level of interest in local cuisine and that tourists would like to be informed about locally made products (Murray & Haraldsdóttir, 2005). It seems that developing strong culinary tourism could enhance and strengthen the image of Skagafjörður region as a quality rural tourism destination.

Local food production and processing is an important prerequisite for developing regional food tourism. The main industries in Skagafjörður are agriculture and fisheries, and several successful food-producing and processing

Figure 5.3 Logo for the culinary project. It references one of the islands on the fjord, which for centuries was called the food chest of Skagafjörður due to its abundant bird life

companies are located in the region. In particular, Skagafjörður takes pride in its innovative fish, meat and dairy processing companies.

Credibility has to be an integral part of the regional branding process. Thus, the culinary project encourages participation from both the food processing as well as the restaurant sector. Substantial time has been invested in building up trust as well as establishing a shared vision and commitment to the project. This has enabled coordination among tourism and food producing/processing businesses in terms of labelling and increased accessibility of local products. This cooperation successfully culminated in the project's logo (Figure 5.3) whereby the participants accepted a unifying identity and resolved to use the logo on menus, packaging, and promotional materials. Restaurants experienced a positive response from customers as dishes labelled with the logo increased sales. The benefits of the project are not as immediate to the food processing companies which are not in direct contact with tourists. However, the project was frequently profiled in the Icelandic media which increased the companies' faith in the project.

This culinary project reveals that opening channels of communication between food companies and restaurateurs is important in developing a reliable rural culinary destination. The next hurdle to clear is to keep up the interest and commitment among the participants, and wider involvement is also required in order to sustain the authenticity of the culinary destination image.

Economically, farm tourism is a rather small branch of both fields of tourism and agriculture. According to the European Commission (1997) marketing development report, farm tourism represents about 2% of the total turnover of the tourism in rural areas and only 0.3% of the turnover from agriculture. Nilsson (2002) has contextualised the role of farm tourism in the wider field of the tourism industry in Sweden. According to him, farm tourism makes up only three overnight visits in 100,000 overnights for the whole country. Still, farm tourism may play an important role for the image of agriculture and a part of the touristic image of Nordic countries (Gössling & Mattsson, 2002). It can also be locally relatively important in economic terms. In Norway, for farms which rent moderate cabins or similar accommodation, the net tourism income represents approximately 15% compared to net agricultural income; but for farms with more developed tourism activities the net income is almost 50% of the net income from farming. The scale and role of farm tourism also varies regionally. In Denmark and Norway about 10% of farms are connected to farm tourism which is much less than in Sweden where about 20% of the farms are estimated to have some form of rural tourism activity (Hjalager, 1996; Rural Tourism International, 2006).

Although tourism may represent a small economic activity for rural areas, many farms involved in tourism are increasingly networking and organising their activities, including marketing, which may help growth and also quality development. In Finland, the Association of Rural Tourism Entrepreneurs has about 270 member farms. In Denmark, the nationwide organisation for farm tourism called *Landsforeningen för Landboturisme* (The National Association for Agritourism) has about 210 members and in Iceland there are around 120 farms within the Icelandic Farm Holidays (IFH) association (Nilsson, 2002; Rural Tourism International, 2006). At the time of writing the website of the Swedish farm stay association Bo på Lantgård reports that there are more than 400 farms available for farm stays in Sweden.

Tourism for Rural Development

The case of Finland: General background

The following case aims to demonstrate the expectations concerning the role of rural tourism and challenges of the rural economic transition process towards tourism by using the Finnish rural tourism development strategy (Rural Policy and Rural Tourism Groups [RPRTG], 2000) as an example. Based on the Finnish case, the issues in rural tourism development are later discussed in the wider Nordic context.

In Finland too, tourism is increasingly viewed as a potential sector of the economy for rural areas (see Komppula, 2004). Based on the future prospects of tourism the Finnish Ministry of Agriculture and Forestry appointed two committees (collaborative groups) which jointly processed the Strategy and Development Programme for Rural Tourism until 2007 (RPRTG, 2000). In the programme rural tourism is seen as a specific part of the tourism industry which draws its development possibilities from the intrinsic resources of 'rurality' and the rural way of living. In other words, not all tourism in rural areas is considered as rural tourism. In addition, the strategy also outlines that the foundation of rural tourism is based on environmental responsibility, health (e.g. in food production) and preservation of the rural cultural heritage which integrates the goals of the programme with the idea of sustainable tourism.

Strategy and development programme for rural tourism

The Strategy and Development Programme has both qualitative and quantitative goals (RPRTG, 2000). The first goals refer to the desired characteristics of future business and marketing skills, products and their quality, education level and environmental issues and planning. The latter goals, which will be discussed here in more detail, focus on the three elements of tourism in rural areas: occupancy rate of accommodation capacity, revenue and employment.

According to the programme the occupancy rate of rural tourism accommodation capacity should be 50% or more by the end of the year 2007. The total turnover of rural tourism business should be over 330 million euros and the employment effect should be 6000 year-round jobs. In practice, based on the situation in the late 1990s and the estimations of the programme, the 50% (or more) occupancy rate would mean an annual 8% increase in tourist numbers. The global trend of alternative tourism development, which often includes rural tourism, nature-based tourism and ecotourism activities, has been previously estimated to be 8–10% yearly (see Fennell, 1999; Scheyvens, 1999). In this respect the trend gives some support to the goal. In contrast to this, however, only in larger cities the use of accommodation capacity (hotel rooms) has been 50% or more.

In 1999 the estimated occupancy rate of rural tourism accommodation capacity (rooms) was about 30% (RPRTG, 2000). The official statistics do not include the category 'rural tourism accommodation' but in Finland their use and seasonal patterns of utilisation can be assumed to be rather similar to the use of holiday villages and camping sites which are included in the statistics. According to Statistics Finland (2000) the occupancy rate in holiday villages was 31.5% and in camping sites 32.1% in 1999. Their use rate has not increased dramatically in recent years; in 2004 the occupancy rate in holiday villages was 34.5% and in camping sites 34.3% (Statistics Finland, 2005a).

In relative terms the 330 million euros revenue goal refers to a 30% annual growth. Similarly, the employment goal of 6000 year-round jobs in rural tourism means almost 30% annual growth by 2007. Unlike the case of occupancy rate these two goals reflecting the relatively long-term 30% annual development are very unrealistic. However, they are not only unrealistic but also illogical and problematic in their relation to the very idea of rural tourism expressed in this specific rural tourism development plan and with the concept of rural tourism as a policy tool.

Cases and issues 5.3: Rural development and tourism entrepreneurship in Finland

RAIJA KOMPPULA

Rural tourism development in Finland

Finnish rural tourists expect to have a peaceful, quiet and rush-free country holiday. The expectations of the service during the holiday included food, sauna and some kind of nature activities (most often swimming, rowing and walking in the forest). For Finns, nature, particularly lake and forest scenery, is a self-evident part of a country holiday (Komppula, 2005). According to a study conducted in 2003, about 37% of 15–74-year-old Finns had spent at least one holiday in last five years in a rented cottage in the countryside. Forty-one percent of the travellers were planning to spend their holiday in a rented cottage in the near future (see Liiketaloudellinen tutkimuskeskus, 2003). Rural tourism customers value the traditional country scenery, a living countryside, nature, opportunities for activities, easy accessibility and personal service. The entrepreneurs, for their part, aim to keep their units fairly small, emphasising personal service, offering services near nature, and emphasising the rural lifestyle. The entrepreneurs' willingness to maintain a small firm size hence seem to support the fulfilment of the wishes of the customers.

The Rural Policy Committee in Finland has been founded by the government and is appointed by the Ministry of the Interior. This committee's task is to harmonise rural development measures and to enhance the efficient use of resources targeted for rural areas. This committee has appointed a theme group on rural tourism in order to promote overall development of rural tourism. The development programme of the Theme Group on Rural Tourism mentions, for example, family and small businesses, local and independent development work and collaboration within the field as its strengths. Its weaknesses include, for example, a lack of efficient and long-term marketing, earlier investments in buildings, poor profitability, the mixed level of services and production-based service culture.

According to the vision of the theme group, rural tourism in 2007 will be an essential and solid part of tourism and rural areas in Finland as a source of livelihood, employment and cultural advocacy. The most important underlying values when developing rural tourism are responsibility towards the environment, health and a high appreciation of the cultural heritage of the countryside. The focal strategic points in order to reach the goals of the programme will be placed on the enhancement of product development work and marketing, the development of the quality control system, an increase in know-how and the targeting of public funding towards these areas of emphasis (see RPRGT, 2000).

There are plenty of actors to develop rural tourism. The Ministry of Trade and Industry bears the main responsibility for tourism policy. Regional policy, which is ruled by the Ministry of the Interior, affects the development of tourism on regional development. According to the guidelines of the rural policy programme of the Ministry of Agriculture and Forestry, the opportunities from tourism for the development of rural areas are enhanced by the promotion of rural, village and nature tourism based on small businesses and network cooperation (see RPRGT, 2000).

Rural tourism entrepreneurs in Finland

Rural tourism entrepreneurship in Finland typically consists of micro family businesses, run by couples, and often also portfolio entrepreneurship. In the small enterprise register of rural areas, there were a total of 3248 businesses that offer tourism and recreation services in the whole country in 2000. These employed a total of 5736 person work years and their turnover adjusted according to the wholesale price index was a total of 410 million Euro. In the capacity report of the Theme Group on Rural Tourism, it was estimated that there are 3600 rural tourism businesses (Markkola, 2003a, 2003b; see Komppula, 2004).

The average age of the rural tourism entrepreneurs is 50.5 years. Approximately 69% of the businesses operate year round, but only one quarter consider tourism as their main form of livelihood. Each enterprise employs approximately 1.5 person work years. The turnover of 44% of the businesses is under the limit of VAT (8400 Euro), only 12% of the businesses have a turnover over 84,000 Euro. Half of the businesses have some kind of a quality control tool at their disposal. The occupancy rate of accommodation services vary considerably according to season: in summer over 50% but during other seasons under 15% (Markkola, 2003a, 2003b). About 90% of the businesses offer accommodation services, 58% food and two-thirds also offer different types of activity services. The year round accommodation capacity in an enterprise is about 16 beds, so it can be estimated that there are approximately 42,000 beds in total in the whole country within rural tourism.

Previous research shows that the motives for starting rural tourism businesses in Finland are in most cases related to existing premises (Komppula, 2004), which make the accommodation or activity services an opportunity to earn extra income and/or employ one or more family members. A major part of the rural tourism businesses start as part time tourism entrepreneurship. The diversification of the businesses is one of the reasons why the growth in tourism businesses is often very slow. If the entrepreneur has to manage several professions at the same time, personal resources to develop all of them are limited. Another barrier to rural tourism firm growth is the entrepreneurs' reluctance to employ full-time workforce outside the family. Entrepreneurs also want to maintain their close personal contact with the customers. Customer satisfaction and the creation of long-term customer relationships is an important indicator of success. Satisfied customers are important marketers of rural tourism, and regular customers will guarantee the growth of turnover in the long run.

A typical rural tourism entrepreneur tends to avoid economic risks and does not want to make rapid growth investments in his/her businesses. Rapid growth expectations might be more likely among young, full-time operators that have bought or inherited the premises of the tourism businesses. Reasons for a low desire for growth are often related to the entrepreneurs' age or state of health: an entrepreneur near retirement does not have the motivation to invest more in the tourism business, especially if there is no possibility for succession or other means of continuity. In many cases the family has already reached a certain standard of living, which the entrepreneurs do not want to risk. In general, success is often measured by affective and subjective measures: reasonable level of income is more important than getting rich, time for family and hobbies is respected, interesting and challenging work and the opportunity to work at home or in the countryside is important for the entrepreneurs.

Growth in turnover is desirable according to most entrepreneurs but without any growth in labour or investment in expansion of capacity. For example, renovation of the accommodation living quarters may increase the occupation rate as well as turnover, but not the capacity itself. But, regardless of which measures of success are used, it can be argued that owners of small rural tourism establishments are in business primarily to earn their living, which calls for at least sufficient profit.

Conclusions

In order to discover the real growth potential of the rural tourism industry in Finland, development activities should be targeted to younger proprietors, who, at the moment, are a clear minority in the rural tourism businesses in Finland. A new approach is needed in the small tourism business support policy and its implementation. Many of the support systems in rural tourism development currently necessitate a high degree of commitment and involvement on the part of small firms. In most cases, financial support is allocated on condition that the company may commit to growth objectives. Nevertheless, many of the lifestyle-oriented owner-managers may be reluctant to make such a commitment and they may be unwilling to participate in development programmes. In the tourism development programmes, the acceptance of 'growth in quality but not in volume' might be the way to balance the divergence between common development policy and private interests.

Reasons and outcomes for the high expectations in rural tourism

Unfortunately there are no updated estimations for rural tourism in 2007 which could be used in evaluation of the goals and success of the programme but, in general, the goals can be assumed to be far too high (see Chapter 3 with respect to the importance of evaluation in tourism planning). The too high expectations also reflect the wider issues and challenges of tourism and its usage as a policy tool for rural development. First, they manifest a certain lack of knowledge and understanding of tourism dynamics and the very nature of tourism. The high expectations of growth and development are implicitly or explicitly an argument based on global scale trends and the significance of tourism. However, tourism is a highly polarised activity and global or even national trends will not become easily, if at all, concrete in a specific local (especially rural) context. As global scale phenomena tourism and its development are very time-space specific issues and the high estimations of alternative or rural tourism development may be based on relatively localised developments and trends, both in the geographical and seasonal sense (see Campbell, 1999).

Thus, generalisations based on these time-space specific tourism developments and individual cases should be drawn carefully. In spite of the fact that multiple studies and policy papers indicate that rural tourism is good business there are also many examples for the opposite situation (e.g. see Fleischer & Pizam, 1997; Hjalager, 1996). In addition, it must be noted that, at least in the Nordic peripheral contexts, high tourism development and tourist numbers have been mainly concerned with larger tourist destinations and resorts. The growth of tourism has also been strongly based on the internationalisation of visitor flows and this has channelled tourism-related economic development mainly to tourist destinations, resorts and cities rather than to rural communities. What makes this especially problematic are the missing operative and policy links: there are no effective tools to integrate developing tourism spaces and seasons, like the rapidly growing Christmas tourism in northern Finland and Sweden, to rural communities and rural tourism activities. Therefore in the Nordic peripheries, as in developing countries, tourism development often contributes to those areas and actors that least need it in broad economic terms (see Müller & Jansson, 2007a).

Second, the development goals in rural tourism are often based on a thin layer of facts and research. As widely discussed in the literature the statistical basis of tourism in general is problematic (see Hansen & Jensen, 1996; Lennon, 2003; Smith, 2004; see also Chapter 1) but those problems are relatively easy compared to the challenges faced with the special sectors and segments of tourism phenomena such as rural tourism. How is it possible to distinguish rural tourism economically, for example, from the wider tourism industry in rural areas or the rural tourism entrepreneur from the 'general' tourism entrepreneur in statistics (see Sharpley & Sharpley, 1997)? In practice this is perhaps impossible, which almost automatically leads developers in rural tourism to use assumptions and estimates. Such figures can be constructed in multiple and 'innovative' ways and if a research or evaluative component is not involved with a development plan or initiative, which is often the case, local estimations may reflect more the developmental hopes than factual present day reality (Hall, 2007d).

Third, and perhaps most importantly, high development goals challenge the connection between rural tourism and sustainability. By definition the concept of rural tourism is an ideological one – at least this is the case in the context of rural community-based tourism development discussions (see Lane, 1994). As an ideological concept it aims to support a rural way of living without changing the countryside too much, or at least the control concerning economic, social and environmental changes should be local or the result of a collaborative decision-making process (see Countryside Commission, 1995; Hardy et al., 2002).

Also, in policy papers, such as the Finnish Strategy and Development Programme for Rural Tourism (RPRTG, 2000), rural tourism aims to sustain local (i.e. rural) economies and protect the environment. Therefore, with strong links to the idea of sustainability the concept of rural tourism can be understood more as a value-based policy tool and process, rather than an idea where 'rurality' is only representing a stage for tourism activities in general (see Hall & Jenkins, 1998;

Lane 1994). From that perspective tourism is seen as a tool for rural community development.

In contrast to this, the high level of rural tourism development goals and the narrow focus only on tourism indicators in growth transforms rural and community development into tourism development. Tourism becomes an end in the development process. Burns (1999) indicates this outcome with the term 'tourism first', in contrast to the term 'development first', referring to the idea of tourism as tool, in which the tourism industry and its needs dominate the development discourses and practices. Conceptually this challenges the connection between rural tourism and sustainability and it may do this in practice as well. The desired goals of rural and community development are not necessarily the same as the tourism industry's goals (see Butler, 1999; Saarinen, 2006; see also Chapter 3).

Summary

In Nordic countries tourism has been widely promoted and used in rural development as a replacement economy for 'traditional' livelihoods based on rural production. The future prospects of tourism, especially, are seen as a potential instrument to control the economic transition process and its social consequences in rural communities. In addition the real or perceived lack of alternatives in rural areas has served almost as an absolute necessity to develop tourism (Hall & Page, 2006; Müller & Jansson, 2007a, 2007b).

Tourism can be a 'saviour' of peripheral rural communities but, in the context of rural tourism as an ideological development tool with strong links to sustainability, tourism should be integrated primarily with regions, communities and their development goals, not the other way around. As has been addressed here, high development goals for rural tourism may separate rural communities and tourism actors from each other. Thus, tourism may grow and increasingly provide employment opportunities but they will not necessarily meet local employment needs. Tourism may also have social, economic and environmental costs in rural areas as far as other local sources of livelihood are concerned, and these costs are often disregarded in tourism development discourses and related practices. This may lower the threshold of dissatisfaction and raise the intensity of possible conflicts between rural communities, traditional economies and tourism industry which is the case in some parts of Norway (Larsen, 2001), for example.

Rural tourism development plans, strategies and other initiatives are important for rural community development and the tourism industry in rural areas but, instead of monitoring tourism development indicators in rural development processes, the indicators of the developmental contribution of tourism should be established and evaluated. That would serve both the long-term development needs of communities and tourism. The relationship between rural and tourism development is not necessarily a 'zero-sum game' but too ambitious and unrealistic goals in tourism may lead to that situation and the separation of rural communities and tourism development. Rural tourism as an ideological concept and policy tool is not automatically sustainable tourism nor sustainable rural development, but it has a stronger connection to the idea of sustainability as a small-scale, long-term and locally controlled economic activity rather than often non-locally controlled, short term and more massive (conventional) tourism development. However, it is probably needed in practice if some of the high development goals are going to be fulfilled.

Demand for sustainability in rural tourism may also have its challenges (Gössling & Mattsson, 2002). In addition to practical problems in achieving sustainability a certain kind of imperative of sustainability in rural tourism development may also direct development more towards social, cultural and heritage issues. This may limit the potential for economic development of rural areas as outlined by Sharpley (2004). Indeed, there is a need for a broader view on the relationship between tourism and the countryside beyond rural tourism as a specific and very limited niche market. However, in the context of rural community development, growing tourism should be integrated into larger rural policy issues and goals. The development indicators in tourism strategies and plans should also reflect widely indicated developmental contribution and the 'promise' of tourism in rural communities.

Discussion questions

(1) What are the reasons for the growth of tourism in rural areas?
(2) What are the positive and negative aspects of developing tourism in rural areas?
(3) Is rural tourism 'just tourism' in rural areas?

Essay questions

(1) Discuss the developmental role of tourism in rural areas.
(2) What are the connections between rural tourism and sustainability and what are possible obstacles to developing sustainable rural tourism?

Key readings

Rural tourism has been one focus area in tourism development and studies in Nordic countries. Internationally there are well cited text books such as *Tourism and Recreation in Rural Areas* edited by Butler *et al.* (1998) and *Rural Tourism and Recreation: Principles to Practice* edited by Roberts and Hall (2001). In the Nordic context there are numerous individual paper including Aronsson (1994), Hjalager (1996), Gössling and Mattsson (2002), Nilsson (2002) and Komppula (2004). In addition Hall and Müller (2004a), Hall and Boyd (2005), Hilding-Rydevik *et al.* (2005), and Müller and Jansson (2007a) include chapters and elements involving rural change and tourism in Nordic rural and peripheral areas.

There are numerous web pages covering rural tourism accommodation opportunities. Many of them (Novasol, DanCenter and Lomarengas) are the same as second home providers (see Chapter 8). Rural tourism development-oriented and educational organisations include both governmental (e.g. Ministries of Forestry and Agriculture and Ministries of Environments in Nordic countries) and international and national non-governmental bodies such as Rural Tourism International and European Rural Tourism Development.

> European Rural Tourism Development: http://info.nortcoll.ac.uk
> Rural Policy Committee Finland: http://www.ruralpolicy.fi/
> Rural Tourism International: http://www.ruraltourisminternational.org
> Bo på Lantgård (Swedish farm stay association): http://www.bopalantgard.org/Valkommen/
> Maatilamajoitus (Finnish Farm stays): http://www.maatilamajoitus.fi/
> > Opplev Gode Norge (Norbooking, Norwegian Farm stays and experience products): http://www.opplevgodenorge.no/

Chapter 6
Nature-based Tourism in Northern Wildernesses

Learning Objectives

After reading this chapter, you should be able to:

- Understand the concepts of nature-based tourism and ecotourism.
- Understand the concept of wilderness in a Nordic context.
- Identify some of the competing concepts and uses of wilderness in the Nordic countries.

Introduction

In the Nordic region nature has been one of the main attractions in tourism and characterise much of the Nordic tourism experience. Although there is a wide variety of ecosystems, habitats and landscape types in the region including coastal environments, mountains, mires, meadows and taiga forests, for example, perhaps the most dominant feature among the images of northern nature and landscape is wilderness. In recent years wildernesses have been increasingly the focus of Nordic tourism development, and the tourism industry has become a very significant user and element of change in wilderness environments.

In particular, the international and growing trend of nature-based tourism and related forms, such as adventure tourism and ecotourism, have influenced the economic, social and also ecological issues in northern wilderness areas. These forms of tourism are also increasingly competing with traditional uses of wilderness areas and this situation has prompted both political and economic discussions on the relevant meanings, uses and values of these areas (Saarinen, 2005). Many of the regional economies that utilise wilderness environments are also facing a situation in which other industries and their impacts and future prospects are compared with a growing tourism industry and its development potential. This has been the case with forestry and tourism in Finland and Sweden and hydro power generation and tourism in Iceland, for example (see Lundmark, 2006; Saastamoinen, 1982; Saethorsdottir, 2004; Vail & Heldt, 2000; Vuorio, 2003). This chapter introduces the basic concepts and evolution of tourism in Nordic wilderness and elaborates the possible impacts of growing tourism in wildernesses environments and other natural areas.

Nature-based Tourism and Wilderness

Nature-based tourism and related terms, such as adventure tourism (see Chapter 10) and ecotourism, are often conceptualised under the umbrella of alternative tourism (AT). According to Krippendorf (1982) alternative forms of tourism refer to tourist activities and practices opposite to mass scale tourism. In addition there should also be an emphasis on the demand for unmodified natural or cultural environments and local needs in development. However, in practice, it is often rather difficult to differentiate the operations and impacts of nature-based tourism from general tourism. For example, the *sun-sea-sand* kind of tourism is typically seen as mass tourism, but those same elements also form a basis for many nature-based tourism activities whose scale may sometimes be relatively close to mass tourism destinations. Many national parks in the Nordic countries can face hundreds of thousands visitors per year, and a major part of their visits may be based on packaged tours to tourist resorts near the parks.

Nature and natural attractions are 'today universally regarded as a source of pleasure' (Wang, 2000: 80), but natural areas and wilderness environments have not always been the focus of positive attention for visitors, nor have they always been a source of positive experiences. According to Short (1991: 6) 'fear of the wilderness was one of the strongest elements in European attitudes to wilderness up to the 19th century' (see Hall, 1992, 1998). Thus, the history of nature as a significant tourist attraction is quite short, in fact. The famous late-18th century Swedish natural scientist Carl von Linné (1969/1889: 68) described Lapland, which is nowadays promoted in terms of positive images of wilderness and a source of vitality, as a hell on earth for which 'the right name should be Styx'. Today, these same 'landscapes of evil' are visited internationally by masses of (nature-based) tourists, simply to experience the natural beauty of the wild landscapes and mountain sceneries (see Chapter 2 with respect to the notion that tourist attractions are socially constructed). A good example of such changes is the extent to which Arctic regions are now seen in a favourable light with respect to tourism (see Case 6.1).

Cases and issues 6.1: Svalbard: Wilderness tourism in the High North

BJØRN P. KALTENBORN

The archipelago of Svalbard is now one of the prime tourist destinations of the Arctic (Kaltenborn, 1996). The High Arctic has traditionally been quite inaccessible to tourism, but this is changing fast. In a world where wilderness in almost any sense of the word is shrinking fast, the growing nature tourism industry is actively looking for places to expand and develop new products. The Polar regions are often marketed as the last wilderness, and in a sense the Antarctic and High Arctic represent a last frontier in wilderness tourism. It is not hard to grasp the appeal of Svalbard: vast expanses of pristine nature, wildlife, a fascinating cultural heritage, remoteness from urban centres, a geographic size that greatly contributes to the impression of wilderness and a well organised nature tourism industry. Tourism has made definite and significant contributions toward the upholding the communities in the island and have formed interesting partnerships with management.

Historically, tourism has played a modest role in developing livelihoods in the Norwegian High Arctic, although visitors started coming to these regions for pleasure purposes more than 100 years ago. Cruise ship traffic actually began as early as the 1870s for the privileged few (Arlov, 1996). Like most communities in sparsely settled areas dependent on primary resources such as fishing, wildlife or mining, modernity has been a tough challenge. At the same time, many of these communities tend to be located in areas that are fragile in terms of ecological conditions, and often also in terms of the social and cultural aspects of livelihoods. With the decline in coal mining and reduced military strategic importance of Svalbard from the 1970s and on, the problem was really one of dealing with unwanted and unforeseen ecological and economic changes at the same time. Unlike what often happens, this challenge has improved the economic development in the islands, professionalised the tourism industry and provided impetus and legal backing for improved tourism management and management planning.

Prior to around 1990 there was virtually no locally based tourism industry in Svalbard. In 1997 Longyearbyen, the largest community in Svalbard, received around 46,000 paid visitor days. In 2004 this number had increased to 77,926 (SSB, 2005). Included in this figure we find business travel as well as courses, conferences and tourism. In 1997 about half of the visitors were tourists, while tourism had risen to 60% in 2004. A look at the cruiseship sector also gives an indication of the development of the industry. In 1996 the ships made stops and went ashore in 52 different locations around the archipelago. In that year 23,846 tourists went on shore for short stops – typically just a few hours. In 2004 this was more than doubled to 140 locations enjoyed by a total of 40,713 passengers. All of these, however, represent overseas cruise tourism. Large numbers of people visit Svalbard as part of an extended tour in the North Atlantic, but they spend very little time on the islands. Hence, both the economic and environmental impact is limited. However, the economic gain is not insignificant even if these people barely spend time on shore. In 2003 around US $300,000 was spent by tourists at the various facilities (e.g. dining, souvenir shops) at the harbour in Longyearbyen (Svalbard Reiseivsråd, 2004). Yet, there is clearly potential for increasing environmental impacts since this part of the industry is difficult to influence and control beyond the actual landings around the coasts, and this segment is increasing.

Locally-based coastal tourism has also been developed in Svalbard since the early 1990s, and this type of tourism is typically combined with longer stops and a range of recreational activities on shore such as hiking, wildlife viewing and mountaineering. In 2001, 8654 people participated in locally arranged coastal cruises, while the number had risen to 13,404 in 2003 (Svalbard Reiselivsråd, 2004).

A number of people travel to Svalbard on their own, that is, without participating on a guided trip. This segment comprises almost all forms of wilderness tourism such as: hiking, skiing, dogsledding, mountaineering, sea kayaking, wildlife viewing and visits to cultural heritage sites. Since the motives of the wilderness users often are complex, a trip to Svalbard may typically involve a combination of activities (Kaltenborn, 1997a, 1999, 2000; Viken, 1995). For many, Svalbard represents the lure of the North, the attraction of the outpost of the ultimate marginal area. Viken (1995) has argued that some people express a need to symbolically conquer Svalbard as a faint echo of the time of the explorers. Be that as it may, there is no doubt that a pristine natural environment, abundant wildlife and all the remains from the formerly active hunting and trapping era, provide core elements of the attractiveness to the wilderness tourist. It has also been shown, as in many other studies of nature tourism, that the average visitor is well educated, well prepared and quite knowledgeable about the Svalbard conditions (Kaltenborn, 1999).

Both the business and the management response to this development has been noteworthy. The tourism industry has clearly matured from a rogue pioneer phase to a sustainable industry in a relatively short time, that is, 15 years and makes a significant contribution to the local employment structure. From 1998 to 2002 there was an almost 50% increase in economic turnover related to tourism in Longyearbyen. In 2002 this totalled approximately US $30 million. The actual earnings due to tourism was around US $8 million in 2002. The number of local jobs (man-years) increased from 59 to 175 (39% increase) during this time period (Svalbard Reiselivsråd, 2004). Furthermore, the tourism industry has developed cooperative forums, guide training, environmental standards and a general mainstreaming of its activities. Environmental authorities have implemented a management plan for tourism covering the entire archipelago, and regulations pertaining to permissions, insurance, activities and travel modes are in place (http://www.sysselmannen.svalbard.no/). Field staff monitor visitor activities and provide information. Yet, environmental management challenges are plentiful. While some evaluations concur that the development is well controlled, little is actually known about the potential environmental impacts.

In general, nature-based tourism refers to forms of tourism that use natural areas and resources in a wild or undeveloped form with the purpose of enjoying natural areas and/or wildlife (Goodwin, 1996). Thus, nature-based tourism does not include strict or actually any limits of use levels or elements of sustainability. In nature-based tourism the focus is on the natural environment which forms a stage and the main attraction and motivation for a variety of (possible) tourist activities. In contrast to the general notion of nature-based tourism many natural area management authorities in Nordic countries use the term *sustainable nature-based tourism* when referring to small-scale tourism and limited forms of tourism activities and impacts (e.g. see Hall, 2006a).

The idea of sustainable nature-based tourism is closely related to the international concept of ecotourism. Fennell conceptualises ecotourism as

> a sustainable form of natural resource-based tourism that focuses primarily on experiencing and learning about nature, and which is ethically managed to be low-impact, non-consumptive, and locally oriented (control, benefits, and scale). It typically occurs in natural areas, and should contribute to the conservation or preservation of such areas. (Fennell, 1999: 43)

Ecotourism is a relatively challenging idea for the tourism industry and tourism development actions. In many cases ecotourism may actually be used in contexts and activities that are not actually referring to 'a sustainable form' but to a general notion of nature-based tourism in which tourism occurs in relatively non-urbanised areas or uses non-built environmental resources. This tendency has created some pessimism among tourism researchers and natural area managers about the term (see Duffy, 2002; Holden, 2003). However, it should be noted that the misuse of the term does not necessarily make the concept and idea of ecotourism useless, but rather raises the need to higher ethics in tourism development and management processes (see Fennell, 2006). Viken (2006) argues, however, that the concept of ecotourism in a Nordic context is perceived as obsolete, since Nordic outdoor traditions are congruent to modern ecotourism ideologies. Hence ecotourism means nothing more than 'business as usual'.

Wilderness, Tourism and Place Ethics

Nature-based tourism and various forms of alternative tourism have been successful in commodifying the wilderness. The term wilderness is a contested idea and it is difficult to find any consistent definition for the concept of wilderness, for as with concepts as a whole, its definitions tend to vary with time and from one culture to another at the same point of time (see Nash, 1982; Short, 1991). Therefore, it is quite understandable that different objectives and values, often mutually contradictory ones, are connected with the use of wilderness in tourism, other economies and activities.

The idea of wilderness can be perceived above all as a value-bound, ethically loaded one, a locus for the examination of questions concerned with the regulation of 'appropriate' land use (e.g. for tourism, nature conservation or forestry). Our notion of wilderness as a place and its character reflects our values and relation to it and also the types and levels of activities which we consider acceptable in the place called wilderness (Saarinen, 2005).

Although it is challenging to find a common ground for the idea of wilderness an attempt has nevertheless been made in legal contexts and in connection with various international and national scale agreements. When preparing a national wilderness law the Finnish Wilderness Committee, for example, defined the basic characteristics of wilderness areas as follows (Erämaakomitean mietintö, 1988: 23):

(1) A wilderness area should comprise a minimum of 15,000 hectares and usually be more than 10 km in width.
(2) The area should be ecologically as diverse as possible and all human action should be adjusted to nature so as not to spoil the wilderness character of the area.
(3) The area should as a rule have no roads.
(4) The landscape should be in a natural state condition and unspoiled. Any structures connected with human activity should merge with the natural landscape.

However, these kinds of elements of wilderness characteristics vary from one country to another. In Sweden the corresponding minimum size is 50,000 hectares and in the United States the minimum size for an official designated wilderness area is 2020 hectares (Hallikainen, 1998), but a common rule for different national and international legislation is that wilderness areas should be in a natural state, have no roads and contain natural fauna and flora (Erämaakomitean mietintö, 1988; Hall, 1992; Saethorsdottir, 2004; Watson & Kajala, 1995).

The first wilderness legislation was prescribed in the United States in 1964 (Nash, 1982). According to the US *Wilderness Act* (United States Public Law, 1964), wilderness is placed outside society and culture: 'A wilderness, in contrast with those areas where man and his own works dominate the landscape, is hereby recognized as an area where the earth and its community of life are untrammelled by man, where man himself is a visitor who does not remain.'

The idea of the Act – 'man himself is a visitor who does not remain' – is somewhat opposite to the basis of the previously referred to Finnish Wilderness Act (Erämaalaki, 1991) which states that designated wilderness areas are founded for: (1) maintaining the wilderness character of the areas; (2) securing Sami (indigenous people) culture and traditional livelihoods; and (3) developing the versatile use of nature and possibilities for the different (economic) uses. Apart from the first somewhat open characterisation, a wilderness is clearly a resource inside the local culture and economy where people are not only visitors but potentially also inhabitants and/or daily users, and it is also defined through its present local use and economic value.

Present wilderness uses, values and images are increasingly defined in seemingly distant global scale processes such as the mass media. In these processes wilderness accommodates new meanings and values, and some of the previous ones may become relics, traces of the past with thinning connotations for new generations. The tourism industry plays an important part in this transformation process. Many traditional wilderness-related activities, such as fishing and hiking are nowadays modified towards adventure treks, snow-mobile and wilderness safaris, for example (Saarinen, 2005).

Generally, human attitudes towards wilderness may be categorised as being either anthropocentric or biocentric in nature (Hall & Page, 2006). The touristic use of wilderness is based on anthropocentric values and views on nature. The emphasis of the biocentric approach is that the value of wilderness emerges in its potential for direct (or in-direct or future) human use and benefits. A basic assumption in anthropocentrism is that wilderness values are principally human-bounded. In contrast to that, biocentrism places emphasis on the preservation of the natural order. In tourism the important distinction between anthropocentrism and biocentrism is the extent to which human benefits of wilderness are seen as dependent on the natural integrity of the wilderness settings (Hendee et al., 1990: 18–19). However, biocentrism may also include connotations with the idea of deep ecology in which wilderness has an intrinsic value without direct or indirect human uses and meanings.

The different values and attitudes towards wilderness can be distinguished as four basic attitudes (Table 6.1): *utilism, humanism, mysticism* and *primitivism* (biocentrism) (Pietarinen, 1987). The most common of these is utilism, denoting 'purely conceived notions of utility' (Pietarinen, 1987: 323). According to Pietarinen

Table 6.1 Four attitudes to wilderness areas

	Objective	*Justification*	*Wilderness image*
Utilism	High standards of social and human well-being by increasing production	Unrestricted right of man to exploit wilderness areas to promote his well-being and production	A source of raw materials and fuels
Humanism	Human perfection and mental balance	Unrestricted right of man to exploit the wilderness to promote his/her perfection	A valuable opportunity that people should develop through their own actions
Mysticism	Unity of man and nature	The highest value of human life is to aim at the sacred state embodied in unspoilt nature	Basically a large spiritual entity
Biocentrism	Safeguarding the inherent value and functions of wilderness areas	All species are equally valuable – man has no special position	A total ecological system with an inherent value of its own

Source: Partly after Pietarinen (1987).

(1996: 290) by utilism he means 'an attitude or way of thinking that nature exists only for the welfare of humankind'. Its substantial characteristics including

(1) End: A high level of welfare for people.
(2) Conception of nature: Nature is a system regulated by causal laws; it provides a huge and valuable source of energy and raw materials.
(3) Legitimisation: Humanity has an unlimited right to use nature for the welfare of people.
(4) Relation to technology: Science and technology are all important, especially technology which helps us improve the effectiveness of production and thereby improves human welfare.
(5) Optimism: All problems related to the welfare of humankind can be solved by promoting science and technology (Pietarinen, 1996: 290).

The approach underlines the unrestricted right of people to exploit the natural environment, and maintains that any excessive exploitation can be compensated for through ever-advancing technological innovations. Humanism, in turn, maintains that the natural environment should promote human development in a variety of ways, not only as a source of raw materials, but also as a means towards attaining ethical, aesthetic and mental equilibrium. In this respect humanism does not attempt to break loose from the human values attached to wilderness but rather makes a conscious effort to emphasise the instrumental character of the 'non-material' values attached to them. Mysticism perceives people as part of a more extensive entity formed by nature, and searches for an experiential unity between people, the natural and the divine. The fourth basic attitude, primitivism, or biocentrism, represents an approach in terms of values which clearly recognise and insist on the inherent worth of the natural environment, maintaining that people have no special rights to exploit nature and that human well-being should not rest on a foundation which causes damage to nature.

Tourism and the Competing Ideas of Northern Wildernesses

Traditional and conserved wilderness

Historically, wilderness (*erämaa* in Finnish, *vildmark* or *äremark* in Swedish) has signified a region outside permanent settlement that has economic importance in terms of hunting and fishing (Lehtinen, 1991). In the medieval system of the hunting economy, the wilderness was considered to have been an integral part of the system of society at that time. In this system, the hunting areas and especially their taxation were often organised according to families and communities. At that time separate wilderness areas provided wealth to the areas of permanent settlement and commercial operators (see Hallikainen, 1998). This kind of traditional Nordic idea of wilderness represents a utilitarian approach, in which the natural environment is regarded as a source of numerous raw materials. Arguments of this kind continue to be supported by objectivising the wilderness only in terms of cubic metres of timber, jobs and sums of money, that is, wilderness environment is not regarded as possessing a value in itself.

Nevertheless, the current ideas of wilderness cannot be solely understood by the past practices of the fishing and hunting economy or their present modes such as forestry serving the needs of distant companies and shareholders. In the Nordic context the legislative protection need of the wilderness became more prominent in the late 20th century. That kind of *conserved wilderness* was, in many parts, based on the Anglo-American concept of wilderness and the corresponding history of using the wilderness. The Western concept of nature is thus mainly formed through conquering wilderness (Hall, 1992; Short, 1991: 5–10). As Nash (1982: xiii) writes in *Wilderness and the American Mind*: 'Civilisation created wilderness'. Saying this, he refers to the construction of the polarisation of nature and culture in north American societies of European origin, and to the clear line drawn between wilderness and organised society.

The conserved wilderness, including the recreational and aesthetic aspects of nature conservation areas, can be regarded as representing mainly a humanistic attitude towards the wilderness, and as far as hiking (walking) is concerned, also a mystic relation to nature, for the motives for this often involve experiencing the natural environment as sacred and seeking unity with it (see Thoreau, 1955). The basic humanistic approach also means that man should make use of technology to ensure that nature can be used more fully as a means of promoting the edifying objectives of humanism. This is what national parks and designated wilderness areas are in a sense all about, for the provision of overnight huts, log walkways, campfire sites and bridges, for example, may be seen as an attempt to give almost everybody a chance to reach and explore an 'untouched and sublime' natural environment. Stretches of wilderness are regarded as offering aesthetic, ethical and educational experiences at both the individual level and at that of humanity as a whole (Martin & Inglis, 1983).

Touristic wilderness

Today wilderness areas are an integral part of modern societies and their institutional and economic relations in the globalised world. Conserved wilderness represents these kind of globalised values and meanings of wilderness but new, and increasingly important, international and global phenomena created by modernisation and, for its part, using, producing and reproducing the ideas of wilderness, is the tourism industry. Tourism is also increasingly challenging and changing the historically constructed ideas of wilderness in the Nordic context.

According to Hall and Page (2006) the growing demand for tourists to experience the wilderness has been based on the changed and more positive attitudes towards the environment in general which have been transmitted to touristic uses of wilderness. In addition, the better access to natural areas has integrated peripheral areas more closely to global tourism markets and also to global ideas and attitudes towards wilderness environments and their uses and values. Thus, the impacts of tourism on wilderness and the surrounding regional structures are not simply directly physical, economic or affecting local employment, for example. Tourism, and especially the related advertising and media (travel, programmes, literature, magazines and tourist brochures) construct images of distant places, destinations and cultures on the basis of our own culture and way of life (see Chapter 2). These created images may affect the practices related to the distant wildernesses, for example, through nature politics and touristic demands.

The *touristic wilderness* is constructed through direct and indirect tourism development, its uses are based on consumption, marketing, visualising natural environments and staging wilderness settings for the purposes of the tourism business. In advertising, positive images, such as experience of freedom and naturalness are connected to the product to be marketed, and eventually as a part of the identity of the consumer: the tourist. As such, the touristic wilderness is a commercialised space. It is a commodity, a resource and product that can be produced, replicated and consumed.

The basis of touristic wilderness represents a process of change affecting wilderness that is simultaneously global, perceived through the eyes of outsiders, and local, perceived through the eyes of those who live there or are frequent users. In tourism, wilderness environments are 'glocal' units that combine local economic benefits and uses with global values and meanings attached to the wilderness. Wilderness emerges as a resource inside the local and a wider economic system and culture. Touristic wilderness is based on the active production, reproduction and recycling of the representation of wild, free, harsh and rugged nature, and their combination in relation to the modern cultures of consumerism and tourism activities.

Touristic wilderness refers to humanistic perspectives but it also has a reverse side. There is no great ideological difference between an institutionalised tourist industry and any other form of industry, in that the former is devoted to exploiting the natural environment under the same terms of a market economy, and often to the maximal possible extent, and as noted by Gössling and Hall (2006b); tourism cannot claim to have any moral high ground. However, as far as conservation issues and their relation to forestry, hydropower contraction or other similar utilisation processes are concerned, tourism has often provided an opportunity to promote the economic benefits of conservation in the form of jobs and incomes (Power, 1996) (see Case 6.2 on wilderness tourism in Iceland). There have also been attempts to steer tourism practices in an environmentally more sustainable direction. Internationally, the tourist industry has reacted to this in the wilderness context in part by introducing ideas such as ecotourism (e.g. Cater, 1994), and nature conservation occupies a prominent position in ecotourism in particular (Boo, 1990; Fennell, 1999; Gössling & Hultman, 2006; Hall & Boyd, 2005).

Cases and issues 6.2: Wilderness tourism in Iceland: Threats and opportunities

ANNA DÓRA SÆÞÓRSDÓTTIR

Iceland is a sparsely populated country, with only 300,000 inhabitants. Most of the population lives in coastal communities. The mainstays of the Icelandic economy are renewable natural resources: rich fishing grounds, hydro and geothermal power and pastureland. The most important is export of fish and fish products, which in 2004 accounted for around 38.5% of total export revenues; aluminium and ferrosilicon accounted for 13.5% and tourism revenues 12.4%.

During the last few decades the Icelandic tourist industry has been selling nature, as is shown by the marketing slogan 'Iceland naturally'. The industry also wants to create the image that Iceland is clean and unspoiled, containing 'unspoiled wilderness' shown by the slogans 'Nature the Way Nature Made It', and 'Pure, Natural, Unspoiled'. Surveys undertaken by the Icelandic Tourism Board show that this is in fact the image tourists have as 76% of the 350,000 international tourists who visit Iceland each year come to experience nature (see also Case 2.3).

The main wilderness areas in Iceland are in the central Highlands that cover about 40% of Iceland. The largest part of the Highlands is wilderness, that is to say relatively unspoiled areas. It is mainly the absence of anything human-made which makes travelling in the wilderness interesting for tourists and they go there to escape the complexity of modern life, experience the authentic and primitive, and strengthen body, mind and spirit.

The Highlands used to be very isolated and getting there and travelling around was a real adventure. Rough landscape with sand, lava and big glaciers and glacial rivers made them almost inaccessible. In the 1970s this gradually began to change as roads were built, mostly in connection with the construction of hydroelectric power plants in the southern part of the Highlands.

The Icelandic Highlands are a big attraction for tourists. More than a quarter of foreign tourists go there. Additionally, the Highlands are popular with Icelandic tourists and every year around 13% of the population stays overnight in the Highlands. From these figures one can estimate that annually at least 130,000 visitors go to the Icelandic Highlands.

Landmannalaugar is located in the southern part of the Icelandic Highlands and is by far the most popular tourist destination there. The main attraction is the wild nature, volcanoes and lava formations, colourful mountains and geothermal areas with natural warm springs where one can bathe. The Icelandic Travel Association built the first mountain hut there in 1951. At that time the area was fairly isolated and getting there was a real adventure; one had to drive through glacial rivers and over rough landscape. Construction of hydroelectric power plants in 1968 and 1978 greatly increased accessibility to the area. In 1970 the area became even more accessible when a road was built from the main mountain road eliminating the need to ford rivers. Now one can drive to Landmannalaugar in an ordinary passenger car. This accessibility, and the fact that it is only a three hour drive from the capital city of Reykjavík, has made Landmannalaugar the most visited Highland area in Iceland. Due to the popularity of the area, its infrastructure has been improved repeatedly, both to provide services and comforts and to protect the vegetation. This highlights the contradiction between wilderness, based on the image of 'unspoiled' nature, and the fact that visitors inevitably spoil the wilderness. It is a difficult task to maintain the image of wilderness in a popular tourist destination as tourism is dependent on a viable infrastructure and only a limited number of visitors can visit an area simultaneously if it is to provide the experience of solitude.

A study conducted in Landmannalaugar showed that visitors are western Europeans, mainly from France, Germany, Switzerland, the Nordic countries and Iceland. A little more than 80% of the travellers were visiting the area for the first time, while some 20% had visited it before (Saethórsdóttir, 2004). A majority of those that answered the survey, some 83%, wanted to come back to Landmannalaugar. This shows that in general the visitors were impressed with the area. Around two-thirds of the visitors stayed overnight in the area and their average stay was 2.8 nights. Around one-third of the visitors were on a day trip and stopped on average for four hours in the area. Most of the tourists travelled in buses in an organised tour and a quarter of the tourists were travelling either with family or friends. The tourists considered Landmannalaugar to be a trekking or hiking area, and more than 90% of them took the opportunity to go for a walk there. On average people spent some seven and a half hours trekking, though the range was large, from a half hour stroll up to a 50 hour trekking tour. Most of the trekkers stayed in the lava fields and the mountains next to the hut.

The study indicated that Landmannalaugar is a symbol of wilderness in the minds of many people; they value it and use it to gain a wilderness experience. Approximately 90% of the visitors felt that unspoiled wilderness was a part of the attraction of the area and considered it the most important experience they had. At the same time they found it important to have well organised camping places, mountain huts, marked walking paths and bridges over small rivers. Other infrastructures such as service centres and hotels, or structures unrelated to tourism, such as power stations, were considered inappropriate in the Highlands.

There has been considerable demand for more and better services at Landmannalaugar. As these demands have been acted upon the site has lost some of its wilderness character. This will continue and as there are many commercial interests willing to act on these demands it is not possible to foresee the final results. Some of the tourists who stay at Landmannalaugar are not really prepared for a wilderness experience and expect services such as a shop where they can buy groceries. Permission has therefore been given to run a 'shop' there. The study indicated that visitors have different opinions about whether the shop should be a part of the area or whether it spoils the wilderness experience.

Most visitors (90%) were satisfied with their visit to Landmannalaugar. However, there was a small number of visitors (8%), who were very dissatisfied with their visit. The main complaint was the weather, poor quality and high prices of facilities and service, and the large number of tourists in the area. The research indicated that solitude is an important part of the wilderness experience of Landmannalaugar. Many had that experience, as 68% thought that the number of tourists was within the limits they thought appropriate, but some

20% perceived crowding and thought there were too many tourists in the area (Saethórsdóttir, 2004).

The results demonstrate that providing recreational opportunities in a wilderness area like Landmannalaugar is a complex issue as pluralism exists in values and interests and there are conflicts of opinion about the development of tourist facilities. The development at Landmannalaugar has resulted in the area becoming less appealing to the most purist tourists, so they are hardly found there anymore. It can be assumed that displacement has occurred; purists who used to visit Landmannalaugar now go to other more isolated Highland areas where there is less tourism and less environmental damage.

Up to now wilderness areas in Iceland have been transformed randomly into recreational areas. This happens, for example, when accessibility suddenly increases as roads are built for hydroelectric power plants. This kind of change of land use should not be allowed to happen randomly. A planning system such as ROS (Recreation Opportunity Spectrum), and the concept of tourism carrying capacity should be valuable in hindering unplanned change in land use. These concepts and frameworks help us understand the importance of diversity in recreational opportunities and how much infrastructure and development is appropriate in each place.

Wilderness tourism in Iceland has other threats than tourism itself. The country has extensive resources of renewable hydro and geothermal energy, and the Icelandic government has given emphasis to further harnessing the abundant energy for power intensive industry. In 2003 the Kárahnjúkar Hydroelectric Project, in the northeastern part of the Highlands started. The power plant will be by far the largest in Iceland and will produce 690 MW of energy, compared to the 273 MW produced by the largest plant today. The electricity will be sold to an aluminium smelter being built in a small fishing village in eastern Iceland. The projects will be finished in 2008. Few projects, if any, have provoked as much discussion and protest in Iceland as Kárahnjúkar power plant and the aluminium smelter. A fierce debate took place between environmentalists and those in favour of the project, among them the National Power Company. A frequent argument was whether the project would benefit or harm the tourism industry. Those that supported the project pointed out that road construction carried out in connection with the construction of the power plant was positive for tourism as it made 'new' destinations accessible. The environmentalists on the other hand drew attention to the fact that 'unspoiled wilderness' upon which wilderness tourism in Iceland is based is limited.

If government policy for further expansion of power intensive industry is realised, dams will be built during the next 20 years on all major glacier rivers in the country and geothermal power plants at some of the biggest geothermal areas. The most suitable locations for the reservoirs that unavoidably accompany hydro power plants in glacier rivers are in the Highlands and some of the largest geothermal areas are located there as well. In the future this will undoubtedly cause further conflicts with wilderness tourism.

Land use issues have not been resolved in Iceland and only a fraction of the energy resources have been utilised so far. There is no doubt that further development of energy resources in Iceland will be a challenging task as interests of other users than those concerned solely with energy will have to be taken into account. The government of Iceland is working on a master plan for the development of hydro and geothermal energy resources. In this master plan a number of proposed power plants will be evaluated and ranked based on economic profitability, efficiency and minimising negative environmental effects. In addition, the impact on tourism and outdoor life, such as fishing and hunting, will be evaluated. If this work is professionally done and not biased towards the interests of the power plants, the strongest stakeholder in the system, the master plan should be a valuable tool for maximising the benefits of utilising the Highlands, for power production, wilderness tourism, grazing, hunting and nature conservation.

In the tourism planning procedure it is extremely important to have a variety of environments that range the whole ROS scale, from primitive and pristine experience to urban and high service experience. In Europe there are few sites left that fit the primitive side of the ROS scale and the vast majority of foreign tourists who come to Iceland do not have the opportunity to enjoy wilderness in their own country. In Iceland there are still some primitive sites. This gives Iceland a great competitive advantage in providing experience of wild nature. Consequently the Highlands are an important resource for tourism and must be handled with care. Their importance will only increase in the future as the wilderness is a resource that is fast disappearing in the world.

The Development of Nature-based Tourism in Northern Wilderness Areas

Tourism is perhaps the latest significant form of economy and mode of consumption using the northern wilderness areas. In this respect national parks are perhaps the most used wild environments in tourism. Most of the national parks are situated in the peripheral areas of Nordic countries (Figure 6.1) but there are increasing needs to safeguard natural environments also in more urbanised areas. The expansion of the nature conservation network is also justified by the needs of tourism and regional development.

Internationally, tourism is considered a highly important economic use of wilderness environments (see Butler *et al.*, 1998; Hall & Page, 2006). This is an essential perspective for the future of the northern wilderness areas, because nature-based tourism and ecotourism in general are considered to be some of the fastest growing sectors of international tourism. Unfortunately, there is no clear knowledge of the scale of tourism in the northern wilderness areas or other wilderness environments outside nature conservation areas. However, it can be assumed that the development of tourism in the national parks corresponds to the

144 Nordic Tourism

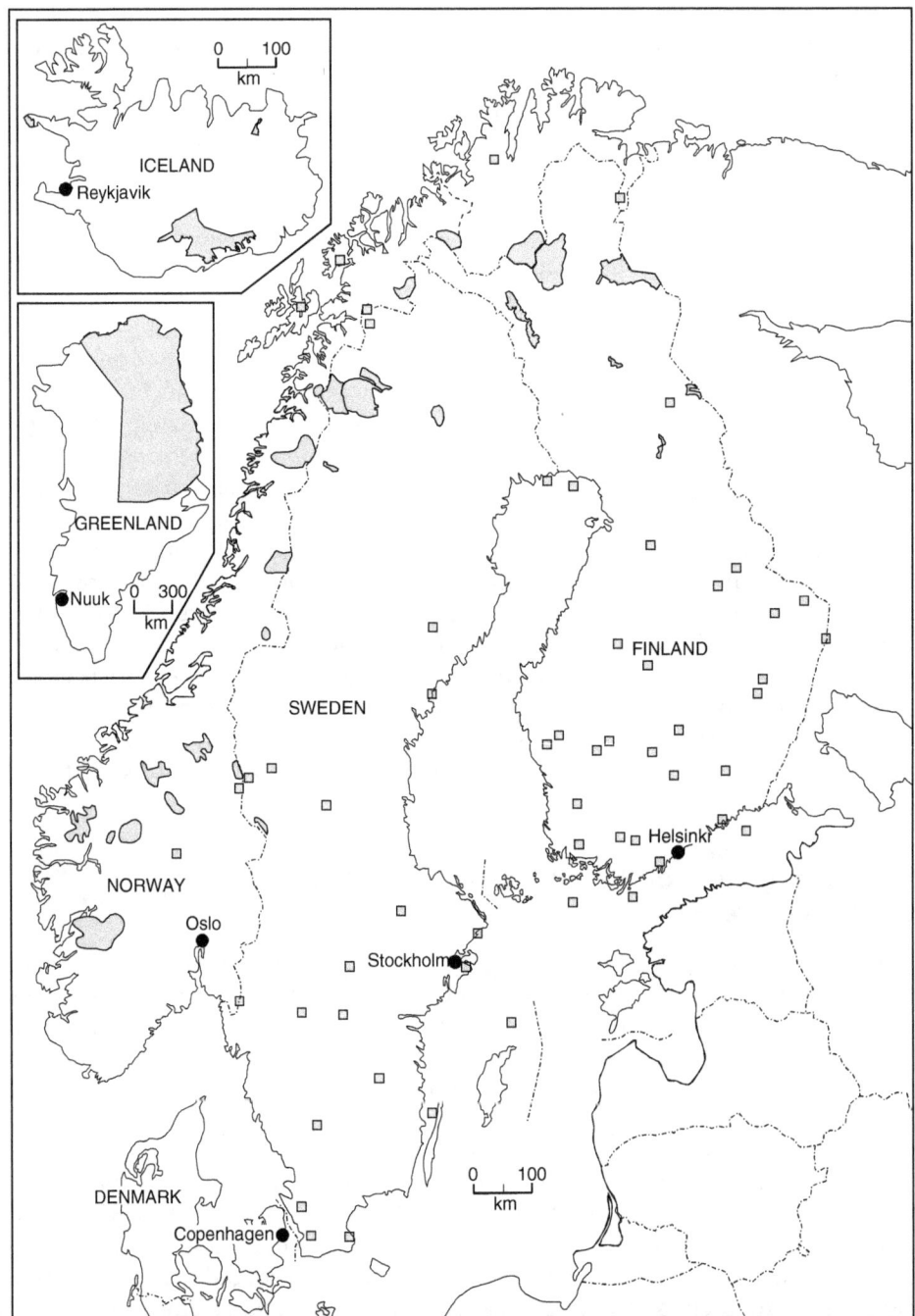

Figure 6.1 National parks in the Nordic countries

Table 6.2 The visitor numbers of selected national parks in Finland in 2000–2006

National park	2000	2001	2002	2003	2004	2005	2006
Lemmenjoki	10,000	10,000	10,000	10,000	10,000	10,000	10,000
Nuuksio	80,000	100,000	100,000	100,000	100,000	110,000	142,000
Oulanka	145,000	143,000	162,000	165,000	173,000	173,500	183,000
Patvinsuo	15,000	15,000	15,000	15,000	20,000	14,000	15,000
Perämeri	10,000	10,000	6500	7200	7200	2500	5500
Petkeljärvi	15,000	15,000	17,000	17,000	17,000	17,500	18,500
Pyhä-Häkki	12,000	11,000	11,000	11,000	11,000	9000	15,500
Päijänne	8000	8000	8000	8000	10,000	12,000	12,000
Archipelago	40,000	40,000	60,000	80,000	80,000	60,000	60,000
Salamajärvi	8000	7000	7000	7000	9000	10,000	12,000
Seitseminen	37,000	37,000	37,000	40,000	40,000	40,000	42,000
Syöte	30,000	30,000	25,000	24,000	34,000	33,500	33,000
Urho Kekkonen	150,000	150,000	150,000	160,000	160,000	165,000	170,000
Visits in all 30–34 NPs	833,000	851,800	1,012,000	1,123,200	1,153,900	1,410,000	1,493,000

Source: Finnish Forest and Park Service.

development of the use of wilderness areas and reflects the development of nature-based tourism in the Nordic context in different kinds of areas with different degrees of access.

Table 6.2 shows the development of the visits in selected national parks in Finland, as an example of the increasing tourism and recreation in wilderness settings that have uniform follow-up statistics for the period of several years. The total average sum of visitors has almost doubled in the period 2000–2006 in the Finnish national parks (NP). The development, however, is not equally distributed among different parks. The development of nature-based tourism has not been as strong in the areas that are far from the main tourist attractions and routes, for example, in Lemmenjoki NP and Perämeri NP. In contrast Oulanka NP, for example, is located near the Ruka tourist resort with almost 20,000 bed units, and Nuuksio NP is situated near Espoo inside the Helsinki metropolitan area.

According to international and Nordic estimates, the annual growth of nature-based tourism would be as high as 8–10% (see Fennell, 1999; Ympäristöministeriö, 2002). However, even lower direct growth of the touristic use of wilderness areas and new forms of tourism activities can significantly influence the use and character of these areas, and the relationships between different forms of use (see Saethorsdottir, 2004; Vuorio, 2003; Watson & Kajala, 1995). In addition to the increasing numbers of tourists in the wilderness settings the nature of tourism has been changing. The northern wilderness areas have traditionally offered touristic opportunities for recreation activities based on nature. In the summer, these have especially been backpacking, hiking, canoeing, boating, fishing and hunting, and skiing in the winter (see Case 6.3 on hikers in Södra Jämtlandsfjällen, Sweden). However, since the mid-1990s snowmobile trekking has become one of the most central and the most visible form of the new nature-based tourism activities (see Vail & Heldt, 2004).

These new specific forms of nature-based tourism are capitalising in particular on the increasing need to experience nature and peripheries by modern urbanised people. However, the development of nature-based tourism in wilderness environments has not only had an effect on the range of tourist products and experiences among modern customers but also has had impacts on the resources and people in peripheries (see Case 6.4). For a long time nature-based tourism has been a useful tool to introduce new use patterns and ideas of nature to peripheries. For example, tourism was integrated to nature conservation from the very early stages of the national park movement (see Butler & Boyd, 2000), while in Nordic countries the aesthetic values of nature for tourism and recreation were deeply involved with the establishment process of the first national parks and nature conservation areas in the early 20th century.

Cases and issues 6.3: Nature-based tourism in mountain areas: Hikers in Södra Jämtlandsfjällen, Sweden

TUOMAS VUORIO and LARS EMMELIN

Effective spatial planning in the mountains presupposes better information on tourism and outdoor recreation than is available at present. However, data on tourism and recreation are not systematically collected at either local or regional level. Nevertheless, municipalities need adequate data on which to base planning. In other sectors the regional administrations are responsible for information support to spatial planning. Thus planning for tourism and recreation infrastructure and resolution of conflicts with other forms of land-use becomes difficult. The lack of information influences perceptions of the existing situation, its problems and opportunities, but also hampers establishing of a common understanding among actors in the planning process.

The case of Södra Jämtlandsfjällen in Sweden describes briefly a study made for a planning process that was aimed at providing information on summer use patterns, especially hiking and camping outside the vicinity of marked trails and mountain huts. The conflicts needing resolution were between tourism, conservation and reindeer herding. Apart from general information on use patterns a number of specific management issues that were to be dealt with in the planning process and subjected to experiments with local management were studied.

Description of the case study area and methods

Södra Jämtlandsfjällen is located in central Sweden, on the border of Norway in the west. The total area is about 2300 km² and consists mostly of bare mountains and forested mountain valleys. The landscape is diverse with the highest peaks over 1700 m above sea level. The Södra Jämtlandsfjällen has the densest network of publicly managed trails in the Swedish mountains, about 500 km of marked summer and winter trails and about 200 km of snowmobile trails. There are three mountain stations and six lodges in the area managed by the Swedish Tourist Association. The whole area is used for reindeer herding.

So called self-registration boxes at entry points were used in the study. The total number of boxes was 21. A total of 15,238 persons registered. Based on non-participation observations this gives an estimated total of 17,301 visitors. A questionnaire was sent to a sample of 2138 with a response rate of 67%. For observation of tents, flight observations were carried out on nine occasions.

Results

Swedish visitors dominate with foreigners accounting for only about 10% of all visitors (mostly from Norway, Germany and Denmark). A large share of the Swedish visitors come from the big towns in southern and central Sweden. The proportion of men and women is about equal. The share of highly educated visitors is relatively high, which is common among the mountain hikers (Vuorio, 2003).

Hiking is the dominant activity among visitors. Camping is important as an activity in its own right. Most of the visitors come in groups of two, most often families. On average Swedish visitors spend 5.6 days in the area. About one-fourth of the visitors were visiting the area for the first time. About 60% of the first time visitors had never done a multiple day hike in the Swedish mountains (Vuorio, 2003). This shows that Södra Jämtlandsfjällen is an important area for Swedish outdoor life and as a recruitment area for mountain tourism.

The use of the area has some clear geographical features. The northwestern trail heads have almost half of the visitors. Other important starting areas are Vålådalen in the northeast and Ljungdalen in the south with about one-fourth's share each. It is obvious that the western parts of the area have more visitors than the eastern parts. The pattern of camping confirms this. A total 649 tents were observed during the nine observation flights that were spread out over the summer season. Most of the tents were close to the marked trails and the mountains huts and lodges. Only about 4% were pitched outside the immediate vicinity of the marked trails. Only 22.6% of the visitors stated beforehand that they were planning to hike outside the marked trails for at least one day.

The most common reasons given for following the marked trails were convenience and security (Vuorio, 2003). Visitors to Södra Jämtlandsfjällen were also identified as less puristic in their relation to the wilderness experience than visitors in other mountain areas in Scandinavia.

Recreational impact is not seen as a problem by the majority of visitors. This is in line with other studies carried out in the Scandinavian mountain region (e.g. Vistad, 2002). It was only a limited number of purists who were disturbed by the presence of other users. This is much less than in many other countries (see Freimund & Cole, 2001; Kearsley et al., 1998). The visitors were satisfied with the quality of the service and establishments, in particular the infrastructure, such as wind shelters, signs, footbridges and bridges.

Information is the most acceptable management action to the majority, especially information on nature conservation and reindeer herding. Technical infringements such as wind generators, snowmobiling, terrain vehicles and power transmission lines are considered as the most important threats to the area by most of the visitors along with crowding and overexploitation. The desired management actions to avoid impacts were nature conservation, footbridges and limited entry to sensitive areas, to control the number of visitors during the sensitive periods, to guide visitors to the marked trails and to prohibit tenting in certain areas.

Stricter restrictions are seen as less desirable by a majority of the visitors. The most negative reactions were caused by the suggestion of new fees for visitors and limitation of the right of common access. This has been a common result in studies carried out in Scandinavia (see Vistad, 2002).

A series of scenario questions were used to study responses to management actions. The number of respondents that would be affected by restrictions with regard to nature conservation and reindeer herding would be quite small. This is also a group that is more puristic in their attitudes towards untouched nature and would probably be more disturbed by some restrictions. However, this group is not the one that aims to heighten the wilderness characteristics of the area.

Cases and issues 6.4: Tourism development, amenity values and conflicting interests in Pyhätunturi National Park, Finland

JARKKO SAARINEN and MATTI VAARA

Nature conservation areas have for a long time been important tourist attractions, and national parks have constituted the most popular of these. When national park systems were initially established in north America and in Nordic countries their use for tourism and recreation was emphasised (see Hall, 1992). However, in more recent years the national parks and other conservation areas

Table 6.3 Forms of tourist use and activities regarded by second home owners or residents as suitable, suitable with reservations, or unsuitable for the Pyhätunturi National Park, Finland

Suitable	Suitable with reservations	Totally unsuitable
Skiing	Adventure programmes	Snowmobiles
Picking berries	Telemark skiing	Dog sleighs
Guided trips	Reindeer sleighs	Mountain biking
Hiking	Slalom skiing	Hang gliding
Hiking with snowshoes		Horse-riding

Source: Saarinen & Vaara (2002).

have been set up primarily on account of their flora and fauna; their primary purpose being to ensure that ecosystems will be managed and used in an appropriate manner according to good conservation practice.

In the present day there is increased interest in the use of national parks and other conservation areas for tourism development purposes. Therefore, it is increasingly important to monitor developments in tourism and its impacts on local conditions (see Williams, D., 2001). Apart from information about the natural environment itself and changes taking place in it, successful planning for use and management calls for a wide range of data on the use made of particular sites, types of visitors, temporal and spatial variations in usage and the attitudes and values adopted by users and changes in these with time. This case example focuses on one particular group of users: persons owning (or living in) second homes close to the Pyhätunturi National Park, northeastern Finland (Saarinen & Vaara, 2002). These people may represent a relatively small group in quantitative terms, but they can be looked on as having a very special relationship to the park, its use and development.

The data were obtained by means of questionnaires sent to the owners or occupants of holiday cottages in the Pyhätunturi tourism region in spring and summer 2000 (Saarinen & Vaara, 2002). A total of 400 questionnaires were sent out, of which 128 were returned, giving a response rate of 32%. There were altogether some 400 holiday cottages in the region.

The owners or occupants of holiday cottages/second homes made a clear distinction between different kinds of tourism activities they saw as appropriate to develop in the National Park in future (Table 6.3). Those tourism activities were regarded highly suitable to develop included most of the Nordic population's favourite outdoor occupations, such as skiing, picking berries and hiking. Also guided trips were judged acceptable, whereas the forms of activities that were thought more suitable for areas outside the conservation area were ones that tend to be favoured only by a minority of people, such as travel by snowmobile, hunting, mountain biking and horse-riding. The tourism activity group between these two ends of the continuum contained activities that only a minority of respondents would exclude from the area, whereas the majority would accept them entirely or with some reservations. These activities are typically ones that are not in direct conflict with the widely accepted forms of nature-based tourism pursued in the area, such as skiing or hiking, and are frequently guided activities or ones that gather large crowds of people together, for example, adventure programmes, reindeer sleigh rides and down hill skiing (on already existing slopes) (Saarinen & Vaara, 2002).

As a tourist destination Pyhätunturi primarily is represented as a ski resort, although action has been taken to develop the region towards a year round resort. Second home owners play an important role in this process. Owners of second homes can be assumed to make repeated and fairly regular use of the tourism region in the summer season, differing in this respect from those tourists who visit the area just once or only occasionally and still mainly in winter time. In addition, these private persons, organisations and companies have invested relatively large sums of money in the area, and it can be assumed that they also have a broad interest in the use and development of the tourism region and possibilities for new activities. Those who have bought second homes here have presumably evaluated the level of attractiveness of the place relative to other comparable destinations on a more profound, long-term basis than the average visitors and customers to a hotel will do when making their decisions.

For the owners of holiday cottages, the Pyhätunturi region may represent a place that has values and meanings attached to it that go beyond those associated with purely tourist attractions, and these values and meaning structures may also influence their attitudes towards the use and the development of the whole tourism region. In this respect, many tourist destinations represent *amenity landscapes* for second home owners (see Hall & Müller, 2004a) (see also Chapter 8). These amenity landscapes carry a positive loading culturally and socially, and normally have values and meanings associated with them that reflect a positive quality of life and life-style (see Godde *et al.*, 2000), possibly including aspects such as closeness to nature, freedom from pollution and contamination, rural qualities, security, health and peace.

Conclusions

During the past two decades the growth potential of nature-based tourism has raised optimistic hopes in the peripheral and rural areas of Nordic countries for regional development. However, it is apparent that tourism and its development prospects in practice are highly polarised. The development possibilities and regionally positive impacts of nature-based tourism are not evenly distributed geographically, economically or socially. In addition, the development of nature-based tourism or other forms of so-called alternative tourism is perhaps a more knowledge-based activity than the development of the industry in general, but the production of appropriate and relevant knowledge has not developed as rapidly as nature-based tourism activities in wild landscapes.

There is, therefore, a need to see nature-based tourism within the broader natural, sociocultural, political and economic systems within which it is embedded and that determine its development. Also, in remote wilderness contexts tourism activities should be seen as an integrated part of the regional and increasingly global tourism industry, which usually needs the same infrastructure as 'average (mass scale) tourism' and which uses the same transportation systems in order to get customers to the peripheries, where the natural tourism resources are mainly located. Global competition for resources in land-use and economic activities are increasingly affecting Nordic wilderness environments and the ideas their uses are based on. In this respect it is rather obvious that tourism, especially nature-based tourism and its new forms, will increase in northern wilderness areas. The patterns of wilderness related tourism will become more international and contrast to the development of tourism and the idea of touristic wilderness, the importance of the traditional northern livelihoods, such as reindeer husbandry, fishing and forestry, will decrease and, in the future, they cannot be used as tools for developing the local economy in the same way as tourism in the Nordic peripheries (see Jóhannesson *et al.*, 2003). This will also locally decrease the importance of the idea of traditional wilderness. There is still synergy and symbiosis between tourism and nature conservation – especially in their relation to the traditional wilderness and questions on how *not* to use the wilderness environment as ethically loaded place. International tourism demand also encourages the adoption of the conserved idea of wilderness with the touristic commoditisation elements of 'the wild'.

Obviously, tourism will not completely replace the traditional or conserved meaning structures and values of wilderness. However, nature-based tourism has brought new and very influential changes to layers of cultural and economic meanings and uses of Nordic wildernesses. Perhaps more than the direct present use of the wildernesses, tourism constructs people's ideas of wilderness through the indirect touristic use, such as marketing and the related images of wilderness as a 'product'. At the same time, it affects people's opinions concerning the suitable uses of wilderness areas and potentially subsequent actions. In future, these processes and related ideas localised in the peripheries will most probably change the wilderness landscapes more permanently than the present scale nature-based tourism directly could ever do.

Discussion questions

(1) What does it mean that wilderness is a contested idea and concept?
(2) What are the main differences between the concepts of nature-based tourism and ecotourism?
(3) Why is the idea of wilderness different in Nordic and Anglo-American contexts? Or are there presently major differences between the cultural contexts and is there one coherent idea of wilderness in Nordic countries?

Essay questions

(1) What issues may have influence on a person's attitude towards wilderness and nature?
(2) What are and will be the main impacts of tourism on wilderness and its traditional uses?

Key readings

Although wilderness is characteristic of Nordic landscapes and tourism images the connection between tourism and wilderness has remained a relatively understudied subject. Most of the research is based on north America; the book *Wilderness Management* provides a good overview of wilderness areas, their meanings, uses and management in the USA (Hendee *et al.*, 1990; see also Hall, 1992 for an Australasian perspective on wilderness). Frost and Hall (2009) provide an account of tourism and national parks in an international perspective, including several chapters that have a Nordic dimension. In the Nordic context Mels (1999) has studied the evolution and constitution of wild landscapes and national parks in Sweden, and Hallikainen (1998) has focused on the historical background and present perceptions of wilderness in Finland. Visitor conflicts in wilderness and national park settings have been studied more widely (e.g. Saarinen, 1998; Saethorsdottir, 2004) but the impacts of tourism on the wilderness environment and its uses would need further analysis in Nordic countries. Gössling and Hultman's (2006) book on ecotourism in Scandinavia provides several chapters that look at ecotourism in the context of the wilderness.

There are several web pages introducing wilderness and nature conservation areas and nature-based tourism in Nordic countries but relatively few aim to combine these issues. In addition to major governmental conservation agency web pages there are numerous project-based sources.

National park management agencies:

Finnish Forest and Park Service (Metsähallitus): www.metsa.fi
Danish Forest and Nature Agency: www.sns.dk/nationalparker/english.htm
Environment and Food Agency of Iceland: http://english.ust.is/about-UST/
Swedish National Environmental Protection Agency: www.naturvardsverket.se/en/In-English/Menu/
Norwegian Directorate for Nature Management: http://english.dirnat.no/
Nature based tourism project: www.naturebasedtourism.net

Chapter 7
Coastal, Marine and Ocean Tourism

Learning Objectives

After reading this chapter, you should be able to:

- Understand the concept of coastal and marine tourism in the Nordic countries.
- Understand the relationship of coastal and marine tourism to ecotourism.
- Appreciate the economic importance of specific forms of coastal and marine tourism in the Nordic countries such as as ferry tourism and cruise ships.

Introduction

Water, as an arena for leisure and tourism, has been largely neglected in the scientific literature, although it is considered a major amenity for tourism even in a Nordic context. Orams (1999) argues that coastal and ocean tourism is the fastest growing segment of tourism in the world. This statement, is, however, hard to confirm since coastal and ocean tourism are difficult to distinguish from other forms of tourism. Figures are therefore not easy to collect. As Hall (2005a) points out marine tourism has not been a self-evident phenomenon. Instead, it is socially constructed and has its roots in aristocratic demand in previous centuries (Löfgren, 1999). Nevertheless, selling sun, sand and sea is also a major component of tourism in the Nordic countries, although climatic conditions counteract developments comparable to those of the Mediterranean.

The Baltic and North Sea are increasingly being converted into spaces for leisure and tourism. Besides individual yachting and cruise ship tourism, ferry travel forms a major attraction along the Norwegian coast and between Denmark, Sweden, Åland, Finland and Estonia. This chapter will illustrate the patterns of marine based tourism and provide insights into the mechanisms and processes that facilitate its development. The chapter also introduces other forms of marine tourism such as whale-watching.

The Concept of Coastal Tourism

The concept of coastal tourism departs from a geographical zone, the coast, and hence it refers to a number of different activities that can be pursued there. These include ventures within the accommodation and restaurant sector, second home development and also various infrastructure developments facilitating vacations in the coastal zone (Bramwell, 2003). Moreover, a number of recreational activities can be added. Swimming, boating, recreational angling and marine based ecotourism are often mentioned as examples of these activities (Miller, 1993). Marine tourism is closely related to the concept of coastal tourism, but also highlights activities outside the more intimate space of the coastal zone. Cruising tourism is thus included in this concept (Hall, 2005a; Orams, 1999).

Sletvold (1999) provides a conceptual overview of the meaning of the coast and its environment. He argues that the coast symbolises openness that contrasts the more closed and fragmented landscapes of interior lands. Moreover, the coast gains meaning from tradition and heritage. The existence of particular lifestyles based on marine activities such as fishing has contributed to maintaining the construction of a typical culture. Historical ports and fishing villages document a history of harvesting marine resources and have often become scenic locales for vacationers (Plate 7.1). In Norway, the fjord landscape has become an attraction of

Plate 7.1 The fishing village on Ulvön within the High Coast World Heritage area in northern Sweden (Photo: D. Müller, 2005)

international significance and is used to symbolise Norway as a nation (Fitje, 1999). Indeed, the fjords had been a primary reason for international tourist development in Norway during the 19th century. Moreover, contemporary landscape and lifestyle tastes mean that property ownership along the coastline also symbolises social status and thus, traditional patterns of settlement and recreation are contested not least by second home development (Marjavaara, 2007a, 2007b; Marjavaara & Müller, 2007; Sletvold, 1999).

Ferry Tourism

The Nordic realm is dominated by the Scandinavian Peninsula. Traditionally this has been seized as opportunity for transporting people and goods along the coastlines and over the sea. Today, land transportation is the preferred mode of transportation and hence, the sea separating the Nordic countries from each other and from mainland Europe is seen as an obstacle. A permanent link between Malmö via Copenhagen and Puttgarden has been initiated by the construction of the Öresund Bridge, which is expected to be followed by the construction of a new bridge during the next 20 years or so closing the gap of the Fehmarn Belt between the German island of Fehmarn and the Danish island of Lolland – a distance of 18 km. Sjalland and the Danish capital Copenhagen are already connected to Jutland by other spectacular bridges. Even in the Bothnian Sea, permanent constructions are being discussed to substitute ferry lines and to allow a permanent connection under difficult winter conditions.

Meanwhile these bridges have become tourist attractions in their own right; even many ferry links have maintained their attractiveness for goods transportation and tourism. The touristic demand is also met by adaptation of the ferry boats. Hence, despite their principal function as a basic means of transportation between the Nordic countries and mainland Europe, many ferries offer entertainment and shopping opportunities catering for a clear touristic demand. Thus, trips are made not only to move from one place to another, but also as leisure activity in its own right. This is also indicated by the scope of marketing among the Swedish ferry lines which amounted to more than €60 million (*PRF*, 2006). Viking Line, a major operator in the Åland Sea, consequently attempted to lure tourists from Sweden to Åland by promising that a land visit was not necessary. Instead you could stay on the boat for the entire time. As noted below such a venture was especially important because of Åland's special tax-free and duty-free status (see also Case 5.1 on tourism in Åland).

In 2005 the Swedish passenger ferries alone carried about 29 million passengers of which more than 12 million were foreign residents (*PRF*, 2006). Altogether 4.3 million cars were transported indicating the amount of tourists using the ferries for cruising rather than transportation to another destination.

Two kinds of ferry connections can be found. Particularly in Norway, fjords and islands require ferries as a part of the national road system. Although many of these ferries have been substituted by tunnels, there is a comprehensive network left offering many scenic transportation opportunities to travelling car tourists. Even in Sweden and Denmark ferries are used to connect the countries' islands with the mainland. Particularly in the Stockholm archipelago, these ferries cater for tourists. The most well-known domestic ferry connection is the Norwegian Hurtigruten operating along the several thousand kilometre coastline from Bergen to Kirkenes at the Russian border. Established in 1893 by government contract to deliver mail, freight and people, travelling with one of the company's boats is now considered a prime attraction in the Nordic countries (see Case 7.1). The success of this ferry has also caused advocacy for similar lines along the Baltic coast (Nilsson, 2005).

Cases and issues 7.1: Hurtigruten

OLA SLETVOLD

The Norwegian Coastal Voyage (NCV) is the tourist product brand of a 112-year-old transport system, 'Hurtigruten'; which is basically an 11 day route between Bergen and Kirkenes with goods, post and passengers, calling at 35 harbours on its way (www.hurtigruten.com). Its long lasting and rising tourism fame supplies its status as a societal and cultural institution. 'Hurtigruten' was an important tool for the integration of the northern regions when building modern Norway. Its machine-like regularity and strict itinerary, coming and going every day in all kinds of weather, in peace and war, made it a routine fact in many peoples' lives. Its status is reflected in the production of a TV docu-soap dramatising life on board the newest ship in 2005. Heavy airport and road infrastructure development since the 1970s reduced the transport legitimacy of the NCV. In itself a sign of modernity the NCV was overrun by accelerated modernisation and passenger numbers went from 500,000 (1976) to 200,000 in 15 years, and the fleet was in strong need of renewal.

A period of political and industrial uncertainty since the mid-1990s has given way to a new era for the NCV. There have been large investments in a total renewal of the fleet to meet demands for efficient operations. The upgrading of tourist facilities is influenced by modern cruise ships design and international ferry architecture. The renewed market strategy is seen in more flexible prices, a breakthrough in the national tourist market and better control of international distribution. The basic NCV product is supplemented with a move towards broader cruise operations, to Spitsbergen, Greenland and Antarctica, Norwegian fjords and the Mediterranean, and special event oriented cruises (Sletvold, 2006).

Passenger numbers in 2002 were 550,000 and the main season (April–September) sees a highly international clientele. The NCV companies divide their customers into three broad categories: experience oriented, mainly round trip tourists, transport oriented port-to-port passengers and conference groups (Sletvold, 2006). The basis of the NCV experience is extraordinary coastal landscapes that lend some truth to the branding claims of 'the most beautiful voyage in the world'. The vessel brings the tourists days and nights of different landscape experiences. On board life is combined with land based attraction visits and over land trips.

All through its history the NCV has served tourists, offering a special voyage towards the midnight sun. Nowadays there are well over 100 cruise ship visits to the North Cape annually, in principle sailing the same seascapes as the NCV. However, there are qualitative differences compared to most cruises: the ships are small (11,000–16,000 tons, 500–600 berths); they have the rhythm and regularity of a working ship; the ships stay quayside for 1–5 hours on average three times a day for the loading and unloading of fish, cars and consumer goods, and for quick walks ashore; tourists co-exist with port-to-port passengers and meetings and conference groups. Compared to mass tourism oriented cruises (Douglas & Douglas, 2004) this is a product that is part of a society, not a self-contained and introvert world of its own.

Website: www.hurtigruten.com

Besides these domestic ferry connections, international ferry networks link the Nordic countries to each other and mainland Europe (Figure 7.1). In certain transport corridors intensive competition between different ferry companies can be found. Major international corridors include the following:

- Northern Jutland (Denmark) – Göteborg (Sweden) and Oslo (Norway)
- Schleswig-Holstein (Germany) – Copenhagen/Sealand (Denmark) – Malmö/Scania (Sweden)
- Stockholm (Sweden) – Åland – Turku, Helsinki (Finland) and Tallinn (Estonia)
- Helsinki (Finland) – Tallinn (Estonia)

Moreover, additional regular connections are available between Jutland and southern Norway and between Mecklenburg-Vorpommern (Germany) and southern Sweden. Additional connections can be found between Poland and Sweden, in the Bothnian Sea between Sweden and Finland and between Norway and the United Kingdom. Iceland and the Faroe Islands are served by a ferry connection from western Norway.

Ferry cruises in some parts of the Baltic Sea have become big business even within the leisure sector. Viking Line, a major player in the Baltic Sea connecting Stockholm with Finnish and Estonian harbours via the Åland islands, transported more than 5.3 million passengers and 412,000 cars during November 2004 and October 2005 (*Viking Line*, 2006). The figures imply an increase of 8% compared to the previous year despite increasing competition caused by new boats more specialised in the cruising component of the journey. Economic margins within the ferry sector are, however, decreasing since the total turnover is decreasing constantly during the 2000s despite increasing number of passengers (*Viking Line*,

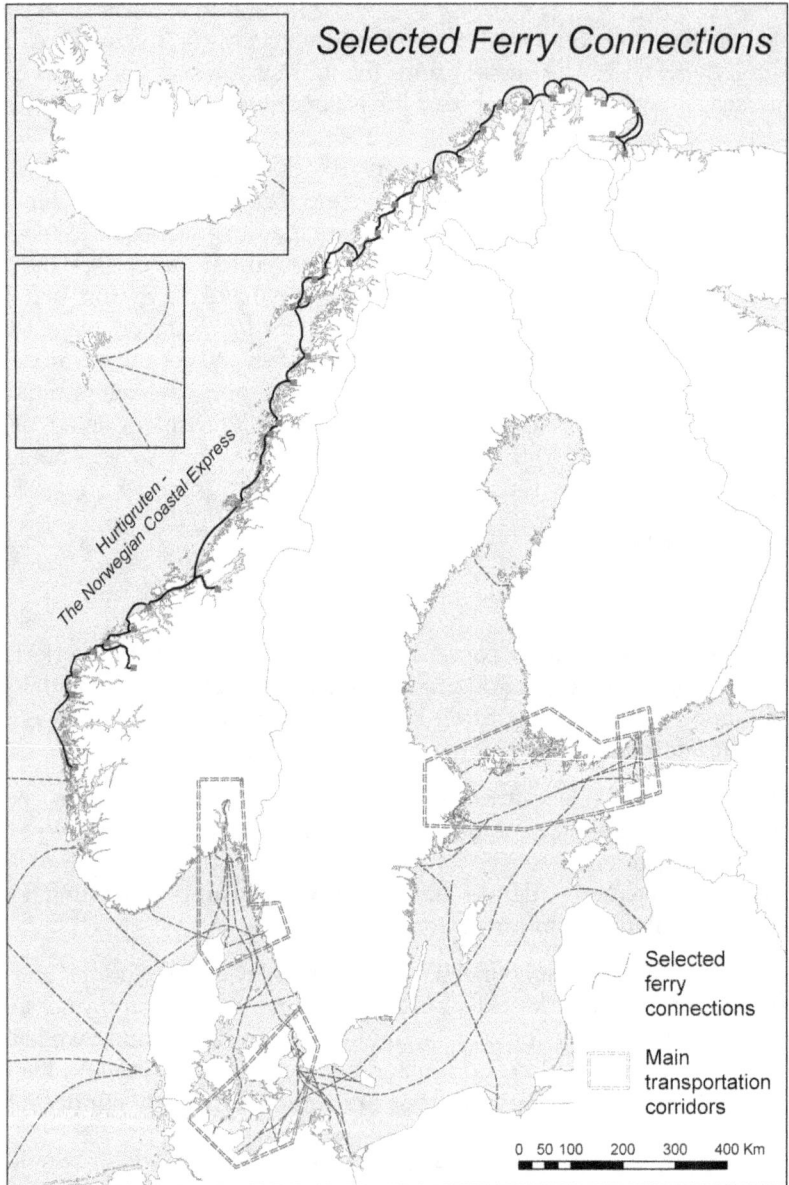

Figure 7.1 Selected ferry connections and transport corridors in the Nordic countries

2006). Short ferry trips taken out of season can be bought for less than €10. Incomes are thus mainly from tax-free shopping which is a major component of the ferry product.

The crucial importance of the shopping component was clearly demonstrated by the abolishment of tax-free regimes at the introduction of the Common European Market in 1995. The ferry line between Umeå and Vaasa in the Bothnian Sea vanished almost immediately since price differences became only marginal. Moreover, economic development in Vaasa also meant that differences in the shopping supplies between Sweden and Finland disappeared. Today, the ferry line is operating with public support. Ticket prices for return transport of a car have increased to more than €300, several times higher than on connections going through the Åland islands which managed to negotiate a special tax status when Finland entered the EU in 1995. According to a report issued by the Ålandic government in 1993 the Åland economy would have shrunk by up to 50% if duty-free shopping had been abruptly discontinued on the routes between Åland, Finland and Sweden (Fagerlund, 1997). Given that it was estimated that over half of the ferry operators' income was derived from tax- and duty-free sales, the future of duty-free sales to travellers was therefore a critical factor in Åland's strategy with respect to Finland's negotiations for EU membership (Fagerlund, 1997). Hence, all ferry lines between the Stockholm area and southern Finland and Estonia respectively have to stop there, which also led to the construction of a new harbour for quick docking only. Ferries operating between EU-member countries lacking tax-free shopping have now to rely on other advantages such as offering a break during the journey without interrupting the transport to the final destination.

A similar example can be provided regarding the connection between Helsinki and Tallinn. Not least cheap Estonian liquor lured Finns to take trips to Tallinn annually, and in 2002 more than 1.9 million adult Finns arrived in Tallinn (Standl, 2004). This situation entailed a comprehensive review of Finnish alcohol policy ending in a dramatic decrease of taxes on alcohol in Finland. This change also affected the ferry boat connection between Sweden and Finland, suddenly depriving them of their comparative economic advantages.

Ferry connections are also of vital importance for economy and society, especially for the autonomous Åland islands where a majority of all tourists reach the island using a ferry link. In 2003 tourists spent about €130 million on ferries owned by Åland shipping companies. This amount corresponds to 55% of all tourism incomes (ÅSUB, 2004b). Altogether the share of tourism is more than a third of Åland's total GNP and hence, the ferry lines facilitate a vital sector of the islands' total economy. Viking Line is providing employment to more than 2100 persons on their fleet. Additionally, staff in administration and other shore-based services comprise more than 700 (*Viking Line*, 2006) (see Case 5.1).

In summary, ferry boats are an important factor in Nordic tourism. Having been traditionally a part of the tourist route and bulk transportation, they recently transformed also into a tourism product of their own. Despite the set-back caused by EU-membership and the consecutive abolishment of tax-free shopping, they remain an important tourism attraction although certain structural adjustments due to intense competition seem to be on their way. For the Åland islands in particular, ferry connections are the mainstay of their economy.

Cruise Tourism

A closely related field of ferry tourism is cruise tourism. As already discussed in Case 7.1, cruises were a common way to explore the north and are thus a least one important root for today's tourism, not least along the Norwegian coast (see also Lundgren, 2003). Besides the Caribbean, the Mediterranean and Alaska, the Nordic area is today the fourth most requested destination for cruise tourism (Dowling, 2006). Since statistics are gathered in the harbours and account for arrivals but not for bought tickets, it is difficult to assess the exact scope of cruising in the Baltic and along the Norwegian coast. In general it should be noted that cruise tourism is a growing market segment in the entire world. Larger and more comfortable vessels are increasingly floating destinations of their own. Indeed, their size actually limits the number of destination ports able to host them (Nilsson et al., 2005).

The Nordic realm has a long tradition of cruise tourism, however, it only accounts for about 10% of the global market (Holloway, 2002). Nevertheless, Cruise Baltic (2006), a network of cruise destinations in the area, claims that cruise tourism has increased by 15% annually since 2000. Altogether about 240,000 tourists cruised the Baltic in 2005 (Nilsson et al., 2005). American visitors dominated the demand for cruises and Germans accounted for more than 20% of the cruise tourism.

Cruise Lines International Association (CLIA) is the world's largest cruise association and is composed of 24 of the major cruise lines serving north America. The 24 cruise lines represent 97% of the cruise capacity marketed from north America. More than 16,000 travel agencies are affiliated with CLIA. In 2007, 14 of the cruise lines were promoting sailing in Scandinavia and North Cape: Azmara Cruises, Costa Cruises, Crystal Cruises, Cunard Line, Holland America Line, Hurtigruten, Norwegian Cruise Line, Oceania Cruises, Orient Cruises, Princess Cruises, Royal Caribbean International, Seabourn Cruise Line, SeaDream Yacht Club and Silverseas Cruises. The 2006, a CLIA survey of the north American market reported that cruisers average 49 years of age, with above average incomes ($104,000 household income per year). They are typically married (83%), have college educations (57%) and are commonly employed full time (57%). On their last cruise they typically sailed with their spouse (79%) for about 6.2 days and spent approximately $1690 per person for their cruise and onboard expenses (not including airfare). More than half (56%) took their first cruise in or after 1999.

Marcusson (2004) identified three main stakeholder groups within the cruise tourism sector. First, the production of the cruise product requires tourists, travel agents and cruise lines including the ships. Second, destinations provide ports and services and facilitate the visit of the cruise ships. Moreover, tourist offices and numerous tour operators are involved offering information and guided land excursions. Finally, the passengers themselves are an important stakeholder group. Nationality and demographic structures influence demand even during the journey.

Since cruise tourists are usually an affluent segment of the market, economic impact is a strong motivation for destinations to compete for cruise ship arrivals. Economic impact entailed by cruise tourism occurs in a number of locations (Dwyer & Forsyth, 1996). Passengers need to travel into the cruise region, usually by air. Moreover, accommodation prior to cruise ship departure, land excursions, visiting tourist sites and shopping, all contribute to the regional economy. However, the tour operator has to pay considerable taxes and fees for port use and service

Coastal, Marine and Ocean Tourism 161

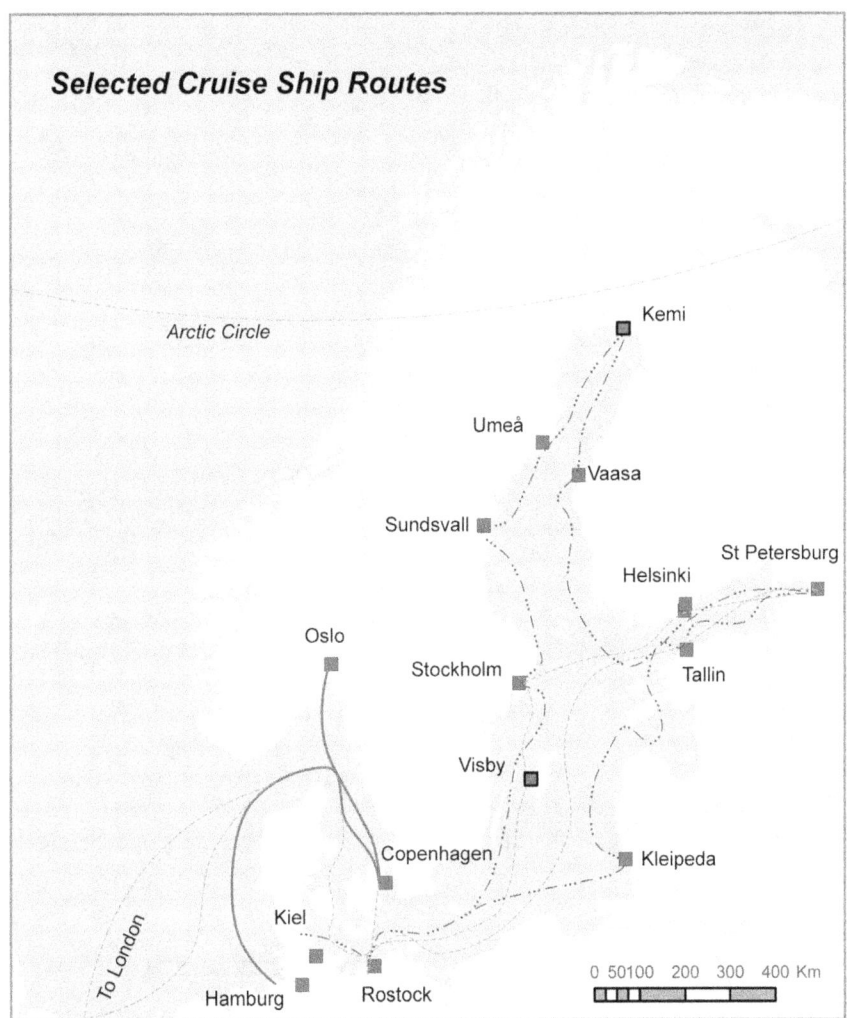

Figure 7.2 Selected cruise ship routes

supply in the ports. Additionally, crewing, ship maintenance and marketing are other areas of expenditure.

The routes mainly cover the southern parts of the Baltic Sea (Figure 7.2). Departures are not always within the Baltic, with Hamburg, Amsterdam and London being important nodes in the cruise networks. Sometimes, even the British Isles are included in the travel programme. Short cruises are on offer starting at about €350 for a four day trip. However, many cruises in the Baltic last up to two weeks and cost, dependent on the programme, more than €2000 per person. In searching for new destinations and products, cruises have now also reached the northern parts of the Bothnian Sea.

162 Nordic Tourism

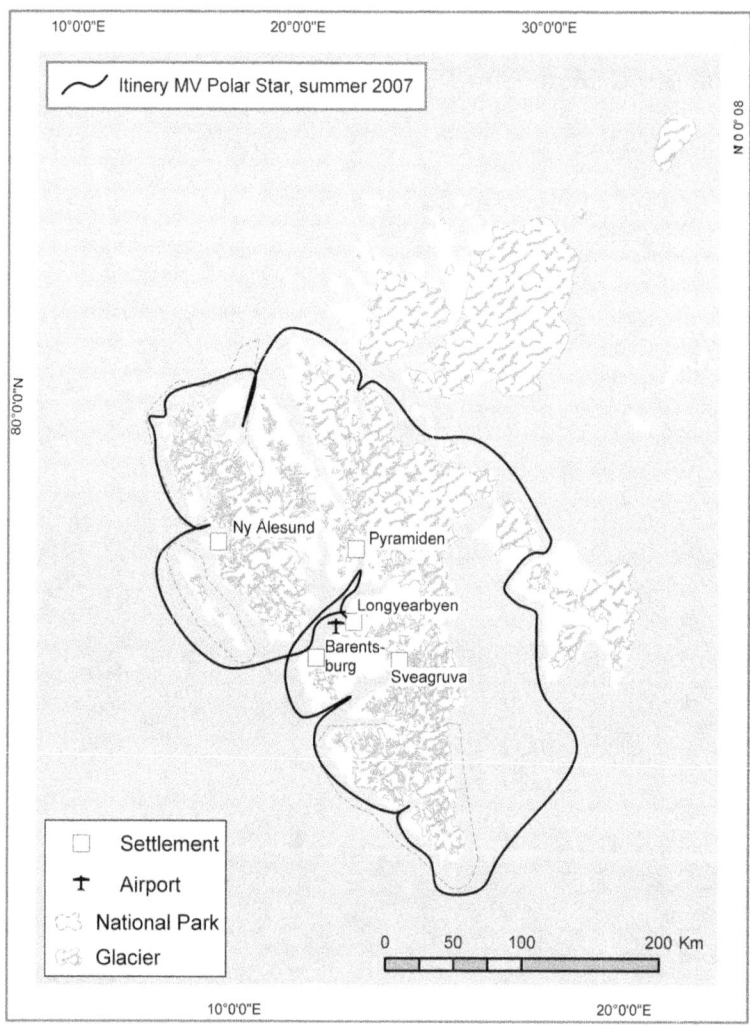

Figure 7.3 MV Polar Star's Svalbard route

Arctic destinations are also part of the Nordic cruise tourism product (Thomson & Thomson, 2006). Svalbard is an exclusive destination offering insights into the area's environment and heritage (Figure 7.3) (see also Case 6.1). Primary attractions include polar bear safaris and whale watching. Prices including airfares are from €5000 in a shared cabin. Even excursions to Barentsburg, the Russian mining settlement, are included in many trips and form one of few external sources of income to this otherwise marginalised community. A milder and shorter touristic Arctic experience is offered in Kemi on the Bothnian coast of Finland by the Sampo icebreaker during the winter months. In 1988 Sampo was modified to serve for tourism and 'ice adventures'. The cruise includes experiencing the icebreaking operation, swimming or walking on the frozen sea and a return trip by snow-mobile (see Grenier, 2004).

In 2003, the number of cruise passengers was over 40,000 and the World Wildlife Foundation (WWF) warned of the implications of further development with respect to following environmental impacts (WWF, 2004);

- oil pollution;
- pollution through waste water;
- pollution through garbage;
- air pollution;
- emissions of ballast water;
- anti-fouling hull paints;
- physical damage to underwater structures;
- wildlife disturbance;
- degradation of vegetation;
- degradation of historical sites;
- degradation of geological sites;
- littering.

Hence, often rigorous codes of conduct are developed to attempt to ensure sustainability of the Arctic environment (Mason, 2007). However, there is considerable debate about the most viable management systems, while the greenhouse gas emissions of cruise ships are also becoming an increasing concern (see Chapter 11).

Marine Tourism and Ecotourism

In recent years marine tourism products have been developed and have made marine environments accessible to a greater number of people than ever before (Hall, 2005a). Although the number of tourists in coastal areas is usually high, the long Nordic coastline also offers opportunities to go further and experience more remote and less crowded marine environments. Traditional products such as fishing (see Case 7.2) are increasingly accompanied by new and more exclusive products. Hence, marine ecotourism in particular, offers an opportunity to provide an exclusiveproduct to a small number of people. Usually the basic product implies transportation to a specific marine environment and interpretation of the local wildlife. Wildlife in particular is at the centre of interest. Besides these commercial offers, there arealso leisure activities such as sea kayaking, allowing for similar experiences in anon-commodified form exercised by people all over the Nordic countries.

Cases and issues 7.2: Marine fishing tourism in Norway

TRUDE BORCH

Norway occupies large parts of the chain of mountains that make up the Scandinavian Peninsula and massive mountains that slope towards the coast dominate the country. This topography makes Norway one of the most sea-oriented countries in Europe. Being a coastal nation, both commercial and recreational fishing have been important to Norwegians. Norway has also been an attractive destination for passionate recreational anglers from abroad. The first fishing tourists visiting the country were the British lords catching salmon in Norwegian rivers. During the last few years however, there has been a substantial growth in the number of marine fishing tourists visiting the country. These tourists are interested in both catching a range of marine species as well as in trophy fishing, mainly for halibut, cod and coalfish. The main fishing takes places close to the coast with boats in the range 5–7m. However, in some cases the fishing tourists rent a larger boat for ocean or deep sea fishing. There are commercial activities built up around the increasing interest in recreational fishing in Norway. Along most parts of the coast there are now tourism companies offering a combination of accommodation, boat rental and gutting, as well as freezer facilities. The largest number of companies offering fishing to tourists are to be found in the southern and central parts of Norway.

Among the 1000 companies offering sea fishing to tourists in Norway, there are great differences in the quality of accommodation facilities, boats, as well as in the facilities for handling the catch. There have recently been substantial improvements to the quality of the Norwegian fishing tourism industry. This is especially the case for the many new companies that are being established to provide services to fishing tourists. This part of the Norwegian tourism industry is growing fast. Seventy percent of the companies that offer fishing as an activity to tourists were established during the period 1990–2000. The tourism industry is welcoming the growth in fishing tourism because it represents a lengthening of the tourism season and because fishing tourists tend to have a high return rate to destinations. The tourism industry defines fishing tourism as an opportunity to utilise the rich fish resources along the Norwegian coast, the knowledge that many Norwegians hold about fish and fishing and the re-use of infrastructure from a declining commercial fisheries industry (boats, quays, buildings). In spite of the natural, cultural and technological advantages that Norway holds for offering fishing as an activity to tourists, there has been a substantial critique from commercial fisheries interests to the growth in this part of tourism. A main component in this is the unresolved management and rights questions in this part of tourism, that is, how to regulate marine, fishing tourism and who has the right to fish resources?

Marine ecotourism implies a number of problems, not least related to the disturbance of wildlife (Garrod & Wilson, 2003). In theory marine ecotourism follows the same principles as other forms of ecotourism. This includes economic viability and sociocultural aspects as well as ecological concerns (Cater, 2003; Cater & Cater,

2007). Hence, marine ecotourism is not only expected to make marine environments accessible to tourists, it also should provide profits to entrepreneurs, revenues for conservation and enhanced coastal livelihoods. This equation is, however, difficult to solve in practice. Cater (2003) argues that marine ecotourism has performed poorly, not least with respect to involvement of the local community. Instead, it is suggested that a lack of insight into the wider context hinders successful development of marine ecotourism (Cater & Cater, 2007). The number of stakeholders appears to be limited, but comprises diverse sectors such as agriculture, research, fisheries, power production, military activities, shipping, water supply and nature conservation, all competing for scarce marine resources, to name just a few.

Moreover, marine ecotourism is embedded in a multi-faceted policy environment (Wilson, 2003). In this context, different sectoral approaches conflict with each other, meanwhile the level of coordination is low. For example, coastal zones are often of primary interest to national defence, which obviously can obstruct conservation as well as development initiatives. The Swedish harbour of Karlskrona and parts of the Blekinge archipelago were, due to their military importance, not open to international visitors for a number of years during the Cold War period. A stranded Russian submarine in the 1980s implied an increase of military interest in the area and led to water bombing of potential alien submarines along the Swedish coast. Here ecological and touristic concerns were obviously sidestepped and security issues prioritised.

A tourist activity which has attracted particular attention is whale watching. At several spots along the Norwegian and Icelandic coast excursions are available allowing for studying whales at a close distance. Hence, the resource whale, which is still harvested by Icelandic, Faroe and Norwegian hunters, is given an alternative value aside that of its consumption as food. This development is not least welcomed by international environmental groups.

An example of this development can be found at Tysfjord in northern Norway (*Tysfjord turistsenter*, 2006). Here research and tourism are organised in parallel. Most tourists choose to purchase a trip with a large excursion boat to the whale grounds. Excursions start with a 30 minute presentation on orca whales, where tourists are informed about whale ecology and behavioural rules to be applied during the boat trip. During the trip, which does not guarantee that whales will actually be spotted, a detour is made to a rock used by sea eagles. Altogether the programme lasts for six hours and costs about €85. For those tourists eager to come even closer, excursions with inflatable boats are offered. These tourists are charged about €85 for a three-hour trip. An additional and more expensive option is to actually snorkel close to the whales. All activities are guided.

The economic impact of whale watching tourism is, however, not limited to observational activity as such. Instead accommodation can be rented for additional payment and tourists are encouraged to stay extra days. The company offers free trips when no whales are spotted during two consecutive days. Moreover, information about the frequency of whale occurrence is provided to lure tourists into the area, and alternative nature-based activities, such as kayaking and moose-safaris are offered close by to allow for tourists to spend more days in the area.

Similarly, in Sweden, several companies organised by *Nature's Best*, the sales arm of the Swedish Ecotourism Society, offer a number of activities that can be characterised as marine ecotourism (*Nature's Best*, 2006). Here, tourists can choose between sailing trips and kayak tours to remote bird islands and national parks. Bosmina, a company in the Haparanda archipelago offers tours into Sandskär national park, where wildlife watching is the main activity (*Bosmina*, 2006). This entrepreneur has a background as school teacher and is also active as consultant within environmental education.

Although the number of products in the marine ecotourism segment is increasing, the overall number is still limited. Moreover, not all products can be characterised as ecotourism. Short cruises along the coastlines to certain attractions may cater for other demands than watching wildlife and experiencing the marine ecosystem. Hence, boat trips can be seen as transportation primarily, or as a way to enjoy being on the water. Often, value is added by food and beverage services.

Boating

Boating is an activity that is strongly embedded in Nordic culture. For example, Norway has more than 400,000 boats (*KNBF*, 2004), meanwhile the number in Sweden amounts to 1.3 million (*Västsvenska Turistrådet*, 2003). Every sixth Swede owns a boat, while figures for Norway and Denmark are every eight and every 125th, respectively. Half of the Swedish population is expected to use a boat at least once a year; meanwhile the average use of boats is 30 days a year. In Finland there are 600 harbours that are managed for recreational boating and which provide a wide array of services. There are also a number of less well-equipped harbours in lake districts, coastal areas and the archipelago. Sievänen *et al.* (2006), in a survey of boating groups and recreational activity choices in Finland, report that almost half (47%) of the population had participated in boating in the course of one year. The most popular type of boating was rowboating, in which almost 40% of the population participated, and 76% of boaters had used a rowboat to some extent. Small motor boating was also a common activity, with 20% of the population and 44% of boaters participating. Canoeing was less popular (5% of the population and 11% of boaters). Boating with larger boats that also allow for cruising possibilities was also popular with a smaller portion of population. Motor cruising was a recreational activity of 7% of population and 13% of boaters. Sail cruising interested 3% of population and 5% of boaters. Boating frequency varied between the different types of boating. The average frequency of participation was 17 times per year for rowboating, 18 for small motor boating, seven for canoeing, 15 for motor cruising and 11 for sail cruising. Nevertheless, figures and knowledge on boat tourism is strongly limited.

Equipment and involvement in boating differ among boaters. Meyer (1999a) states, for example, that owners of larger boats allowing for overnight stays on board are more involved in their hobby than those with smaller boats. Also, access to sailing boats is influencing active involvement positively. In Denmark this results in more than 500,000 overnight stays in guest harbours, equivalent to about 1.5 million individual overnight stays (Statistics Denmark, 2006). Moreover, almost 50% of all overnight stays were owing to foreign boat tourists, primarily from Germany and Sweden.

Local authorities acknowledge the importance of boat tourism by, for example, establishing guest harbours all along the Nordic coastlines. However, natural harbours complete the total supply and thus, accurate statistics of boat tourism are difficult to accomplish. In Finland, boat tourism has been encouraged via special waterways for boats being established, guaranteeing a safe journey along the country's coastline (*Veneily*, 2007).

On the Åland islands, strategically located for boat tourism in the Baltic Sea, a total of almost 30,000 boat nights was registered in 2006, corresponding to about 90,000 tourist nights. The majority of boats in the guest harbours were registered in Finland (65%) followed by Sweden (27%) and Germany (5%). The season for boat tourism is however short; 75% of all overnight stays were noted for July (ÅSUB, 2006).

In a study on boat tourism in western Sweden, it was shown that most boat tourists stayed for about 10 nights in the area often sailing around in the archipelago (*Västsvenska Turistrådet*, 2003). It was also revealed that boat tourism had an impact on many parts of the economy since money was spent on dining and pleasure (39%), retail (43%) and accommodation (16%). In Sweden, the Göta Canal that links Göteborg on the North Sea with Stockholm on the Baltic, is regarded as one of the world's most impressive boat journeys and is a substantial tourist attraction as well as a safe route in its own right. However, its overall contribution to the tourism economy, although substantial, is unknown. Similarly, the economic and other related impacts of canal based tourism from Finland to Russia via the Saimaa Canal are under studied.

Planning for Marine and Ocean Tourism

Planning has rarely included marine areas, at least not in the same regulating way as on land. However, the establishment of marine national parks, nature reserves and bird sanctuaries has highlighted the need of also regulating marine tourism. However, besides assigning waterways to larger boats and limiting visitation temporarily to areas such as bird sanctuaries, planning appears to be rather rudimentary.

Nevertheless, recently the number of marine sanctuaries, nature reserves and national parks in the Nordic countries has increased. Hence, the demand for planning action has also increased. In this context current planning often applies a conservationist perspective. Accordingly certain areas are considered more valuable than others owing usually to the occurrence of threatened species or heritage. As has been shown for the Luleå archipelago this is also used for informal zoning (Ankre, 2005). However, a lack of data on marine tourism actually implies that attempts to influence the spatial patterns of marine tourism fail to accomplish the desired results.

Moreover, as demonstrated by Meyer (1999b) crowding in boating areas may be considered a problem, not least with regards to the experience of the boating trip. He showed that the existence of norms among the boat users increased their awareness of social and ecological problems associated with boating. Considering the increasing pressure on marine areas from diverse economic activities, recreation and nature preservation makes planning a mandatory task for the future.

Coastal Tourism

Although located far north, the coastal zone is of vital importance to tourism almost all over the Nordic realm. In particular, beaches in Denmark and along the Baltic Sea offer opportunities for sea bathing. Historically sea baths became popular already during the 19th century (Löfgren, 1999). Since then important changes have occurred entailing first an increasing interest in seaside vacations, but later also in growing competition with warmer destinations in Europe and elsewhere. However, some coastal destinations have maintained their position as major domestic destinations; meanwhile, others mainly cater for the needs of local recreation seekers.

Traditionally, boarding houses and camp grounds were the major vacation domiciles for tourists. Hotels accommodated travellers later on and even privately-owned cottages contributed to the total supply of tourist beds. Most cottages in Denmark and Sweden are located at the seaside and also in Norway and Finland coastal locations play an important role (see Chapter 8).

An example of early coastal tourism can be found in Hanko in the southwest of Finland. Already in 1874 when the town was formally founded, a spa with warm bath and sea bath was planned and established. Various amusement and leisure facilities such as restaurants, tennis courts, gyms and riding courts were added to contribute to the attractiveness of the new resort. Also, boats were offered for rent. Well-off families from nearby Turku ordered summer houses close to the spa and the surrounding park as appropriate and representative domiciles for societal life during the summer months. This behaviour was subsequently copied by the growing bourgeoisie entailing the development of new quarters in Hanko. Even today the town is an important destination for domestic tourism in Finland. However, it is not necessarily the opportunities for bathing in the sea that attracts visitors to Hanko. Instead it is the remains of the early tourism development that distinguish Hanko from other seaside areas and contribute to its attractiveness. Similar places can be found also in the other Nordic countries, often in the close vicinity to the main cities.

The Nordic seaside resorts are in general not very big. Instead they are characterised by their small scale and often their heritage as fishing villages. Hence, seaside vacations often combine sun-bathing and visits to heritage sites and environments, implying challenges to community infrastructure, property market and service supply. Moreover, in Sweden irritation about tourism is particularly prominent during the midsummer weekend when partying is relocated to major beach resorts on islands such as Öland and Gotland.

A Swedish example of coastal tourism development is Tylösand outside Halmstad in southwest Sweden. Even here tourism started early. In particular, the growing interest in camping, gymnastics and fitness during the early decades of the 20th century lured a growing number of tourists to the Halland coast and its sandy beaches. Tents were the major form of accommodation and later on they were accompanied by an increasing number of second homes in the vicinity of the beach (Rasmusson, 2004). Many properties in the immediate neighbourhood of the beach are now converted into exclusive real-estate; meanwhile, second homes can be found in less attractive areas beside the beach. A major camp ground offers a cheaper alternative accommodation. The establishment of a hotel on the beach widened the accommodation supply and made Tylösand a major sea bath in Sweden. Today, the hotel is one of the most prominent and high class establishments in Sweden, attracting the country's jet set to the area. Altogether more than 200 rooms, 20 meeting rooms, a conference venue for 700 visitors and several bars and restaurants cater for the touristic demand. Also, the hotel is known for its art collection and spa opportunities. During the summer, concerts with well-known bands are offered as well as after-beach entertainment. Besides the first class hotel and its services, two golf courses directly at the beach also attract tourists outside the bathing season. The purchase of the hotel by music star Per Gessle, known from the Swedish band Roxette, contributed to further increase interest in the resort luring tourists owing to nostalgic reasons.

Similar development can be found even further north. In Piteå on the Bothnian coast of Sweden, the resort of Pite Havsbad caters for tourist demand from northern Sweden, Norway and Finland. On the Finnish side of the coast, the Nallikari resort in Oulu offers (sun), sea and sand, mainly for Norwegian and Finnish guests but increasingly also to Russian visitors. In both resorts, camp grounds, a large conference hotel with an indoor spa that also offers activities for children, and a long sandy beach form the main attractions. Considering the quickly changing temperatures and weather conditions during the short summer in the high north, the indoor spa guarantees opportunities for bathing independent of weather. Recently, Norwegian visitors discovered the 'Nordic Riviera', lured by various entertainments and shopping opportunities in the area, as well as the bathing infrastructure. The colder North Sea water does not allow for similar development in Norway. Summer festivals such as the Piteå Fotboll Summer Games and a street festival create other highlights. Out of season the hotel is used as conference venue. The construction of a large meeting hall also allows for hosting national and international artists.

In Denmark, coastal tourism is to be found almost everywhere and is a major attraction in Danish tourism supply. In particular, along Jutland's west coast beaches attract visitors during the summer months. Second homes are the major form of accommodation offering a base not only for going swimming, but also for excursions to other sights in Denmark (Haldrup, 2004; Tress, 2002). Since second home ownership in Denmark is restricted to inhabitants of Denmark, rental homes are frequent, not least because they have traditionally been seen as attractive investments. Also in the metropolitan area of Copenhagen, beaches attract not least the local population. Amager Beach Park is an artificially constructed beach landscape on the outskirts of Copenhagen (Amager Strandpark, 2007). The beach offers not just bathing opportunities, but access for handicapped visitors, piers, a lagoon, a marina close by and a variety of activities and events.

The short Nordic summer season also requires alternative activity options outside the main season. The exposed location of much of the Danish coast also creates excellent conditions for wind surfing, and in the area of the mud flats coast in the south, bird watching is promoted.

Future Research

Research on marine, ocean and coastal tourism in the Nordic countries is extremely limited, probably not least because of the nature-based tourism profile promoted abroad. Nevertheless, intra-regional tourism is increasingly including international tourists interested in cruising and yachting in the area making the Nordic seas an arena for international tourism. This implies an almost unlimited supply of research issues to be addressed in the future.

Discussion questions

(1) What are the major forms of marine, ocean and coastal tourism in the Nordic countries?
(2) What are the reasons for recent internationalisation of marine and ocean tourism?
(3) What are the main impacts of marine, ocean and coastal tourism and how do they differ for various geographical contexts?

Essay questions

(1) Select a cruiseline and a particular cruise. Map the itinery and think where and on what tourists can spend their money during the trip. By using the web resources of the harbour municipalities you can also design a programme for the tourists during their landings.
(2) Discuss the impacts of boating and yachting on the marine ecosystems. What management solutions are possible?

Key readings

Regarding cruise tourism, Dowling (2006) published a collection of international cases; meanwhile Orams (1999) provided a state of the art overview of marine tourism. Garrod and Wilson (2003) offer a recent compilation of papers on marine ecotourism, while Cater and Cater (2007) provide a useful integrated overview of some of the key issues in the field of marine ecotourism. Although related more to lake tourism, a number of chapters in Hall and Härkönen (2006) also provide information on coastal and marine ecotourism tourism activities, especially in Finland. More specific and comprehensive studies on marine, ocean and coastal tourism in the Nordic countries are unfortunately not available.

Besides the web pages provided by the national tourism authorities (see Chapter 1), the following web pages offer interesting insights:

Cruise Lines International Association: http://www.cruising.org/
Hurtigruten: www.hurtigrut
Tallink: www.tallink.com
Viking Line: www.vikingline.fi
Hotell Tylösand: www.tylosand.se
Amager Strand Park: www.amager-strand.dk
Hanko: www.hanko.fi

Chapter 8
Second Homes in the Nordic Countries

Learning Objectives

After reading this chapter, you should be able to:

- Understand the second home concept.
- Understand the importance of second homes within Nordic culture and tourism.
- Identify some of the planning issues that second home development bring to a community.

Introduction

Second homes are part of Nordic heritage and are a major feature of Nordic tourism. Nowhere in the world is second home ownership that common as in the Nordic countries. However, second homes are not only important as domiciles for vacation and recreation, they also form important parts of the Nordic folklore and are thus reproduced in, for example, popular coffee table books and tourism promotion. However, second home tourism has also been contested; for instance, during the 1980s international charter tourism had been a competitive alternative to the mainly domestic second home tourism. During the 1990s second home tourism experienced rejuvenation. Today Nordic second homes are seized as opportunities for regional development (Flognfeldt, 2004; Jansson & Müller, 2003) and a potential avenue for improved transfer of funds from central government to municipalities; but they also stand accused of causing rural displacement (Marjavaara, 2007a, 2007b). Hence second homes have become an issue for many authorities all over the Nordic realm.

Although a significant part of Nordic tourism, there has been a broader debate as to whether second homes are really part of tourism at all. Cohen (1974) argued that second home owners are marginal tourists since there is no aspect of novelty involved in travelling to a personally owned cottage. Ontologically, second home mobility as tourism was also questioned by Johnston (2006) who argues that the very different set of practices employed by second home owners in fact distinguishes them from tourists. Destinations based mainly on second homes have thus left their tourism life cycle.

In contrast, Jaakson (1986) argued that the lack of novelty is hardly special for second home tourists, but comprises many other forms of tourism, too. He also noticed that the scope of second home tourism is significant, and that ignoring it would mean that the existence of large parts of the tourism infrastructure would be unexplained, as would the rise and the development of destinations. This was also confirmed by Strapp (1988) who clearly showed the role of second homes in stabilising destination development by supplying steady demand for various tourism services.

Second homes have increasingly been included in tourism research (e.g. Lew *et al.*, 2004) and hence, this chapter provides an overview of second home tourism in the Nordic countries. This is done by sketching the roots of second home tourism and its development until today. Moreover, current issues in second home tourism are reviewed. Finally, future research topics are presented and discussed.

Second homes are, in the context of this chapter, always considered as being a rural phenomenon (Plates 8.1 and 8.2). Sometimes it is argued, however, that

Plate 8.1 Second homes in the Swedish mountain resort of Tärnaby

174 Nordic Tourism

Plate 8.2 Second homes in Iceland (Picture: Müller, 2005)

increasing mobility for many households implies a multiple place attachment (Kaltenborn, 1998; McIntyre *et al.*, 2006). Hence, second homes may be inherited and passed on through generations are truly permanent homes; meanwhile, official primary residences represent temporary shelters only. Urban second homes certainly exist; they are however not the subject of this study.

The Rejuvenation of Second Homes

Second home tourism was an early form of tourism in the Nordic countries and hence, during the 1970s, it was considered as being a major force for rural change (Coppock, 1977). Probably because of the dramatic increase of international tourism, research interest in second homes went into hibernation and hence, only a few studies addressed second home tourism, often focusing on the impacts on the destination communities (Hall & Müller, 2004a, 2004b).

During the late 1990s second homes re-entered the academic agenda. Müller (2002) proposes four main explanations for this development. First, demographically aging societies imply a greater share of population able to allocate their time independent from the location of primary residence. This development obviously favours second home tourism. Second, economic restructuring entailed a concentration of capital and population to urban centres leaving existing housing in the countryside obsolete and consequently creating a resource for second home tourism. Moreover, economic restructuring also meant a transformation of industrial economies into service-dominated economies. This change also implied changes in working life. More flexible working conditions allow for taking short breaks and for distance working. Third, the development of effective transport and communications provide the preconditions for more intensive use of second homes. Decreasing costs for transportation allow for more frequent visits even to remote second homes; meanwhile mobile phones and ICT-development secures the connectedness to centres during visits to second homes. Fourth, European integration and the subsequent abolishment of national property market protection meant that Sweden, Norway and Finland became destinations for purchase by foreign nationals, not least German second home owners (Müller, 1999); although Denmark managed to restrict foreign property ownership in negotiations with the European Community in 1973 (Tress, 2002). A similar solution was chosen for the autonomous Åland islands. Here property ownership is for the most part restricted to inhabitants of local origin. Recently, contrasting property markets within the Nordic realm entailed a flow of Norwegian second home owners to Sweden (Berglund, 2005). Kaltenborn (1998) also argues that globalisation and the rapid changes associated with it, made people long for the local, secure and well-known. Second homes can give a sense of stability in a quickly changing world.

Against this background new questions are put forward, and earlier issues have caused renewed interest. Second homes are increasingly seen in the context of mobility and are located at the intersection of tourism and migration (Hall, 2005b; Williams & Hall, 2002).

Second Home Heritage

In the Anglo-American literature it is argued that a touristic interest in the countryside requires the process of urbanisation (Bunce, 1994). Urbanisation created social structures and populations with experiences of urban life that needed to idealise the countryside. It also created a political economy redefining the urban–rural interdependence. Moreover, urbanisation created an urban-based intellectual and cultural climate that led to the development of societal values that allowed for the idealisation of the rural landscape and the romanticising of countryside life (see also Chapter 5).

Although the use of several houses over the year could be found in agrarian society and among the nobility, even in the Nordic countries, second home tourism took off in latter part of the 19th century when, and in, places where urbanisation increased (Löfgren, 1999; Nordin, 1993b; Tress, 2002). Artists were especially important as protagonists for discovering rural and particularly seaside amenities. Scheduled steamboat connection entailed the development of second home areas close to the landing stages catering for the demand of the upper bourgeoisie (Hansen, 1969; Ljungdahl, 1938). These mansions or *Grosshandlarvillor* as they were called in the Stockholm archipelago, and sets of behaviour associated with them, were soon copied by the emerging urban middle class who saw rural life as a part of a physically and mentally rewarding lifestyle that was also imbued with nationalistic romanticism. In Sweden, the cabin movement *Sportstugerörelsen*, caused the construction of many simple and rustique cabins on the outskirts of the now rapidly growing urban areas that facilitated the desired lifestyles (Pihl Atmer, 1998).

The most significant expansion of second home tourism occurred after the world wars and was mainly facilitated by increased car ownership and government social tourism programmes aimed at offering domestic countryside recreation to large parts of the population. Hence, during the 1960s, 1970s and the early 1980s, second home construction boomed and added cottages to locations on the urban outskirts and amenity-rich areas all over the Nordic countries (Nordin, 1993b; Arnesen et al., 2002; Müller, 2004a, 2004b; Tress, 2002). Besides these purpose-built cottages, second home stock was completed by converting homes that previously served as primary residences, but had become obsolete owing to rural restructuring and out-migration (Müller et al., 2004). Hence, second homes are ubiquitous in the Nordic countries (Figure 8.1). About 24 million people share roughly 1.5 million second homes, that is, there are about 16 persons per second home (Table 8.1). About 50% of all Nordic households have access to second homes, although as shown for Finland the actual number having access may even be higher (Sievänen et al., 2006).

This is also obvious in popular culture and not least tourism promotion where second homes are reproduced frequently. *Stuga*, *mökki*, *sommerhu*s and *hytte* are not only Nordic words for recreational buildings, they also carry an important cultural meaning representing summer, links to the countryside, the good life and family history to name just a few (Jansson & Müller, 2003; Kaltenborn, 1997a, 1997b, 2002; Löfgren, 1999). Hence, staying at the cottage is an important part of national folklore; it is also a special part of family life, and thus in tourism it is promoted as a means of experiencing the Nordic way of life (see Case 8.1).

Cases and issues 8.1: Inhabiting the second-home

MICHAEL HALDRUP

In his book on the history of vacationing Orvar Löfgren recalls his own childhood memories of second-home vacationing in Sweden in the 1950s:

> Packing the car was a tense experience, with a lot of bargaining and pleading about what was necessary to bring or not. 'Do you really need this for the summer?' my stressed Dad would constantly ask the rest of us. (In the end you had to smuggle some of it.) Finally overloaded and usually late, the car took off with four kids, a dog, and number of potted plants precariously balanced on the top of the bags. The eight-hour drive from Stockholm to the west coast had its own microrituals and traditions, from the moment one of my sisters always chose to get carsick to the ceremonial lunch at the same old restaurant, preferably with the same menu as last June. We had to keep up the traditions of the drive, in detail. 'Hey, we always stop here!' There were constant references to landmarks passed or things that had

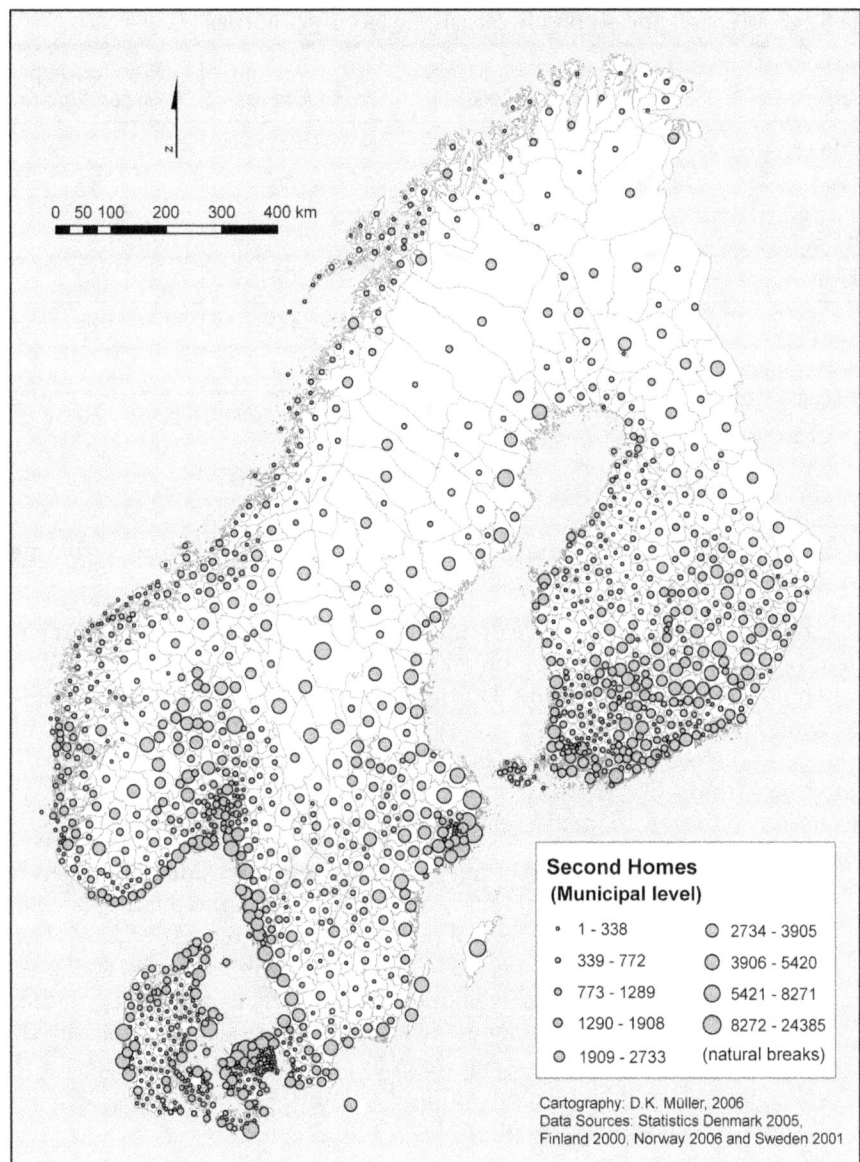

Figure 8.1 Second homes in the Nordic countries

changed since last summer. We were driving back in time, to the territories of all those early summers. Approaching the coast, we got ready for the standard game: who could spot the ocean first. When we were nearly there Dad had to stop the car to let us kids race down the road to the summer

Table 8.1 Population and second homes in the Nordic countries

Country	Population (2006)	Second homes	Population per second home
Denmark	5,427,400	202,500	27
Finland	5,255,600	450,600	12
Iceland	304,400	10,450	29
Norway	4,640,200	379,200	12
Sweden	9,047,800	469,900	19
Nordic countries	24,371,000	1,502,200	16

Sources: Statistics Denmark (2005, 2006); Statistics Finland (2000, 2006); Statistics Norway (2006); Statistics Sweden (2003, 2006), Statistics Iceland (2006).

> house. Just driving into the grounds wasn't enough of a return to summer. (Löfgren, 1999: 134)

Second-home vacationing is a dominant form of tourism in Norden. Yet it is often thought of, spoken of, assumed trivial. In Denmark around 10% of all households own a second home. This figure is however slightly misleading as everyone knows someone in possession of a 'summerhouse', and thus has intimate knowledge and memories of the ritualised performances of frozen family life, continuity, stability and coherence that epitomises second-home vacationing. Löfgren concludes his reflections on childhood summers:

> The dreamscapes of all the cottages, beach houses, lakeside lodges, second homes, have an enormous mental energy condensed into them. Even if the closest we will ever get to the real thing is a weekly rental of a Santa Cruz beach house, a time-share condo, a trailer campsite, or visits to the cottages of friends or relatives, we hope to reenact the classic cottage culture scenario of the perfect summer getaway. Or if we ourselves never had 'cottage summers', the least we can do is to provide that experience for the kids – they should have good summer memories, which means a feeling of ritualized continuity. (Löfgren, 1999: 150)

The second-home takes a central yet rather tacit role in our collective tourist memory. It offers a 'home away from home' (Jaakson, 1986), a refuge from the 'tyranny of time' that increasingly occupies our everyday lives (Franklin, 2003). The 'frozen' times and spaces of the second-home are worlds apart from the thrills and excitement of much contemporary tourism.

Diaries written by German inhabitants of rented second-homes on the Danish North Sea Coast stages the second home as a introvert place 'outside' the maelstrom of daily experiences (Haldrup, 2004). It is a place to 'build up from civilizational constraints', a place providing 'time for us and time to do things with our children'. Some have presented the second home as a place antithetical to 'noisy and fuzzy tourist places of the South', others described how they took pains maintaining the staircases and heating supply system of their (rented) second home. Photographs collected with the diaries not only showed intimate scenes from families everyday life (Haldrup & Larsen, 2003), but also how they rearranged furniture and decorated the spaces of the second home with wild flowers, mussels and stones, thus making it into a home of their own.

According to the British anthropologist Tim Ingold much recent social theory and analysis have failed by exaggerating how people ascribe meaning to things, places, practices and images, instead of engaging with the ways they are inscribed (also materially) in the world. Humans, he argues, inhabit the world from their birth onwards, they use it and their capabilities (language, tool use and so forth) are products of this active use (Ingold, 2000: 185–187). This 'dwelling perspective', as Ingold coins his approach, furthermore implies that memory and sociality are incorporated into places and landscapes. He captures this embodiment of memory in the double notions of landscape and 'taskscape'. 'Just as the landscape is an array of related features so – by analogy – the taskscape is an array of related activities' (Ingold, 2000: 195). Likewise 'Landscape as a whole must (...) be understood as a taskscape in its embodied form' (Ingold, 2000: 198).

Viewed from such a dwelling perspective the obsession with home making in the performance of second-home tourism makes perfect sense. Like this inhabiting, the second-home is no less of a ritualised performance than any other tourist performance. Approaching second-home vacationing from a dwelling perspective illustrates how central apparently trivial materialities are for constructing, staging and performing tourism. Instead of ignoring them, mundane materialities such as dwellings, technological artifacts (such as cars and cameras) and material surfaces of the landscape are central in the way people inhabit the world also as tourists (Haldrup & Larsen, 2006). Cars are used not only as a means of transport, but for extending the domestic spaces of families. Likewise cameras are used for materialising and eternalising 'frozen memories' of moments worthy for memory, and as means to construct and tell families memory stories (Haldrup & Larsen, 2003). Landscapes are not only objects of passive visual consumption, but material surfaces to be engaged with, used and activated.

As one family describes it, the moment they encounter the grounds of their rented second home, a train of recollections of past holiday moments and practices are triggered:

> We arrived at [this place] where we have worked out that we were 12–15 years ago (...). Here we found a lovely exclusive house, with small dunes of its own, just outside you can view the sea. The first road of course took us to the beach. Off with the shoes, and then off we went until we found a wonderful landscape of castles in the sand.

Second-home vacationing is intricately bound up with projects of home making. Not only is the rented house domesticated and transformed into 'our' house, so are the surrounding spaces. The coastal landscape of dunes and coastline is encountered as a 'taskscape' already inscribed with the path to take and the rituals to perform epitomised by the encounter with 'a wonderful landscape of castles in the sand'; traces of past holiday performances to take possession of, reconstruct or oust.

Motives for Second Home Tourism

Motives for second home use are similar to motives for tourism in general. Jaakson (1986) provided an outline of different motives in the Canadian context. However, Nordic studies (Jansson & Müller, 2003; Kaltenborn, 1998) confirm the validity of Jaakson's listing even in the regional context. He lists the following motives:

- *Inversion*: Second homes offer a contrast to everyday life.
- *Back-to-nature*: Second homes offer countryside and nature experiences absent in many owners' urban lives.
- *Identity*: Second homes provide a geographical node for identity formation based on leisure rather than work.
- *Surety*: Second homes are locales for gathering family and friends. Hence, they represent owners' ambition in life and are hence sometimes called *Lyckebo* ('Home of Happiness') and similar optimistic names.
- *Continuity*: Second homes are stable nodes in their owners' lives that are kept sometimes during their entire lives and even passed on to the next generation.
- *Work*: Second homes are places for creative work such as gardening and interior design.
- *Elitism*: Second home ownership can function as symbolic consumption showing their owners' social and economic status.

In addition, at least one more motive can be mentioned. Second home construction, particularly in the mountain resorts, was seen as a promising investment especially during the 1970s. Investment costs were expected to be covered by rental incomes, an expectation that unfortunately did not always turn out to be realistic.

The increasing number of foreign second home owners, originating particularly from Germany and other central European countries, carry similar motives as the domestic owners for their second home ownership in the north (Müller, 1999). However, differences in property prices explain international second home tourism too, as well as a wish to leave the home country behind.

Second Home Patterns

Early academic interest in second homes focused on patterns and diffusion processes (Aldskogius, 1968, 1969). Several studies illustrated the second home hinterlands of urban centres (Hägerstrand, 1954; Hansson & Medin, 1954;

Linkoaho, 1962; Sund, 1949; Svensson, 1954; Wiik, 1958), and others charted the origins of second home owners in certain destinations (Finnveden, 1969). These studies usually aimed at illustrating patterns and distance decay. Moreover, second homes became increasingly relevant for planning (Anonson, 1954; Svalastog, 1981).

As shown in Figure 8.1, second homes are not evenly distributed in the Nordic countries. There are in general three different factors influencing the geography of second homes. First, rural restructuring has caused rural depopulation and hence, a share of the rural housing stock was converted into second homes. Thus, the historical geography of settlements, as well as the geography of economic change, influence current second home patterns intimately. Second, urbanisation during the 20th century created new belts of accessible and purpose built second homes around the growing metropolitan centres that compose the majority of the second home stock in the Nordic countries. Third, the geography of amenity-rich areas directed second home flows to coast and mountain areas. In particular, the two previous factors are highly interrelated since the construction of purpose-built homes is concentrated in amenity rich areas. Moreover, recently there has been a polarisation of second home tourism towards amenity-rich areas mirroring a growing interest for second homes as leisure commodities rather than nodes in a genealogical context (Müller, 2004a).

A majority of second homes were built after the world wars and hence, most of the second home stock is purpose-built. Second home distributions reflect current population patterns and thus, the image of a converted farm now used as second home is the exception rather than the rule. It is also obvious that second homes concentrate around certain types of landscape. In Denmark amenity-rich areas are almost exclusively found in seaside locations, meanwhile in Sweden and particularly in Norway second home locations in the mountains are attractive (Figure 8.2)

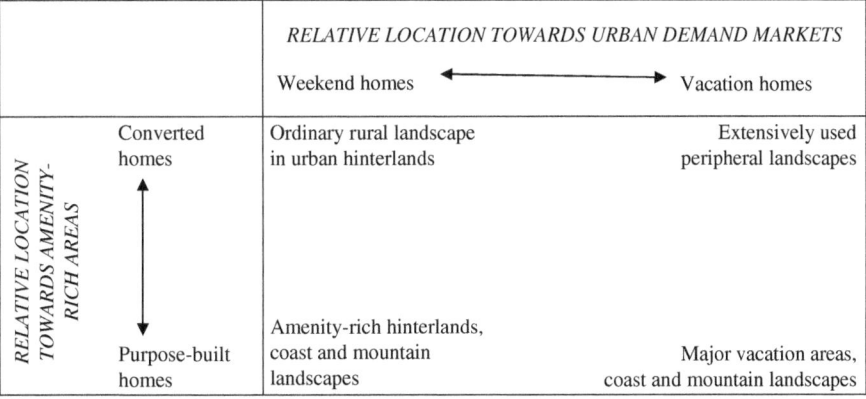

Figure 8.2 Second home landscapes and their characteristics
Source: After Müller *et al.* (2004: 16).

(Müller, 2004a, 2004b; Tress, 2002; Case 8.2). In Finland lakeside destinations compete successfully with second homes along the coastline and the ski resorts in the north of the country (see Case 8.3). In Iceland second homes can also be found in inland locations close to rivers and streams (Figure 8.3).

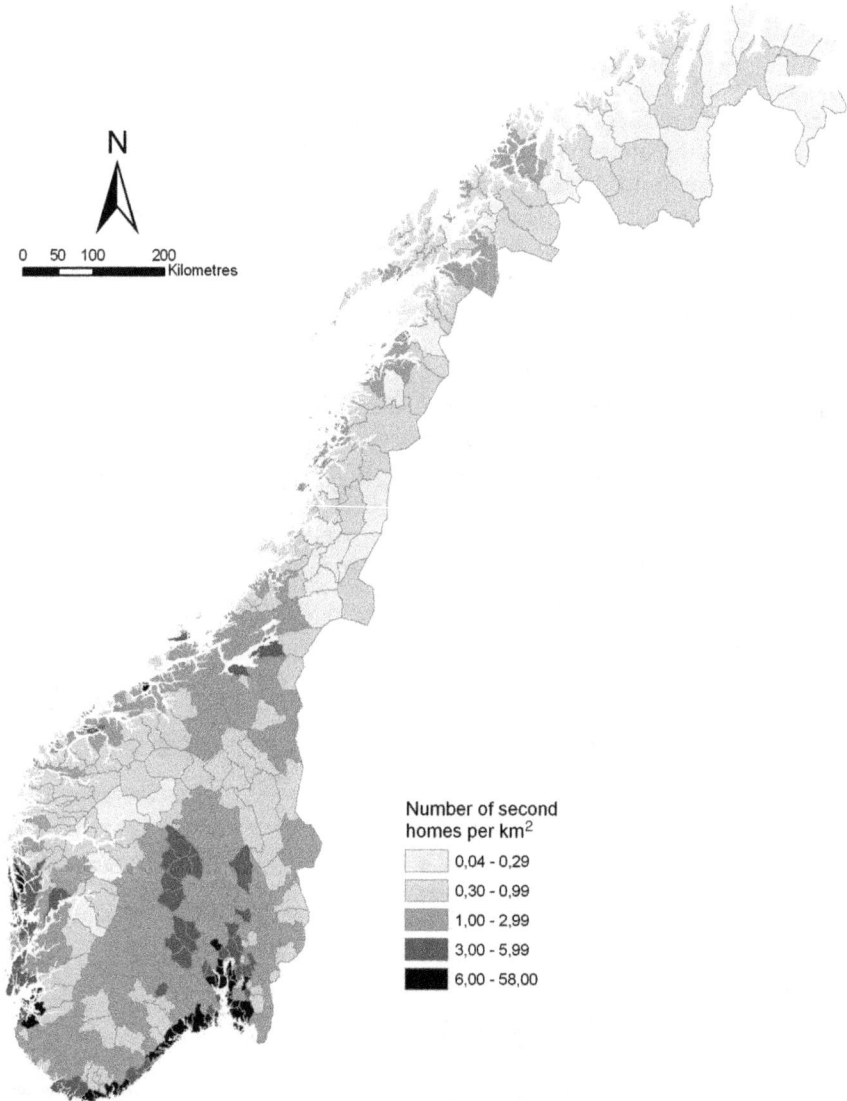

Figure 8.3 Density of second homes. Number of second homes per km² land per municipality, January 2005 (statistics: SSB)

Cases and issues 8.2: Second home tourism in Norway

BIRGITTA ERICSSON and KJELL OVERVÅG

An important part of leisure and tourism

Second homes have played an important role in Norwegians' holiday life for several decades. Today there are 375,000 second homes in Norway, and more than 60% of the population (aged 29–79 years) either own or have regular access to a second home. Access to second homes in Norway is then, in contrary to many non-Nordic countries, widely distributed among people. Such a wide distribution entails, however, significant differences in technical standards, accessibility and usage of second homes. Approximately one-third of all second homes have a technical standard equalling the residential home (Statistisk Sentralbyrå, Teigland, 2000).

About one-third of all Norwegians' domestic vacation trips are accommodated in a second home, either borrowed, rented or owned by the user (Statistisk Sentralbyrå, 2004). Norwegian leisure tourism in general is highly seasonal, and the use of second homes follows that pattern. Reasons are timing of school holidays and general holiday to periods when the Norwegian climate is favourable. With the main holiday period in summertime, second home usage also peaks in the summer season, although there are differences in the height of the peak season between seaside and mountain resorts.

A new feature regarding second homes is their increasing significance in destination development as rental accommodation, especially noticeable in alpine winter resorts. Although this has been a major form of accommodation in, for example, the Alps for a long time, many Norwegians have not until recently let their second homes to strangers. This is also due to lack of market appeal (e.g. standard and recreation facilities), location and accessibility.

Letting second homes could be in the joint interest of resort and second-home owners. The resort gets returning guests without major investments in accommodation capacity and owners of second homes a contribution to 'balancing the sheet'. A combination of owners' own usage and letting second homes could generate a significant number of guest nights and expenditures of benefit to the resort. This strategy is, however, only applicable to a limited number of the most attractive, well located and developed alpine resorts. The market situation in Norway is very much different and more extensive than in more densely populated countries, especially neighbouring the Alps.

Uneven spatial distribution and development

Although one can find second homes spread all over the country they are geographically unevenly distributed. Most second homes have sprawled from major urbanisations in Norway, especially from the capital (Oslo), and are today clustered in (a) attractive seaside areas and (b) in mountain areas, both within a couple of hours travelling time from home (Figure 8.3).

Figure 8.4 Change in number of second homes 1970–2005 by municipality (statistics: SSB)

Since 1970 the number of second homes in Norway has nearly doubled, while the population has increased by 18% and the number of households by 50% in the same period. More recently, growth has been especially strong in valley and mountain regions (e.g. Trysil, Gudbrandsdalen, Valdres, Hallingdalen) in the hinterland of the Oslo urban area (Figure 8.4). This development is generally triggered by a strong national economy and considerable growth in personal wealth of Norwegians. The significant increase in second homes in mountain areas has its origins in push as well as pull components. Push components are mainly lack of available land and rocketing prices in attractive, accessible coastal regions. On the pull side is a significant increase in practising alpine skiing. Furthermore, more widespread opportunities for flexible organisation of work and family life represent a framework for extended use of second homes.

The rise in number of second homes has been accompanied by increases in size, standard and usage, especially from the mid-1980s. In 1985, the average size of new-build second homes was 60 m^2; in 2003 it was 79 m^2. Technical standards have risen and imitate urban-like structures and infrastructure (roads, electricity, water supply, sewage). In sum this tends towards clustering of second homes, especially around major tourist destinations (Ericsson *et al.*, 2005).

Different motivations for ownership

Recreation and leisure are main reasons why people own and use second homes. But just as there are different kinds of second homes, there are different motivations for second home ownership. Knowledge about such motivation is necessary to understand this market, and important when planning for and developing second homes. Based on author surveys, five dimensions for second home ownership can be identified:

- modern outdoor recreation motivations;
- traditional outdoor recreation motivations;
- specialised outdoor recreation motivations;
- second home owners with roots in the area;
- flexible professional second home owners.

It is important to also note that owners' motivations will differ in different areas. This is mainly due to differences in landscape amenities and the motivation of (potential) owners.

A debated issue

Second homes have generated a public debate in Norway in recent years. One is the international theme of costs of environmental protection versus benefits of economic growth concerning land use issues. Prominent themes discussed are whether landscape aesthetics decline, what impacts second homes have on nature and wildlife, and the degree of restrictions on public access to attractive landscapes (especially coastlines) versus private property rights.

From a tourist industry perspective, there are differing perspectives on adequate land allocation in resort development and zoning of accommodation forms. Main questions concern whether second homes occupy valuable areas in tourist destinations at the expense of commercial accommodation, which could be expected to generate more spending tourists to a destination than does low occupancy rates on 'cold beds' in private second homes.

Cases and issues 8.3: Second homes in Finland

MERVI HILTUNEN

The history of second homes in Finland traces back to the beginning of the 19th century and can be divided into four periods. The first period was characterised by societal elitism and seasonal summer villas located in coastal towns and nearby islands along the Gulfs of Finland and Bothnia. During the second period between the two world wars, seasonal second homes spread towards inland lakeshores and became common also among urban bourgeoisie. The third period can be described as the beginning of second home tourism with

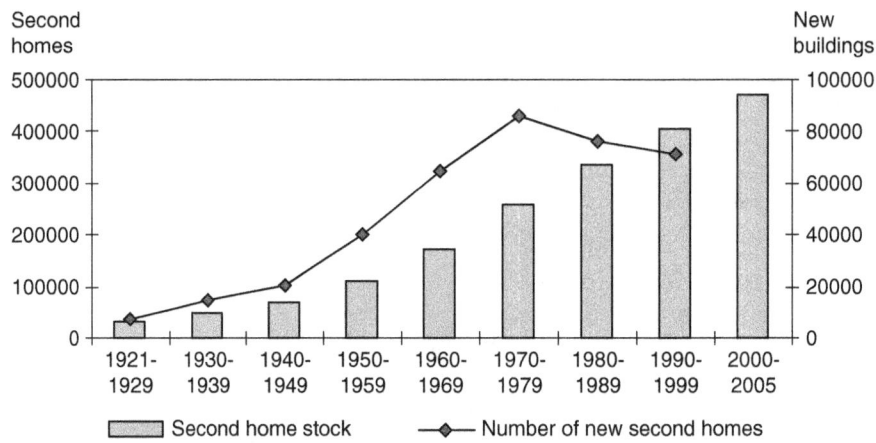

Figure 8.5 Second home development in Finland 1920–2005
Data sources: Statistics Finland (2005a, 2007).

expansive motorised mobility between home and second home. Cottage renting began in the 1950s and it still forms the basis for the rural tourism industry. The construction of new, often rather simple summer cottages was extensive especially during the rapid urbanisation of the 1960s and 1970s. Hence having a second home changed steadily from recreation of the gentry into an activity of common people (Nygård, 2003; Vuori, 1966). Today second home tourism is still growing, even though the building volume of new second homes has been diminishing during the last few decades (see Figure 8.5).

The present and thus the fourth period of second home tourism from 1990s onwards is characterised by governmental management (Jokinen, 2002). Second home development is today actively controlled by planning and building regulations yet also enthusiastically promoted in rural policy programmes. People increasingly want to use cottages throughout the year and have amenities of home life at second homes. Still, the common expression *Mökki* (cottage) and the ultimate motives for second home tourism seem to remain (Periäinen, 2006; Sievänen *et al.*, 2007). Privacy and outdoor activities in natural, quiet surroundings by the waterside keep on attracting urban dwellers to rural second homes. Sauna bathing belongs profoundly to Finnish culture and leisure, and is also an integral part of traditional cottage customs.

Second homes are widespread on the Finnish coast and archipelago, in the Lake District, and in tourism centres of rural and peripheral areas, especially in Lapland (Tuulentie, 2007) (Figure 8.6). By the end of 2005 there were nearly 475,000 buildings statistically counted as second homes and 800,000 Finns (15% of total population) belonged to second homeowner households (Statistics Finland, 2007). However, many of the second homes are used by friends and relatives and consequently around 2 million Finns have access to a cottage.

Second Homes in the Nordic Countries 187

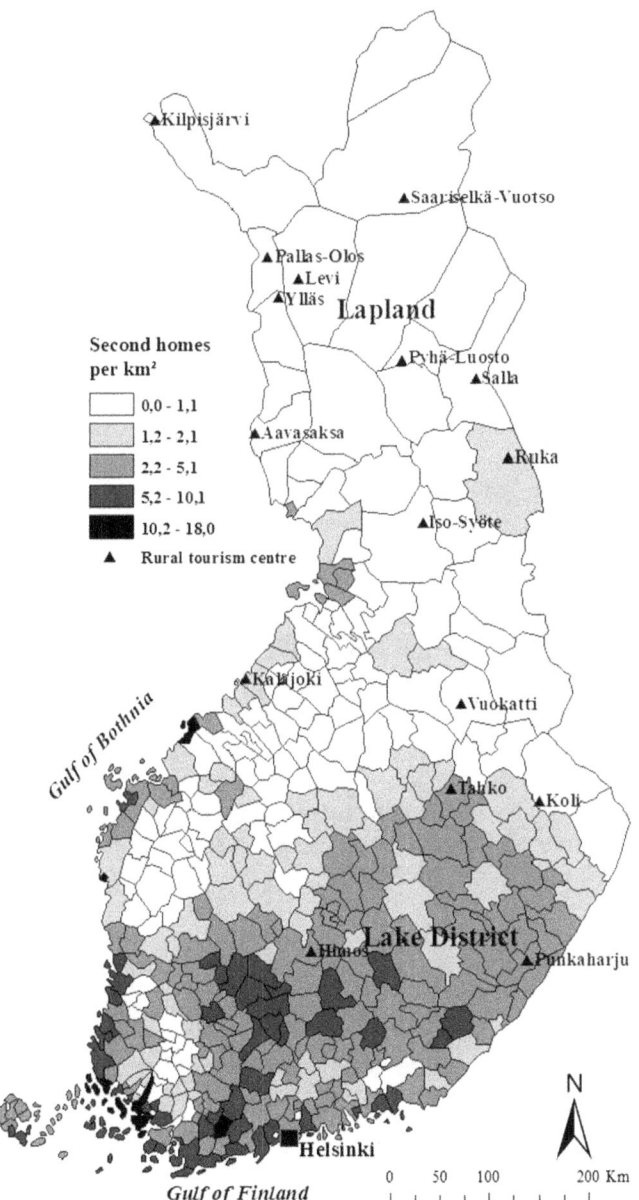

Figure 8.6 Distribution of second homes in Finland, municipal level
Note: Cartography: Katja Ristolainen & Mervi J. Hiltunen (2007)
Sources: National Land Survey of Finland (2006), Statistics Finland (2007); Vuoristo (2002).

There are over 12,000 rental cottages (Statistics Finland, 2006) and, in addition, timeshare cottages have become ever more popular in recent years.

Most of the second home owners belong to the post-war baby boomers of the late 1940s and early 1950s. This generation will soon retire and have increased free time. What does the near future of Finnish second home tourism look like? The following empirical findings are based on a questionnaire survey (referred to as the Lakeland study) carried out in June 2004 (Pitkänen & Kokki, 2005; see also Hiltunen, 2007; Pitkänen, 2008). The survey ($n = 1096$, response rate 45.5%) was conducted among second home owners who live in the metropolitan region of Helsinki and possess a second home in the Lake District (southeastern Finland). Respondents of the Lakeland study are actively commuting to a second home (see Table 8.2) and represent typical Finnish second home owners who are middle-aged couples, relatively wealthy, live in urban areas and have roots in the countryside.

More than one-third (36%) of the respondents intend to spend more time at their second home within the next 5–10 years. However, only 4% would like to move permanently to their second home even though most of the respondents (70%) have family roots in the Lake District. It appears that second home owners wish to enjoy both urban and rural life which unfortunately means a growing environmental burden in the form of commuting and energy use (Hiltunen, 2007). Distance working is increasingly a part of cottage life, although a majority (67%) of the working respondents want to keep a second home solely for leisure. Nevertheless, one-third of them would like to do remote work but only 1.5% do it regularly, although 20% do so occasionally. It looks like the near future of second home tourism in Finland is ensured. It seems, however, that cottage life is gaining new forms and more features of modern home life.

Table 8.2 Some average values of the 2004 Lakeland study compared to national means of second home tourism

	Age of owner (years)	Days at 2nd home per year	Distance from home	Trips to 2nd home per year	Travel by car	Floor space of 2nd home	Wired electricity	Shore site
Lakeland study 2004	57	71	270 km	15	96%	58 m^2	71%	90%
National average 2003/2004	59	72	107 km	9	82%	50 m^2	70%	84%

Sources: Pitkänen and Kokki (2005); Statistics Finland (2005a, 2005b); Second Home Barometer (2004).

Second homes are used in various ways by their owners. Often dependent on their relative location towards the owner's primary home, they are used as weekend homes or vacation homes. Figure 8.7 illustrates this for a number of destinations in northern Sweden and Finland. The second home owners in the ski resort of Tärnaby have their primary residences rather distant; meanwhile owners in Robertsfors and Korsholm (Mustasaari) have their primary residence in the nearby urban centres Umeå and Vaasa, respectively.

In general, second homes are used intensively (Flognfeldt, 2004; Jansson & Müller, 2003; Kaltenborn 1997a). In their study of second home tourism in northern Sweden and Finland, Jansson and Müller recorded that second home owners in the ski resort of Tärnaby used their properties about 20–30 nights a year. In areas more characterised by weekend homes these figures increased up to 80 nights a year.

These figures do not account for use by others than the owners' families. Second homes are however also used intensively for rental. Besides private often unregistered rentals, some larger agencies arrange contacts between tourists and second home providers. In some cases, as for instance the Åland islands, second homes form a major product for the entire destination (Case 8.4). However, statistics and information on the rental sector are scarce. In the Swedish tourism statistics for 2002 it is estimated that second homes are used for 42 million overnight stays or 30% of all tourism overnight stays (Swedish Tourism Authority, 2003).

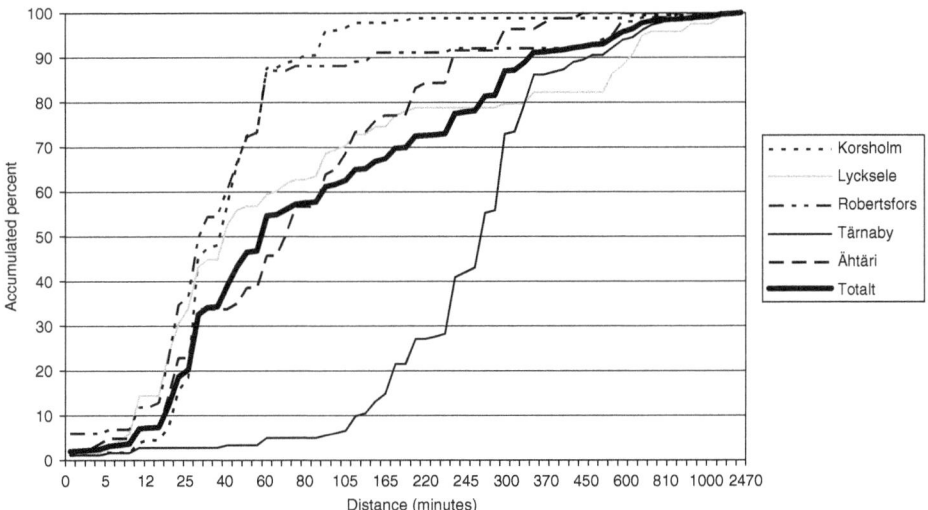

Figure 8.7 Time distance between permanent home and second home for a selection of destinations in Sweden and Finland
Source: Jansson and Müller (2003: 72).

Cases and issues 8.4: Cottage holidays in Åland

SAMU MÄKELÄ

Rental cottages are of great significance for Åland's tourism business. The absolute number of overnight guests in cottages is the largest of any accommodation sector and in general visitors renting cottages stay longer than guests in other accommodation (ÅSUB, 2005). Furthermore, renting a cottage is a flexible and easier option for bigger parties, like families, and it fits in well with the Åland's tourism image and with the provincial plans for tourism. In addition, letting cottages is for some an important extra or main source of income.

Challenges to tourism and cottage rental business in Åland
Today Åland's tourism business faces many challenges:

(1) the increasing competition from the Baltic Sea area;
(2) the lowering of alcohol taxation on the mainland decreases the strategic benefit of tax free sales on board and will probably raise the prices of ferry tickets;
(3) the lack of strategic insight and therefore also slow and insufficient investments in tourism; and
(4) the necessity to increase the number of cruise line passengers staying overnight in Åland.

The biggest challenge in the cottage rental business is to make it profitable enough. Often the profits are not adequate after the building and maintenance investments are deducted and the entrepreneur's work is taken into account. In principle it is possible to increase the profits by raising the rents or prolonging the season. Increasing the rent might work, if the visitor feels that the gained value of the stay increases as well. However, a condition for this is finding the right target group, raising quality and/or adding extra benefits.

According to Lindström and Salminen (2004), about 75% of the rental cottages were built before or during the 1980s. The same study showed that over 60% of the cottage entrepreneurs were over 50 years old and most of the cottages were booked through bigger agencies (Eckerö Line and Viking Line). This means that most of the cottages need to be renovated and modernised. However, many of the owners might feel they are too old to make the required technical and functional investments and many of the owners do not have the means to create a successful business idea or good marketing plans.

Focus areas of operations development
In the cottage rental business, as well as in tourism business in general, the trend is to improve cooperation as well as to better the service and quality. There are fixed criteria for rental cottages classification according to their

standard of equipment, and in the future the entrepreneurs will need to provide some kind of a quality system. According to the general quality concept the disabled should also have better access to facilities and attractions, the operating hours of shops and sights should be better adjusted, information more efficient, transportation better and the know-how of the people in the travel and service industry should be developed.

Furthermore, special courses and information packages need to be made available to cottage entrepreneurs to increase their skills. A well-equipped cottage is not enough, the service must be friendly and knowledgeable. Unique packages will be provided to meet the specific needs and wishes of various visitors. This development will be possible by increasing cooperation among the entrepreneurs, learning service skills and the Åland government's investments in developing nature and cultural attractions. Also, visitors to Åland will notice that new measures to preserve nature will be taken in all fields of travel business.

Åland's tourism business strategies 2003–2010 document (Ålands landskapsregering/Näringsavdelningen, 2004) states the following as the most important target groups: families, pensioners, golfers, athletic clubs and conference and business clients. Most of the visitors and cottage renters are still assumed to come from neighbouring areas; the Stockholm and Mälardalen area and southern Finland. Thus Ålands Turistförbund (the local DMO) faces the most important challenge of creating an appropriate image to attract the target groups.

Second Home Impact and Planning

Although second homes appear to be everywhere in the Nordic countries, related issues differ between geographical locations rather than the Nordic countries. In particular, the location of second home areas in relation to major urban demand markets entails certain problems; meanwhile the location in relation to amenity landscapes causes a different set of problems and opportunities. Müller *et al.* (2004) suggest a matrix representing second home landscapes and their characteristics in terms of second home stock composition and use (see Figure 8.2). Peripheral second home areas dominated by converted homes and seasonal use suffer from depopulation and hence second homes mark decline and disappearing regions (Müller, 2004a). In contrast peripheral locations with a high number of purpose-built second homes can be found in attractive seaside locations or adjacent to alpine facilities. In an urban context second homes close to amenity-rich areas tend to be purpose-built and to be contested by urbanisation and subsequent conversion into permanent homes (Marjavaara, 2007a, 2007b; see also Case 8.5). More ordinary rural surroundings do not usually attract major second home interest.

Cases and issues 8.5: Second home tourism and displacement in the Stockholm archipelago

ROGER MARJAVAARA

The County of Stockholm is the most populous region in Sweden. Consequently, the county and especially its archipelago, contains some of the most attractive and expensive second homes in Sweden (Figure 8.8). High demand for second homes causes some conflicts of interest between permanent residents and second home owners. One issue of conflict is how this demand affects future possibilities for rural communities. It is argued that high demand for second home leads to a displacement of permanent residents through price inflation and shortage of dwellings (Gallent & Tewdwr-Jones, 2000; Gallent *et al.*, 2003;

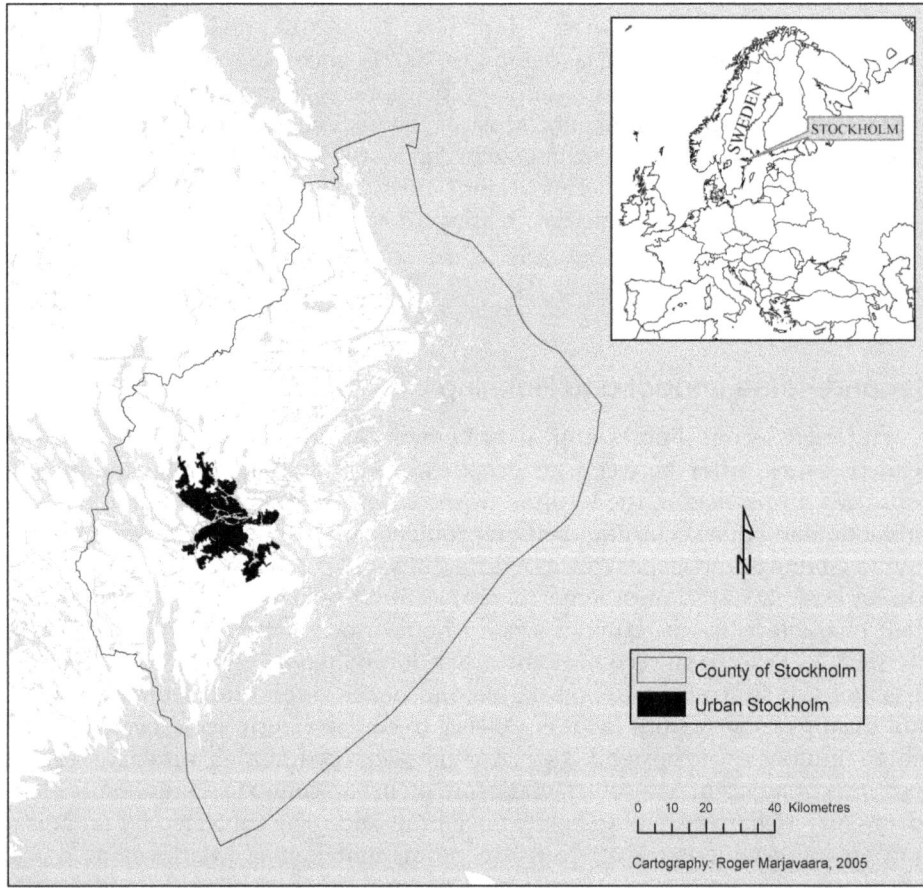

Figure 8.8 County of Stockholm

Skärgårdarnas Riksförbund, 2002). Rising property values and property taxes increase the cost of living for permanent residents, making it difficult for them to stay. However, others argue that the depopulation and economic decline in some attractive rural second home destinations is simply a result of the restructuring of the rural labour market, causing unemployment and out-migration (Keen & Hall, 2004; Müller *et al.*, 2004; Nordin, 1993a). This results in empty dwellings or potential second homes. Hence, second home buyers are only filling the gap caused by rural out-migration.

In recent years the number of second homes has decreased significantly in the Stockholm archipelago, in favour of permanent homes. The archipelago is not in a state of depopulation, but rather repopulation. Today, many former second homes have become dwellings for permanent use. This conversion is often made by young families who want to live close to nature and are attracted by relatively low prices. There is also a growing group of people who are retired or for other reasons do not need to commute to work every day. This group use their second home as a permanent dwelling. In the Stockholm archipelago, permanent homes contribute more to local price inflation than second homes do. Permanent homes' share of the total property value is increasing and the property values of permanent homes are higher than those of second homes. Consequently, any displacement of permanent residents would mainly be caused by increased demand for permanent homes, rather than by demand for second homes. Still, second home owners are blamed for rural decline. This is partly due to the characteristics of second home tourism.

Second home tourism is visible and a convenient scapegoat, compared to less tangible causes of rural decline (Gallent *et al.*, 2005). In the public debate, second home owners are often targeted as holders of alien values not suited to the local rural community. The paradox is evident. Local communities struggle to survive, but see outsiders as a threat to the preservation of the traditional community. The negative development is blamed on second home owners, an easy scapegoat. According to advocates of the displacement effect, the Stockholm archipelago has all the necessary attributes to be exposed to a displacement of permanent residents. Even so, no evidence of displacement can be traced at a regional level.

The economic effects of second home tourism were addressed by Bohlin (1982) who established a relationship between increasing distance and increasing spending in the destination community because of longer visits. The economic dimension of second home tourism was revisited by Jansson and Müller (2003) who stressed the importance of second home tourism scrutinising spending patterns in second home areas in Västerbotten and Österbotten. They also showed that second home tourism particularly entailed benefits for local suppliers of durable and convenience goods, although this varies with location. Nordin's

(1993a) study regarding the role of second homes for rural economic change in the Stockholm archipelago indicated that improved infrastructure also implied increasing competition for archipelago entrepreneurs. However, the increasing number of second homes also provide an opportunity for new entrepreneurship.

Another economic impact is usually expected regarding the housing market. In particular, foreign second home ownership is blamed as the cause of local inflation in the property market (Müller, 1999). This notion is also related to the idea that second home demand causes displacement of permanent residents (see Case 8.5). Aronsson (1993) scrutinised the social impacts of second home tourism illustrating the conflict over recreational resources between second home owners and permanent residents in Smögen on the Swedish West Coast where second homes increasingly convert the village into a seasonal community. This was not seen positively by most permanent residents and even today's Norwegian demand in the area is considered as a competition for property and other amenities (Berglund, 2005). In contrast, Flognfeldt (2004) argues that in the case of Norway second home owners often are influential ambassadors for their second home community.

Environmental concerns are also increasingly articulated towards an increasing second home tourism. The construction of new second homes close to shorelines especially has caused considerable debate in Sweden (Müller, 2004); meanwhile in Norway aesthetical and environmental reasons have been put forward to argue for a limitation of second home development in the mountain regions (Hecock, 1993; Langdalen, 1980).

Doubtless, however, is that second homes form an increasing problem for planners. Varying institutional contexts in the Nordic countries entail various problems. In Finland where second home owners pay local taxes, second homes are seen more positively than in, for example, Sweden where second homes are mainly associated with costs and inflation. Moreover, second home tourism causes a seasonal population distribution that is not measured or registered in official population counts (Müller & Hall, 2003). This implies increasing problems related to the provision of infrastructure and social services.

Future Research Issues

Today second home tourism is undergoing rapid change. New mobilities entailed by the changes mentioned above trigger new questions to be addressed in the future. Obviously, experiences differ between countries even within the Nordic realm which is also indicated by Hall and Müller (2004b) who chose to characterise second homes as being 'between elite landscapes and common ground'. In particular, the following fields of investigation can be identified:

- *Statistical issues*: Second homes are measured in various ways. Common, however, is a high degree of uncertainty regarding number and use of properties. Foreign property ownership especially is greatly unknown.
- *Scapes and flows*: Patterns of second home ownership are largely unknown, particularly regarding foreign second home ownership. This includes

geographical patterns as well as social and economic patterns. Moreover, it is unclear what constitutes the creation of scapes and the size and directions of flows.
- *Mobility strategies*: It can be expected that second home ownership is part of a comprehensive life course strategy where second homes are purchased to secure a spatial centre in life or a place to retire. However, knowledge regarding this is rudimentary.
- *Multiple and transnational identities*: Second home ownership may entail that people are at home in multiple places. These places may be located in different countries and hence the creation of multiple and transnational identities may be directly linked to second home ownership.
- *Impacts*: Second homes change communities not only physically, but also socially and culturally. It is sometimes argued that second homes displace traditional populations by causing a local inflation in property and other consumer prices. Hence, the economic impact of second homes on property markets and regional economies has to be investigated for different geographical contexts.
- *Governance and planning*: Second homes are often neglected in planning simply because their owners are statistically invisible to local administration and policy. This certainly differs between various countries. However, problems in supply of community services to second home owners raise these questions all over Europe.
- *Second home democracy*: If second homes are true permanent or alternate homes, questions regarding citizen rights become urgent. How can second home owners participate in local communities? Is there reason to rethink common democratic practices such as voting systems and taxation?

Answering these questions requires comprehensive research. This issue addresses some of these questions. However, considering the current changes in demand and transportation, there will be reason to revisit second homes in the Nordic countries in the future.

Discussion questions

(1) What are the roots of Nordic second home tourism?
(2) What are the reasons for recent interest in second home tourism?
(3) What are the main impacts of second home tourism and how do they differ for various geographical contexts?

Essay questions

(1) Discuss commonalities and differences of second home tourism (a) between the Nordic countries and other parts of the world; (b) between the Nordic countries.
(2) Analyse the opportunities and risks of using second home tourism as a tool for regional development.

Key readings

Coppock (1977) provided a first major overview of second home tourism where several still valid issues have been raised. Hall and Müller (2004a) addressed second home tourism at the turn of the century covering recent changes related to globalisation, economic change and ageing societies. In this overview a couple of chapters review the situation particularly in the Nordic countries. Another source of recent Nordic second home research can be found in a special issue of the *Scandinavian Journal of Hospitality and Tourism* (2007: 3).

The number of web pages covering a variety of cottage supplies for rent and sale is vast and hence the listed pages only give a glimpse. Novasol, DanCenter and Lomarengas are major agencies specialising in cottage rentals. Nordic-Estate and Schweden-Immobilien are examples for real-estate agencies specialising in sales for the German market.

Novasol: www.novasol.com
DanCenter: www.dancenter.com
Lomarengas, Finnish Country Holidays: www.lomarengas.fi
Nordic E states: www.nordic-estate.com
Schweden Immobilien: www.schweden-immobilien.de

Chapter 9
Culture and Tourism

Learning Objectives

After reading this chapter, you should be able to:

- Understand the concepts of cultural and heritage tourism.
- Understand the role and importance of culture in tourism.
- Identify some of the key issues and challenges in the relationship between tourism and local cultures.

Introduction: The Culture-Tourism Nexus

The relationship between tourism and culture is plentiful and complex and '... culture(s) is/are mobilised *for* tourists and read *by* tourists within particular settings' (Robinson & Smith, 2006: 1). This means that culture can serve as a tourist attraction, but also that the tourist experience as such is culture. In this chapter the focus is on culture as a tourist attraction. However, first a short introduction is given regarding the cultural nature of tourism, mainly reviewing Nordic experiences.

Tourism is perceived as culture itself (Robinson & Smith, 2006; Urry, 1995). This is discussed in MacCannell's (1976) seminal work on *The Tourist*, but also highlighted by, for example, Urry (1990, 1995) focusing attention on tourism as a consumption of place through, not least, symbols and signs. It is argued that this culture of being a tourist creates individual and intimate meaning and relational space (Crouch, 1999; Robinson & Smith, 2006). Hence, personal experiences and encounters outweigh the often promoted emptiness of destinations spreading national stereotypes and narratives all over the market. Thus the touristic consumption of culture is not only about the physical visit to touristic sites, it also is about encounters between tourists with various cultural backgrounds and the cultural expression at issue, producing diverging meanings and experiences (Therkelsen, 2003). This is not incompatible with the notion of a production of tourism based on culture, but indicates the wide spectrum that has to be thought of regarding the relationship between tourism and culture.

Recently, the notion of being 'on the move' and its plentiful meanings has been addressed frequently (e.g. Hall, 2005a, 2005b; Larsen, 2001; Meethan, 2001; Urry, 2000). Here, focus is on the experience of mobility per se. In the Nordic context Jacobsen (2004) investigates the motives of foreign motorised tourists and depicts them as individualistic and romance seeking through moving towards solitude and self-experience in relation to the Nordic scenery. Similarly, Haldrup (2004) analyses the practice of second home living among German tourists in Denmark by scrutinising the performance and experience of several modes of movement; inhabiting, navigating and drifting. (For a broader context of second homes and mobility see Hall & Müller, 2004a; and Chapter 8.)

Besides these studies of tourism as culture, culture is increasingly seen as a resource for tourism (Richards, 1996a, 1996b, 2001; Robinson & Smith, 2006). It is argued that this development is no coincidence, but a consequence of post-modernity de-differentiating previously separate sectors such as economy and culture. Moreover, the financial crisis of public households, not least during the 1990s, called for alternative often marked-based solutions to maintain cultural sites and activities. The consequences of globalisation add to this by forcing regions into a global competition for mobile capital and people in a global marketplace (Hall, 2005a; Harvey, 1989; Kotler *et al.*, 1993). In this context globalisation also contributes to a homogenisation of place, not least because of the global spread of brands and information. Hence, culture is used to highlight and utilise the local for presenting a distinct image in a global market place. This is also true for the Nordic realm, which often, not least in marketing, is stereotyped as Europe's last wilderness. However, a recent review of German guide books on Sweden revealed that almost half of their content related to culture (Zillinger, 2006). Even the Nordic tourism marketing organisations have realised this and promote culture as an attractive feature of their supply of tourism product.

Cultural Tourism Supply

Given the context of culture noted above, the supply of cultural tourism products is thus broad (Prentice, 1994; Timothy & Boyd, 2003). The increasing acceptance of the landscape as a product of culture implies, however, that a delimitation of cultural tourism is problematic. This is also true for the related concept of heritage tourism that almost can be seen as equivalent to culture tourism. Hall and McArthur (1998) state that heritage is what society wants to keep from the past, or as Ashworth and Tunbridge (1999: 105) put it, it is the 'contemporary use of the past'. Accordingly, heritage tourism is the touristic use of the past which places local history and traditions in a present-day context by spatialising history and localising culture to certain tourist attractions. Since a modern rock concert can also be interpreted as symbol of a music heritage rooted in traditions and dependent on previous achievements, it is difficult to distinguish the two concepts from each other. Hence, the following list of attractions represent types of the current supply and presents Nordic examples:

- *Natural history attractions*: Laponia World Heritage, the white chalk Cliffs of Møn, Landmannalaugur, Punkaharju, Geiranger fjord.
- *Scientific attractions*: Universeum (Göteborg); Nobel Museum (Stockholm), Tycho Brahe Museum (Ven), Kon-Tiki Museum (Oslo).
- *Primary production facilities*: farms and food production, Falun Copper Mine.
- *Craft and manufacturing centres*: Kosta Glasworks, Verla Groundwood and Board Mill World Heritage (Jaala and Valkeala).
- *Transportation attractions*: Flåm-railway line, Norwegian Road Museum (Lillehammer), Göta Canal.
- *Sociocultural attractions*: Birka, Viking Ship Museum (Oslo), Alta rock carvings, Bomba House (Nurmes).
- *Attractions associated with historic people*: Trollhaugen (Edvard Grieg's home in Bergen), Ainola (Jean Sibelius's home in Järvenpää).
- *Performing arts attractions*: Dalhalla.
- *Pleasure gardens*: Vigeland.
- *Theme parks*: Skansen, Maihaugen.
- *Galleries*: Louisiana, Barony of Rosendal, Retretti Art Centre.
- *Festivals and pageants*: Roskilde Festival, Air Guitar World Championship (Oulu), Savonlinna Opera Festival.
- *Stately and ancestral homes*: Drottningholm Castle (Stockholm), þingvellir.
- *Religious attractions*: cathedrals, Norwegian stav churches, Ukonkivi (Lake Inari).
- *Military attractions*: Suomenlinna Castle (Helsinki), Bommarsund Castle (Åland), Akerhus (Oslo).
- *Towns and townscapes*: Visby, Røros, Rauma.
- *Villages and hamlets*: Lysekil.
- *Countryside and treasured landscapes*: Österlen (Eastern Scania).
- *Seaside resorts and seascapes*: Lofoten Islands, Hailuoto.
- *Regions:* The Fjord-land, Åland Islands, Lapland, Gotland.

This supply can be classified along different scales. Richards (2001) proposes a typology based on the scales of 'form' and 'function'. In this context form refers to whether culture is presented as objects usually representing the past, or as process with focus on present articulations of culture. Function refers to the proposed impact of a cultural tourism attraction. Here, museums and monuments aim mainly at education; meanwhile art and music festivals focus on entertainment.

Moreover, it is possible to apply a temporal scale since not all attractions are accessible all the time. Cultural events are popular ways of delimiting the cultural supply in time and concentrating it in space (Müller & Pettersson, 2006). This also allows control of the nature of supply. Cultural tourism supply featuring landscape and the built environment can usually be consumed throughout the entire year particularly in the Nordic countries, where the right to roam guarantees almost unlimited access. An economic scale discerns commodified and non-commodified supplies. Certain attractions require entrance fees and offer diverse souvenirs; meanwhile others are displayed in a non-commodifed way.

Demand for Heritage Tourism

According to MacCannell (1976) the search for authenticity is a major motive for tourism. The concept of authenticity is, however, strongly contested within the scientific literature (Fawcett & Cormack, 2001; Timothy & Boyd, 2003). It is thus reasonable to ask what authenticity in tourism actually is all about (Wang, 1999). Current knowledge of the past is incomplete and selective, and consequently questions regarding the authenticity of heritage are always subjected to societal negotiations (Timothy & Boyd, 2003). Tourists do, therefore, always experience a distorted past only influenced by demand, economic and business processes, and political pressures (Timothy & Boyd, 2003). Lowenthal (1985) thus claims that a comprehensive authentic heritage never can be achieved or experienced owing to a lack of knowledge about the past. Hence, heritage places and attractions are often sanitised and idealised (Timothy & Boyd, 2003) and thus an experienced authenticity is always staged (MacCannell, 1976). Still, tourists do not necessarily experience apparently staged experiences as inauthentic. It is obvious that even staged attractions such as theme parks manage to attract considerable amounts of tourists. McIntosh and Prentice (1999) therefore propose to allow people to create their own subjective experiences of 'authenticity'.

Hughes (2002) offers a more recent assessment of cultural tourism and proposes the following framework for analysis. To culture-core tourists the cultural attraction is a primary motive for travelling to a certain destination. For a far greater population, however, cultural attractions are a secondary motive. Instead, cultural attractions are consumed when visiting friends and relatives or maybe as part of a business trip. This is analogous to Chen's (1998) notion of two major motivations for visiting heritage sites; seeking new knowledge and other more personal benefits comprising relaxation and sight-seeing. Timothy and Boyd (2003) also list nostalgia among the motivations relevant for heritage tourists.

Poria *et al.* (2003) are critical towards many approaches to defining heritage tourists. They reject the idea that all visitors at heritage sites can be considered heritage tourists. Instead they focus on the core of heritage and its meaning to the individual visitor as a discerning element. Accordingly they segment visitors to a heritage site into four groups:

- tourists unaware of the heritage attributes at the tourist site;
- tourists aware of the heritage attributes, but motivated by other factors;
- tourists motivated by the heritage attributes, but not personally involved in the displayed heritage;
- tourists motivated by the heritage attributes and personally involved in the displayed heritage.

Only the final group qualifies as heritage tourists. This definitional approach is a result of applying motivation as major element in defining a special-interest segment of tourism. However, from a business perspective this demand-side approach is of minor importance (Smith, 1998), although it may be significant with respect to the definition of heritage market segments. Also, the approach is not dynamic, that is, it does not open up a tourist's opportunity to discover their 'own heritage' through visits to a heritage site.

Although visits to heritage sites can be part of many tourist trips, research suggests that many heritage tourists are more educated than the average population. Richards (1996b) states that 80% of all heritage tourists in Europe had passed through a tertiary education programme. Internationally this is an analogue to the socioeconomic status of the visitors (Timothy & Boyd, 2003). However, in the Nordic countries, with a flat income structure and relatively small differences between the income groups, this is not necessarily true. The Swedish statistics on living conditions (ULF) reveal that white collar workers are more likely to participate in cultural activities than blue collar workers. This does not however correspond with income. Moreover, it is clear that women are more frequent consumers of so called high culture than men. Also, elder persons demand cultural activities to a far greater extent than younger persons, although school students are frequent visitors to cultural events, too. This is quite representative for the international situation (Timothy & Boyd, 2003).

Cultural attractions are also of great importance for various groups. Not least school groups are often, within the framework of their curriculum, frequent visitors of cultural sites. Even groups related to church, pensioner clubs and other special interests are likely to visit cultural sites and attractions (Timothy & Boyd, 2003).

Culture and Heritage for Regional Development

The role of heritage is also changing. Previously it has been exclusively considered as an asset for reinforcement of national pride and identity. Starting in the 1970s, culture has been more particularly seen as a tool for local development. In Denmark, public sector support of cultural festivals and institutions marked this change (Bayliss, 2004). This interest accelerated during the 1980s and 1990s in the Nordic states because of the expected role of cultural industries for regional development and growth (Power & Hallencreutz, 2005). This change has also been noted by Aldskogius (1993) who reviewed summer music festivals in Sweden. He recorded 410 events during the summer of 1993. In comparison, music events occurred only seldom during the early 1960s. With regards to cultural industries, also comprising libraries, print media and so on, Power (2003) records a significant increase in employment during the late 1990s making them an important segment of the labour markets in the Nordic countries, often reinforcing ideas of culture as integral to the development of creative cities and regions (see Chapter 4).

Cultural events are also used to promote tourism places as part of broader imaging and place competition strategies. Bærenholdt and Haldrup (2006) reviewed the efforts in Roskilde to stimulate economic development, relating to three major tourist attractions, that is, the Roskilde Cathedral, the Viking Ship Museum and the Roskilde Festival, focusing pop and rock music, which has entailed a number of spin-off activities within the music and heritage industry. With respect to the latter two attractions they conclude that the embeddedness is seldom local, apart from intensive networks with the local municipality. Instead, networks facilitating the success of the museum and the festival and their spin-off activities are Scandinavian and European. A similar development was observed in Hultsfred, Sweden, where a music industry cluster was developed around the annual 'Hultsfred festival' (Power & Hallencreutz, 2005).

The recent interest in cultural tourism as a tool for regional development is also facilitated by the support provided within the framework of the European Union's structural funds (Richards, 2001). Qualified projects are expected to contribute to economic and social change in the receiving regions often characterised as marginalised or, at least, contested by globalisation and economic restructuring. Hence, the availability of EU-funding substantially increased the number of tourism projects related to cultural tourism in rural and peripheral areas eligible for this funding. In this context focus has also been on the touristic development of already existing sites. Adding commercial value to these sites has also highlighted the role of interpretation (see Case 9.1).

Cases and issues 9.1: Interpretation in Finland

JAAKKO SUVANTOLA

Around the world there are examples of regional or thematic strategies for interpreting natural and cultural heritage. The creators of the strategies range from interpretation related NGOs to local governments and central government agencies. The term 'interpretation' appears in many of the plans explicitly. However, both the term and its meaning has remained absent in Finnish tourism strategies. Metsähallitus (the Finnish Forest Service that, for example, manages national parks as well as forest reserves) has moved towards interpretation by trialling their Nature Services' guiding material (Lehtonen, 2006). As a result they realised nature guiding personnel's need for practical guidelines for the process of creating guiding. While using the term 'guiding communication', they effectively mean interpretation of nature. Naturally, there are other organisations that do interpretation as part of their work, such as the Geological Survey of Finland and National Board of Antiquities. Neither of these, however, have a formal interpretation strategy. However, there is growing interest in the interpretation concept.

In 2006, a course in interpretation was organised for professionals, including workers from visitor centres, nature conservation authorities, museum people, tourism development project personnel and teachers of tourism. None of the participants were aware of the term interpretation (which was not in the name of the course). During the course the participants had to come up with an interpretative plan for the case they selected, often the very place where they were working. Many issues present in the interpretation process were familiar to them, but not as a part of a comprehensive interpretative plan. Furthermore, the meaning of a theme in interpretation was a new concept for the participants. Nobody had been thinking that their object of interpretation could, or should, have a theme or clear message in order to become more interesting for the audience. One comment included:

> When the visitor centre was built, I do not think we had thought what our message to the visitor is and what we want them to remember. The exhibition is content in providing mere information about the nature of the north.

The insights of the participants indicated that attention to interpretative principles has the potential to improve visitor experiences at the destinations and attractions in which they were working.

Many attempts to start cultural tourism have failed owing to diverse reasons often related to peripheral locations (Hall, 2007d; Müller & Jansson, 2007a, 2007b). Lack of human capital and political power are only two of these reasons. Westin and Paju (2003) point at geographically diverging preconditions for developing cultural tourist attractions. Peripheral sites lack visitation since demand for cultural tourist attractions is declining with distance from urban centres; meanwhile, cultural sites close to urban areas may suffer from crowding and degradation (Figure 9.1).

Concerning these disadvantages, Molin *et al.* (2007) question whether it is always the availability of a recognised heritage object that, facilitates the development of small-scaled cultural tourism. Instead they argue that, particularly in peripheral areas, good business ideas and engaged entrepreneurs indeed can create heritage that also has to be acknowledged by authorities responsible for the heritage sector.

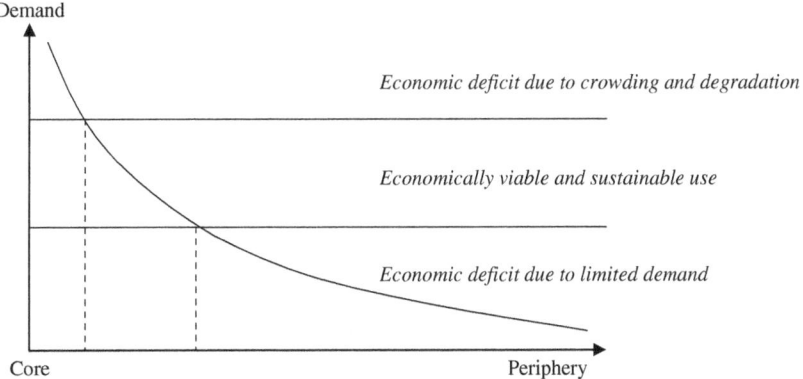

Figure 9.1 Demand for heritage tourism
Source: Adapted from Westin and Paju (2003)

Besides economic benefits, cultural tourism is also seen as a tool to generate local pride and contribute to identity (see Case 9.2) (Schouten, 2007). However, cultural tourists are also important customers since they tend to visit the cultural attraction more frequently than more distant visitors (Timothy & Boyd, 2003).

Cases and issues 9.2: The controversies of a heritage tourism site

ARVID VIKEN

This case study presents local perceptions of the Manor of Rosendal Barony as a modern cultural centre and a tourist attraction. The manor was the centre of the one and only barony in Norway, established when Norway was a Danish colony (1387–1814). In recent years the Barony has achieved a significant success both as a centre for the performing arts and with tourism. The data on which this case study are based are drawn from an investigation into the relationship between tourist attractions and the surrounding communities (*cf.* Viken, 2004). Among the research questions were commercial tourism as a parameter for social and cultural change in the local community. The Barony and Rosendal was chosen as a case due to its significant success and publicity recent years. The method applied in the research project was focus group interviews (*cf.* Morgan, 1988), supplemented by individual in-depth interviews.

The Rosendal Barony

After their marriage in 1658, Ludvig Rosenkrantz, a Danish noble, and Karen Mowat built a manor house on the Hatteberg Estate; it was a wedding present from the bride's parents. The manor was completed in 1665, and the estate was elevated to a barony in 1678. The title of 'Baron' allowed the holder exemption from paying taxes to the King, gave him property rights to churches, right to appoint vicars, supervision of the judicial system and of military conscription. Among the obligations were that the Barony's property could not be sold or mortgaged. From the mid-18th century the estate was classified as a hereditary property, meaning that the possessions of the Baron at that time would remain in his family as long as it existed. One of the most fascinating personages throughout the history of the Barony is Marcus Gerhard Hoff-Rosenkrone, an intellectual and an entrepreneur, who formally was not a baron, as he was born in 1823, two years after the aristocracy had been abolished in Norway. After his death the Barony became more and more of an anachronism, also for its owners, and the Barony was handed over to University of Oslo in 1927, according to its original statutes if no inheritors existed or wanted to take over. After some decades with a low profile, and primarily functioning as a farm, the Barony is today a museum and tourist attraction, and a place for the performing arts.

The model for today's management is Marcus Gerhard Hoff-Rosenkrone, the estate holder from the second part of the 1800s. Among other things he took part in public affairs, contributed to modernisation of farming in the region, and created a park around the manor. He was also a benefactor of artists and intellectuals and his home was a cultural centre (Hopstock & Tscudi Madsen, 1969). Many of those who visited the Barony during his time played important roles in Norwegian cultural life, and included musicians, composers, painters and authors, many of them reckoned as the most prominent artists that Norway ever fostered. Most of them also played a part in the constitution of a Norwegian national identity, and the Barony – despite its colonial roots – was one of the places where they met and discussed the creation of a new national state and a split with Sweden with which in 1814 Norway had been forced to create a union.

Today the Barony is again a cultural centre. Norway's cultural and artistic elite have been invited to perform or exhibit and the public have flocked to the Barony, filling the till with money, and the place has attracted high media attention. Since the mid-1990s an English theatre troop specialising in Shakespeare has visited the Barony every summer. The cultural figures practice their art and reside at the manor, but their presence is not only artistic; sometimes it also has a political dimension, as will be discussed later on. Hoff-Rosenkrone's spirit reigns, but not entirely; modernisation of agriculture that was one of his major interests is not a priority for today's management. The experimental farm established by Hoff-Rosenkrone has been abolished. However, the new regime has devoted the Barony to another industry; tourism.

Today, the Barony is a significant tourist attraction. It has enjoyed great success after it began to market itself. The number of visitors has increased from a few thousand in the early 1980s to 40,000–50,000 today. The building and its surrounding rose garden and park make the place suitable for sightseeing tourism. There are also facilities for meetings and seminars, including 26 bedrooms and dining facilities. In the summer, tourism's semiotic cover is drawn over the place: car and ticket queues, bus groups, happy guides, ice cream-eating and photographing people in a frame of opening hours, ticket counter, café and souvenir shop. These signs are not especially obtrusive, but as in other places do something to the spirit of the place (*cf.* Hubbard & Lilley, 2000).

The Barony as cultural heritage

The concept of cultural heritage can involve a variety of cultural expressions (*cf.* Herbert, 1997; Lowenthal, 1997). A relatively common division is between material and immaterial cultural heritage. In its simplest form material heritage involves cultural monuments, such as stone-age settlements, Viking graves, castles or collections of artefacts from particular cultures or periods – often in institutions named as museums and under a conservation regime. The Barony Rosendal is an example. However, in all societies there are also social remnants from the past; immaterial heritage including history, traditions, customs, legends, myths and other expressions and results of human life and relationships. History and traditions have given rise to many museums and tourist attractions. Due to its immaterial character, the narrative sides are always fundamental to such institutions. Somehow history must be told, and traditions often have to be demonstrated and given an explanation to be understood.

Another distinction is between what Kirschenblatt-Gimlett (1998) characterises as in situ cultural heritage, which relates its own history because it is situated in its natural place in a original or reconstructed cultural environment, and in context cultural heritage, which are representations torn out of its environment, but in a professionally documented way. There are a variety of methods for presenting context, including labels, information placards and other visual and audio-aids. Obviously, compromise solutions exist, and in situ-expositions will normally use some of the informing and contextualising techniques, such as information plaques and guides. The in situ–in context dimension can be combined with the material–immaterial dimension giving four different types of cultural heritage presentations: (1) in situ material, (2) in situ immaterial, (3) in context material and (4) in context immaterial.

Most institutions will to some extent have elements from all four categories. The Barony Rosendal has many of the in situ characteristics, but except for guiding the in context presentations (categories 3 and 4) these are modest. The Barony is primarily an in situ material heritage (category 1), but some category 2 heritage elements also exist; traditions have been revived, among them that of arranging a Christmas party for the children of the parish, bread baking and not least that of forging links between artists and the place. However, the most interesting aspect of the cultural heritage at the Barony is undoubtedly the immaterial in situ aspect that can be termed revitalising of life forms, the revitalisation of the spirit and work of Hoff-Rosenkrone. Instead of staging the past (e.g. performing Viking life) elements of the activities such as they were 130 years ago are taken into today's life form. Unfortunately, heritage as a copy of past life forms is not without antagonisms.

The revitalisation of a former life form also means revitalisation of old patterns of relations between the Barony and the surrounding community. The division between the 'classy set' associated with the Barony – mainly people from other parts of the country – and 'ordinary' local people, has been invigorated and envisaged. There is a group of people that comprises cultural managers, artists, academics and fund donors – and the managers of the Barony, that use the Barony as their hub. It is perhaps unsurprising, therefore, that in local parlance the present manager of the Barony and her allies are spoken of as 'the Baroness and her court'. Justified or not, the people of Rosendal feel that a new division between people on the inside of the Barony and those on the outside has been established in the wake of the revitalisation of Hoff-Rosenkrone's spirit.

The conservation of the Barony

The Rosendal Manor was put under a conservation order in the 1920s. Around 1990, the *Cultural Heritage Act* was changed and provides the basis for comprehensive conservation, so that the whole site is considered to be a cultural monument and the core of a larger whole or context that is conserved, practically the whole village. Besides the protection imposed by central authorities, the municipality is by law obliged to make an area regulation plan for the adjacent areas that take into account the cultural treasure nearby. In the mountains behind there is a national park (opened 2005: Folgefonna National Park) also covering a huge glacier, and a 'landscape conservation area', between the national park and the parish. These conservation pronouncements permanently alter the conditions for tree felling, grazing, hunting, fishing, outdoor recreation, tourism and other industrial development opportunities.

During the last decade several local projects have been suspended due to the conservation regime; for example, a rifle range behind the Barony, construction of houses and facilities on the farms in the neighbourhood and a planned expansion of the local graveyard. On the other side, the neighbouring farmers have contested a road the Barony had planned instead of the one that goes through the Barony courtyard. This conflict reached a nadir before a staging of a Shakespeare piece in the courtyard of the manor, when the police had to be called to hinder a planned demonstration with tractors that would have disturbed the show. However, the most well known conflict concerns a planned hydroelectric development of the Hatteberg watercourse. This conflict appears to have been mistakenly fought out over the Hatteberg Waterfall which, in its time before the Barony park became established, was visible from the manor house. A remote waterfall would have been channelled through pipes if the plan had been realised, but does not affect the Hatteberg Waterfall. 'After having passed through the turbines the water had to find its way to the sea, as before', it is maintained locally. The Barony and a number of their companions – several of them national celebrities – have opposed the development. Locally, it is felt provocative that national singers, actors, authors and researchers appear in media in defence of the watercourse: 'It is irritating to listen to these people, who after a three hours visit act as spokesmen for the protectionists in media', it was maintained, and 'these people do not know the local needs and our everyday life' and 'they even came up with false evidence', referring to mixing up the waterfalls involved.

As most informants saw it, conservation policies and rules are barriers to dynamic local development. It is, among other things, encumbering renewals in agriculture, and building of new housing areas in the parish, and 'the areas suited for new industrial activities are occupied in the name of conservation'. Local residents appear to have little hope that conservation and park management will provide the parish with many jobs, or result in tourism being significantly strengthened. People feel that the brakes have been put on local development.

The disapproval and ambivalence referred to above can be linked to the strategies today's management have used to establish the Barony as a national cultural institution rather than a local one. Alliances with the national conservation authorities and national celebrities have been prioritised before local involvement and acceptance. Formally speaking, the alliance acts correctly towards the municipal authorities and even towards the local community that has been informed through letters and meetings. But there is a perception locally that 'the alliance does not communicate, it informs; and they do not negotiate, just act'. It was maintained that with more flexibility and diplomacy, the relationship could have been much better.

Conclusion

This case study has presented the Rosendal Barony, where many of the lifestyles from 130 years earlier are revived within the frames of a modern cultural institution and a tourist attraction. The revitalisation also seems to incorporate the old relationships between the Barony and the parish, relationships loaded with class differences, and antagonisms between the posh and ordinary people. In local eyes, an already encompassing conservation regime has been enlarged through the attention and support the Barony has got from the national cultural and intellectual elite, and from the media. However, conservation as such, and the tourism that accompanies the conservation, does not give enough jobs and development energy to compensate for the restrictions that are set on the local community. Thus, the result is increased resentment between the Barony and local people, which in fact is an antagonism between national heritage concern and local development.

Baroniet Rosendal: http://www.baroniet.no/en/index.php

World Heritage

A way of marking heritage was established in 1972. The international treaty Convention Concerning the Protection of the World Cultural and Natural Heritage was adopted by UNESCO. The mission of the Convention is to spread out the idea of protection of heritage to all countries and to assist them in conserving listed objects. Moreover, it forwards the idea of local involvement (UNESCO, 2007).

To be awarded world heritage status it is required to be nominated by the national government. Moreover, at least one of the following criteria formulated by UNESCO (2007) has to be met. To represent a masterpiece of human creative genius:

(i) to exhibit an important interchange of human values, over a span of time or within a cultural area of the world, on developments in architecture or technology, monumental arts, town-planning or landscape design;

(ii) to bear unique or at least exceptional testimony to a cultural tradition or to a civilisation which is living or which has disappeared;

(iii) to be an outstanding example of a type of building, architectural or technological ensemble or landscape which illustrates (a) significant stage(s) in human history;

(iv) to be an outstanding example of a traditional human settlement, land-use, or sea-use which is representative of a culture, or human interaction with the environment especially when it has become vulnerable under the impact of irreversible change;

(v) to be directly or tangibly associated with events or living traditions, with ideas, or with beliefs, with artistic and literary works of outstanding universal significance;

(vi) to contain superlative natural phenomenon or areas of exceptional natural beauty and aesthetic importance;

(vii) to be outstanding examples representing major stages of earth's history, including the record of life, significant on-going geological processes in the development of landforms, or significant geomorphic or physiographic features;

(viii) to be outstanding examples representing significant on-going ecological and biological processes in the evolution and development of terrestrial, fresh water, coastal and marine ecosystems and communities of plants and animals;

(ix) to contain the most important and significant natural habitats for in situ conservation of biological diversity, including those threatened species of outstanding universal value from the point of view of science and conservation;

(x) to contain the most important and significant natural habitats for in situ conservation of biological diversity, including those containing threatened species of outstanding universal value from the point of view of science or conservation.

If criteria (i) to (vi) are met, the heritage site is classified cultural heritage, meanwhile meeting criteria (vii) to (x) qualifies the site for being a natural heritage. Sites can also be a so-called mixed heritage site, that is, that they meet criteria from both groups.

In the Nordic countries altogether 30 heritage sites can be found, of which 26 are classified as cultural heritage, three as natural heritage, meanwhile the Laponia area is considered a mixed site (Figure 9.2). The majority of the sites are selected due to criteria (iv) (18) and (iii) (12). Other criteria apply in few cases, meanwhile criterion (x) has not been applied to any Nordic site.

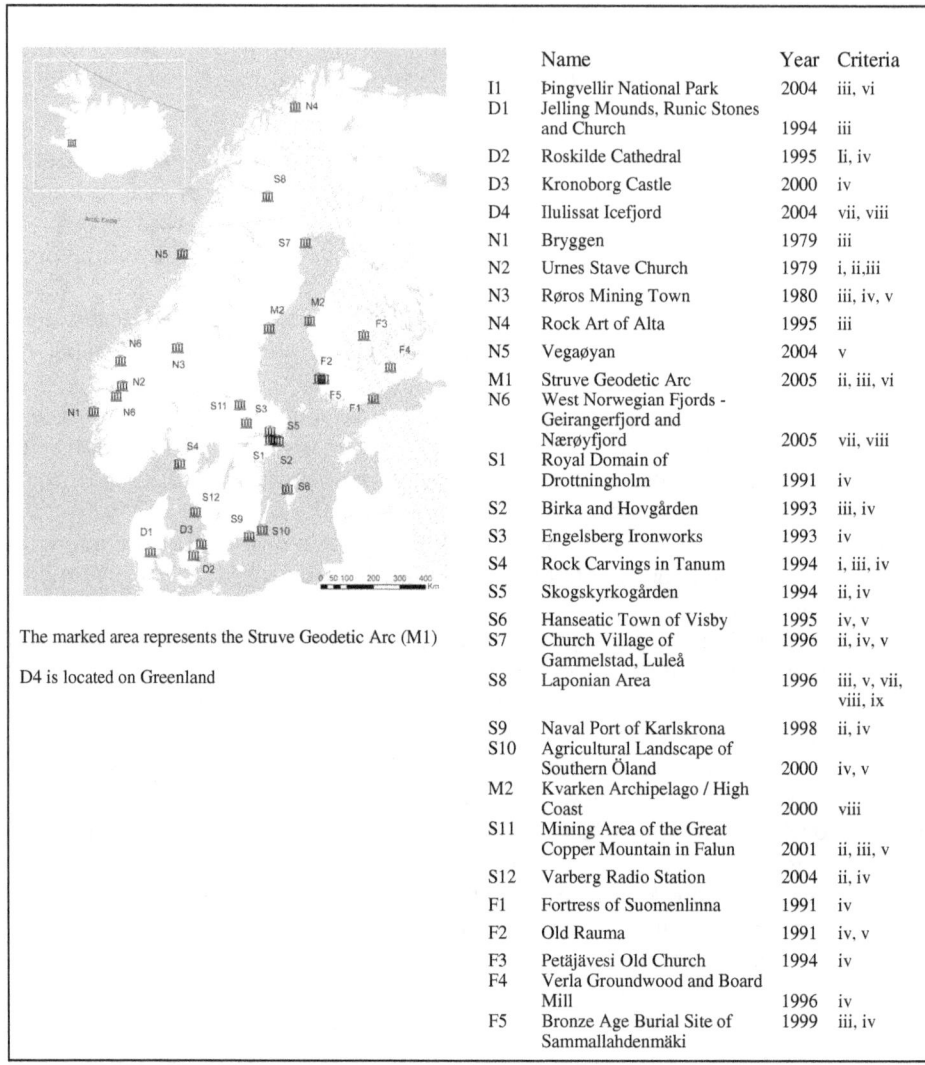

Figure 9.2 Nordic world heritage sites

The objects awarded World Heritage status, differ very much in their ability to attract tourists. This is partly because of their principally national significance; since national governments nominate the candidates they are identified as heritage from a national point of view. Moreover, certain areas lack public appeal and offer not much to see; for example, Struve's Geodetic Arc, lacks visible traces in the landscape and is identified as World Heritage because of its scientific significance.

Museums

Museums are a major supply factor in cultural tourism. Originally developed as institutions engaged in protecting and showing cultural artifacts, they have recently increasingly developed into places that also are expected to attract tourists and provide leisure experiences to local residents (Timothy & Boyd, 2003). Traditionally, museums could be found in heritage buildings carrying historical significance. Now, modern museums are sometimes constructed with remarkable architecture as landmarks to position the local community as a tourist destination.

In the Nordic welfare state museums also traditionally played an important role for public education. The importance of museums as tourist attractions is however clearly demonstrated in Table 9.1. For example, the figures for museum visitation on the Åland islands reveal that the number of visitors to museums is five times higher than the number of inhabitants. This marks the appeal of museums to tourists. The table shows, however, that the situation differs radically between the Nordic countries, leaving Finland in the group of the remote and less-visited island societies of Faroe and Greenland.

A particular set of museums closely related to the Nordic countries are Folk Life Museums. The open-air museum Skansen in Stockholm established in 1891 was designed as a showcase displaying buildings representative for various landscapes of the country. Shortly after, similar museums were constructed in Norway and Denmark. Nowadays, these museums are turned into living experience parks, where buildings, at least during the tourist season, are inhabited by museum staff in historical clothing. The phenomenon spread out over the entire Nordic realm and hence, often small-scale outdoor museums representing local buildings and folklore can be found over the entire area (see Case 9.3). Not all of these museums are however major tourist attractions, although they are usually promoted as such.

Table 9.1 Museums and tourism in the Nordic countries, 2006

Country	Population	Museums	Visitors (1000)	Visitors per 100 inhabitants	Guest nights (1000)	Guest nights per 100 inhabitants
Denmark	5,447,084	262	10,175	187	14,324	263
Faroe Islands	48,350	4	30	62	106	219
Greenland	56,648	20	22	39	245	432
Finland	5,276,955	165	4,340	83	14,552	276
Åland	26,923	6	135	504	173	643
Iceland	307,672	157	1,414	482	1,728	562
Norway	4,681,134	188	9,061	197	17,773	380
Sweden	9,113,257	228	20,139	223	30,163	331

Source: Nordic Statistical Yearbook (2007).

Instead, they cater for a local need to sustain a more local and regional heritage, and are thus places for folkloristic celebrations such as the Swedish midsummer.

Cases and issues 9.3: The Maihaugen folk museum at Lillehammer

THOR FLOGNFELDT JR

Maihaugen, previously named the Sandvig Collection, was founded in 1886 and moved to the present site in Lillehammer in 1904. This is the largest outdoor folk museum in Norway and also houses the national museum collection of crafts and the national postal history (and stamp) collection of the Norwegian Postal Services. In addition, Maihaugen has the responsibility of managing two Nobel Literature Laureates' homes; Bjørnstjerne Bjørnson's Aulestad and Sigrid Unseth's Bjerkebek. Since 2006 the Norwegian Olympic museum, a Hall of Fame institution, is also within the care of Maihaugen.

The collections at Maihaugen can be divided into three parts:

(1) The outdoor collections with several complete farms and a summer farm building, a stave church from Garmo in Lom and several other rural buildings. This collection is designed as being located from down in the valleys up to the high mountains.
(2) The 1900-houses collection with one typical villa from each decade from 1900 onwards. This is closely connected to the town streets of Lillehammer – a collection of several restored buildings from the town centre. Many of those in active use during the summer – a general store, a bakery and a post office plus some other stores and the regional archives. Even an old railroad station is present.
(3) The crafts and postal collection is located in the main building which also contains the Concert Hall built for the 1994 Olympics and is regarded as having some of the best acoustics in Norway. The research offices and the National Crafts Library and Archives are also in this building as is the modern exhibition area with the permanent 'Norway development exhibition', along with space for temporary exhibitions.

Maihaugen web site (English): http://www.maihaugen.no/default.aspx?id=250)

Increasingly, ever larger environments are protected and turned into museum-like institutions. This is not least true for early industrial plants and mining communities as exemplified in Plate 9.1. The island of Norrbyskär in the Kvarken area was originally designed as utopian community around a major saw-mill. After abandonment the available housing is used as second homes, meanwhile the saw mill has been turned into a museum also offering a child-scale version of the historical buildings.

Plate 9.1 Guided tour on Norrbyskär Island

Festivals

Cultural festivals are increasingly used in the competition for mobile people and capital. This is done on different geographical scales and hallmark events of international calibre are accompanied by of more regional or even local significance (see Case 9.4). Festivals are clearly defined in time and space and thus they are able to attract visitors rather easily since date and content are often known to the visitors beforehand. Hence, festivals guarantee access to cultural expressions that otherwise occur more irregularly or not publicly (Pettersson, 2003).

Cases and issues 9.4: Emergent Vikings: Changing character of a micro-festival in Iceland

GUNNAR THÓR JÓHANNESSON

Introduction

In recent years micro tourism festivals have mushroomed in Iceland. Most towns and many villages have their own 'local' festival one weekend during the summer. I intend to give insight into how the character of one of those festivals is currently changing from being a kind of collective family gathering into a themed and potential tourist event. The festival in question is called Dýrafjarðardagar or Days of Dýrafjörður. It is held in a village called Þingeyri, which is located on the Westfjord peninsula. What separates this festival from most others is that there is a good chance for the visitor to meet Vikings. The discussion will focus on some of the practices underlying the emergent Viking character of the Days of Dýrafjörður festival.

Background

Þingeyri has about 300 inhabitants. The local economy has traditionally been based on fisheries. Major reorganisation of the fishery system in Iceland and introduction of new technology in the fisheries since the mid-1980s has put many fishery based localities under stress. People have coped with these changes in various ways, among other things through tourism development projects (Skaptadóttir & Jóhannesson, 2004).

One such project, the Gísla Saga project, is being carried out in Þingeyri. The coupling of the Gísla Saga project to the Days of Dýrafjörður has meant a revival of the festival but has also changed its character or outlook. It used to be an event that primarily drew those that had some connection to the place back for a visit. While Days of Dýrafjörður still has a flavour of an extended family gathering, it is now beginning to be themed in a specific way as a possible tourist event.

Suddenly, the visitor might run into people dressed up as Vikings. Although Days of Dýrafjörður has not become a 'pure' Viking festival, significant changes of the outlook of the festival are taking place. This process of change is interesting as it raises questions about how tourism projects interfere with societal ordering. Not long ago it would have been considered a sign of mild disorder to show up in Þingeyri dressed as a Viking. Now it is accepted and even praised. Below, I will cast light on some of the practices that have made it possible to become a Viking in Þingeyri.

How to become a Viking

There are two interrelated things I want to highlight. First, networked, material practices are of crucial importance for any change of the festival outlook to come about. Second, it follows that the change takes time and effort and may be described as a learning process. A clarifying example follows:

> It is no coincidence that the Gísla Saga project is carried out in Þingeyri. Gísla Saga is one of the more famous Icelandic Sagas and its scene is mostly in the area around Þingeyri. The project started in 2003 with a course in sewing Viking clothes. Apart from creating the material prerequisite for dressing up as Vikings the course had two interwoven effects. First, it created a material commitment to the project as people repeatedly came together for the sewing work. Second, the sewing work provided a forum for discussions about the project and possible venues for tourism development in the area. The course thus created an interactive learning space about Viking culture and the potentials of tourism in the area.

One of the ideas coming out the discussions was to build outdoor Viking-style festival facilities, which could be used for the village festival. This idea was the first step in relating the Gísla Saga project to the Days of Dýrafjörður festival. The facilities were called the Viking Ring. The first part of it was built in the spring 2004 and was finished just in time for the festival. The Ring was then extended in the summer 2005 and the plan is to finish its building and an adjacent Viking-style playground in 2006.

Finding one's sea legs as a Viking
At sea, finding one's sea legs means to get comfortable with the rocking movements of the world. Literally it is about recovering from seasickness, which is basically a learning process (Pálsson, 1995). In the present context, the metaphor of finding one's sea legs may be used to describe how people gain in confidence and get comfortable with the role and theme of Vikings in Þingeyri.

The account above, is in terms of Actor–Network theory, an example of translation (Latour, 1999), that is, the movement to and materialisation of the Gísla Saga project in the locality. As such, it involved a translation of the project into the usual practice of Days of Dýrafjörður and concomitantly the outlook of the festival changed. New elements were translated into its assemblage. The material effects of the translations, that is, the clothes and the Viking Ring, provide the material basics for the changing character of the festival. However, learning, or finding one's sea legs as Vikings, is also part of the translation. Without it, the 'theming' of the festival is not likely to be successful. Every translation is a precarious process. It involves power and politics at every stage. As a heterogeneous network, the Days of Dýrafjörður festival is being negotiated. It is not known how this story will end. Probably it will not end in any simple or pure manner. However, the material practices and the learning process they have initiated add to the emergent character of Days of Dýrafjörður as a Viking festival.

Many festivals are themed and offer numerous activities related to the theme. Popular festivals address, for example, the following cultural expressions:

- music (Savonlinna Opera Festival, Roskilde festival, Hultsfred festival, Molde Jazz Festival);
- arts (Österlen Art Circuit);
- theatre (Odense Theater Festival);
- films (Stockholm Film Festival, Sodankylä Film Festival);
- history and heritage (Visby Medieval Week, Jokkmokk Winter Festival);
- food (Öland Harvest Festival);
- sports (Gothia Cup, Extrem Sports Week Voss).

Not all festivals are however themed, but celebrate important dates like the Easter Festival in Kautokeino, or are in themselves a part of the heritage industry, such as the Winter Festival in Jokkmokk (see Case 9.5). Other festivals are odder, such as the annual Air Guitar World Championship in Oulu, Finland. Here, the concept of a music festival is blurred with urban fairs often organised to entertain the local population as well as to attract visitors to urban areas not least during the summer months when competition with other more established tourism destinations increases. Most of these fairs also feature live music. However, they are not dedicated to certain genres as for example, Hultsfred, Roskilde, Molde and Pori, and attract thus different groups of visitors as those special interest events.

Cases and issues 9.5: The Sami winter festival in Jokkmokk, Sweden

ROBERT PETTERSSON

Introduction

Some events should not attract tourists. The winter festival in Swedish Jokkmokk is one of them. The small town of Jokkmokk, with a peripheral location north of the Arctic Circle (Figure 9.3), offers the mid-winter festival visitors rather harsh conditions with snow, cold weather and a short period of daylight. The peripheral location implies long travel time and high travel expenses. Furthermore, the area has only a limited tourism infrastructure with a lack of travel alternatives, food service and accommodation.

However, for the last 400 years the winter festival in Jokkmokk has attracted an increasing number of visitors. Despite the conditions, Jokkmokk is 'the place to be' during three days in the beginning of February every winter. The number of inhabitants in the small town of Jokkmokk is about 3000, a number which is redoubled about 10–15 times during the market days. The festival is seen as an opportunity to conquer the wilderness, meet with other people and above all to meet with the exotic indigenous culture of the Sami people.

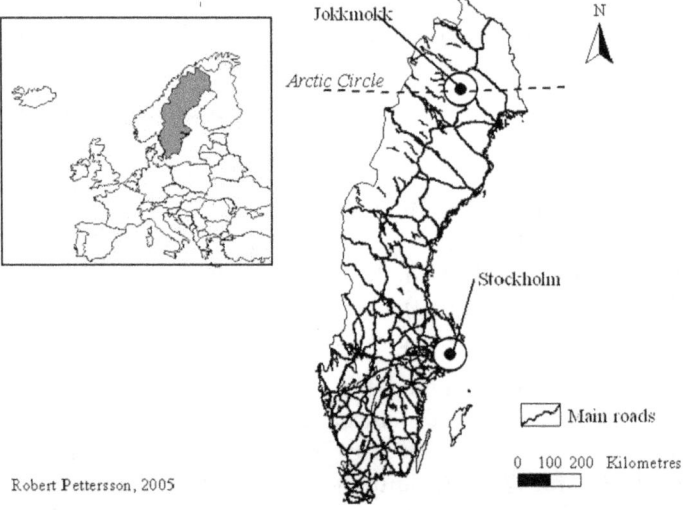

Robert Pettersson, 2005

Figure 9.3 The location of Jokkmokk

This case study gives a brief presentation of the winter festival in Jokkmokk. First some glimpses of Sami tourism and the festival's historical development are given, then it discusses to what extent the festival truly is a Sami event.

Sami tourism

Historically, Sami in northern Scandinavia have been occupied with reindeer breeding. During recent decades there has been a decline in reindeer business activity and the modern reindeer breeder is struggling to make ends meet. Today reindeer farming is carried out using helicopters, motorbikes, snowmobiles and trucks aiming for large-scale food production. Because of this situation many Sami have started looking for new occupations, and tourism is often mentioned as an alternative that offers the entrepreneurs an option to stay in a sparsely populated area. In the Swedish parts of *Sápmi*, as the Sami call their homeland in the very north of Scandinavia, there are an increasing number of individual Sami tourism entrepreneurs, museums, cultural events, outdoor cultural sites and places where Sami handicrafts are sold (Pettersson, 2004). However, there are risks in developing tourism connected to indigenous cultures (Butler & Hinch, 1996; Price, 1996; Viken, 1997). Because of its concentration in time and space, indigenous culture as a part of an event, like the winter festival in Jokkmokk, implies particular risks.

The festival's historical development

In 1605 the Swedish King, Charles IX, established the marketplace in Jokkmokk in the heart of what was called the 'nomadic Laplander's territory'. It was established because the Swedish state wanted to increase their control over the trade between Sami, tradesmen and people passing by. This was also an opportunity for the Swedish state to collect taxes and purchase hides and skins. Moreover, the late medieval period of Sweden, was the time when Christianity was spreading in northern Scandinavia. The market days therefore became an occasion when the Church could meet the Sami congregation that was usually spread over large areas in the reindeer pasture. When people came together during market days there was also a visiting judge who pronounced verdicts in legal cases that had occurred during the year.

A major alteration in the history of the festival occurred in the 1960s. This was when the organisation of the event was transferred to the local government, and the tourism era really began. An expansion of tourism arrangements together with a growing trade fair contributed to an increasing number of tourists visiting the festival. Furthermore, starting in the 1950s, increased car ownership has made it easier for many people to make a visit.

The traditional trade fair is, and has always been, an important part of the festival in Jokkmokk. Research done, focusing the development of the winter festival in Jokkmokk (Pettersson, 2003) shows that the number of traders and market stalls has increased during the last decades to stabilise at about 200 tradesmen distributed along roughly 500 market stalls each year. A majority of the tradesmen come from northern Sweden, and one out of five has Sami-related merchandise for sale. All kinds of merchandise are sold at the trade fair, but there is a dominance of handicraft products.

Festival leaflets from recent decades show that the festival has much to offer besides the trade fair. The number of tradesmen, as well as the number of activities, has increased during the last few decades. The largest increase is in the number of exhibitions and shows, but also the number of outdoor activities and seminars has increased. The festival activities tend to extend into the neighbouring villages to Jokkmokk. Consequently the festival influences a larger area than before.

What is so Sami about it?

Tourist experience of the indigenous culture is often limited to staged culture in museums, exhibitions and festivals. So what and where is indigenous at the festival in Jokkmokk? Studies indicate that the indigenous culture presented at the winter festival is highly staged, although backstage experiences are available to the Sami and to the tourists who show a special interest (Müller & Pettersson, 2005).

To answer the question as to 'where' the Sami is, Müller and Pettersson (2005) define three separate areas of the festival in Jokkmokk. The first area is the *commercial area* where the traditional trade fair is found. The second one is the *activity area*, with the activities offered at the lakeside in Jokkmokk. The third and last festival area is named the *cultural area* and it includes the Sami museum and the Sami educational institution.

The Sami representation differs in these three areas. Much of the Sami influences in the festival are found in and among the market stalls in the commercial trading area. The most common Sami attributes in this area are Sami handicrafts and the colourful Sami costumes (Plate 9.2), worn both by tradesmen and by Sami visitors to the festival. In the activity area at the lakeside, the Sami culture is often a part of the attraction. Here the most common Sami symbols are reindeers and Sami tents. In the culture area, the museum presents Sami culture in many ways, although in a rather static, historical way. In addition to the exhibitions there are also seminars and lectures during the festival. At the Sami educational institution there are many Sami craftsmen selling genuine Sami handicraft.

A media survey that was undertaken shows that reports from the festival were common, especially in the local newspapers and in newspapers from northern Sweden. Altogether almost 100 cuttings were collected during the weeks before and after the festival. The cuttings included many photographs, and the photographs from the media survey present a more exotic image than the image experienced at the festival site.

Interviews show that the festival is enjoyed by the Sami and is very important to them. However, the Sami have the feeling that Sami culture has to some extent been pushed aside, away from the centre of the festival. The interviews

Plate 9.2 Two Sami attributes at the winter festival: The Sami dress and the reindeer (Photo: Robert Pettersson)

also show that the Sami (insiders) experience the festival in a different way than the tourists (outsiders).

Concluding remarks

The festival in Jokkmokk has, thanks to the continuous addition of new attractions, been able to retain a rather high level of popularity despite its peripheral location at the Arctic Circle in northern Sweden. After 400 years, the Sami culture and its role as a meeting place still makes the winter festival in Jokkmokk unique. However, other activities have become an increasingly important part of the festival. A conclusion to be drawn is that if the festival can retain its Sami characteristics and continue to add new attractions, it will most certainly remain popular, both with the Sami and tourists.

The festival offers a broad variety of aspects of Sami heritage that are appealing to various groups of visitors. It is probably this co-existence of more or less staged, authorised and unauthorised representations of Sami heritage that makes the festival attractive. In the end the festival is an arena where indigenous heritage is displayed, tested, contested and re-negotiated by Sami, tourists, tradesmen, planners and public servants. The festival is thus not only an important indigenous attraction, it is also an annual meeting point and display for the redefinition of indigenous heritage.

Jokkmokk winter market page on the Jokkmokk tourism website: http://www.jokkmokksmarknad.com/

Festivals have different roots explaining their establishment. Partly they have historical roots and have been commodified more recently (Pettersson, 2003), and partly they are the result of local tourism strategies aimed at reimaging tourist destinations or taking advantage of the attractiveness of destinations during shoulder seasons. For example, Voss in Norway tries to position itself as the European centre for extreme sports by arranging a week celebrating the various interests of this subculture (Gimothy et al., 2006). In contrast Gotland draws on its medieval heritage by arranging a medieval week involving numerous role players to extend the tourist season by one week. In certain city destinations such as Göteborg cultural events are part of a comprehensive strategy to position the destination as a specialised locale for events entailing a steady inflow of tourists.

Sami Tourism

A specifically Nordic attraction is formed by the indigenous people of the Sami inhabiting the northern parts of Norway, Sweden, Finland and parts of the Russian Kola peninsula. *Sapmi*, the Sami homeland, is today split by national borders. The total number of Sami in the four countries is approximately 60,000. However, differences are considerable. Besides several languages, even different national legislations contribute to keep the Sami apart. Only a few individuals are directly linked to the traditional economy of reindeer herding (Pettersson, 2004; Viken, 2006; Viken & Müller, 2006).

Indigenous populations are frequently used in tourism promotion and marketing (Butler & Hinch, 2007). This is also true for the Sami people in northern Europe (Olsen, 2006; Pettersson & Viken, 2007). In the area, sometimes described as Europe's last wilderness, the Sami and their culture are epitomised as a main asset for an increasing tourism industry. Previously this has caused problems and irritation among the Sami (Saarinen, 1999). Nevertheless, tourism development is also seen a potential solution to problems troubling the Sami society offering new sources of income and future employment (Pettersson & Viken, 2007). This is of major importance considering the challenges facing traditional Sami reindeer herding. Besides problems of profitability and market access, rejuvenated interest for resource extraction, large-scale nature conservation projects and climate change threaten the Sami grazing lands often considered of central importance to the maintenance of the Sami culture.

Internal struggle about reindeer herding is common among the Sami. For example, in Sweden the *Sametinget*, the Sami parliament, has the obligation to handle these conflicts and to develop strategies for future development. In this context tourism has been pointed out as an alternative source of income and an important means for achieving a number of goals (Müller & Pettersson, 2006: 60):

- increased employment;
- increased income;
- broader economic base;
- decreased antagonism (by increased transfer of knowledge);
- improved infrastructure;

- increased possibilities for Sami to stay in Sápmi;
- increased tourism control (reduce disturbance of grazing reindeers).

In Sweden in 1999 about 45 Sami enterprises were offering different supplies such as visits to a Sami camp, homestays, hikes with reindeers, hunting and fishing trips, boat trips, Sami food, reindeer gatherings and visits to Sami craftsmen (Müller & Pettersson, 2001). Most attractions were clearly staged and targeted primarily to an international market. Tourism is not only encouraged by the Sami and Swedish authorities. Sami tourism entrepreneurs themselves have attempted several times to create networks for production and marketing of their tourism supplies. Some of these attempts have failed recently mainly because of the geographical isolation of many of the entrepreneurs involved. On reviewing these attempts Müller and Pettersson (2006: 60) have suggested that Sami tourism should be developed in the following way:

- shall be run in small-scale;
- shall be run with regard to the nature and the culture;
- shall be built on the traditional Sami heritage, including reindeer herding;
- shall be authentic and genuine;
- shall be run by professional staff, initiated in Sami culture;
- incomes shall stay in the local area.

In most of the studies addressing indigenous tourism development it is assumed that: (1) indigenous peoples want to engage in tourism because it provides a set of opportunities; and (2) that tourism development causes threats to indigenous culture and thus has to be managed properly (Butler & Hinch, 1996, 2007; Ryan & Aicken, 2005). This is also true for the Sami (Pettersson & Viken, 2007). So far Sami tourism has not fulfilled the high expectations. Nevertheless, problems related to the commodification of the Sami have already occurred, not least in relation to representations available at the Santa Claus Tourist Center in Rovaniemi, Finland, where Sami culture is mixed with Anglo-American Christmas traditions and is highly commercialised. Hence, sometimes limitations to indigenous tourism may be required to guarantee the culture of the indigenous people.

In this context Zeppel (1998) identifies four indigenous strategies of exercising control: (1) spatial limitations regulate access to land and sacred places; (2) activity limitations define certain tourist activities as wanted and ban others as unwanted; (3) temporal limitations delimit the time for tourist access; and (4) cultural limitations on the dissemination of rituals and other cultural knowledge.

To accomplish these limitations is however difficult since, for example, property rights on visual representations of indigenous peoples are not guaranteed to the indigenous community. An unauthorised exploitation of an indigenous theme may thus occur even without the agreement of indigenous populations. For example, non-indigenous tourism entrepreneurs in Finnish Lapland are using unauthorised representations of the Sami depicting the indigenous group as drunk and dirty (Saarinen, 1999, 2001). Moreover, even within the indigenous community different opinions on how to develop tourism are present (Lyngnes & Viken,

1998; Pettersson & Viken, 2007). Ultimately, indigenous societies may be mainly modern (Pettersson & Viken, 2007; Tuulentie, 2006; Viken & Müller, 2006) and hence, individual decision-making can only be counteracted, but not legally hindered, by community representatives as long as they are within the legal framework of the surrounding society.

The future development of Sami tourism is tightly connected to the development of reindeer herding since many Sami tourism entrepreneurs are considering reindeer herding as their primary economy (Nordin, 2007). It provides important cultural benefits and is important for the Sami identity. A decline in reindeer herding would however force Sami into other industries. In this context tourism appears to be a viable and attractive solution.

Nordic Heritage?

Increasingly, the notion of Nordic heritage can be contested. All Nordic countries are highly involved in globalisation and thus, influences from all over the world can be found. Hence, heritage is usually conceptualised as a national venture representing the history of a certain people within a certain territory. In a mobile world, such a perception becomes obsolete. For example, in Sweden about 20% of the population has a foreign descendant in their own or their parents' generation. To what extent is Swedish heritage even their heritage?

In reality, heritage has become mobile too, particularly with regards to customs and folklore that today can be found far away from their origins. Hence, a recently raised Buddha-statue in Fredrika, far out in the northern boreal forests may

Plate 9.3 A Buddha-statue in Fredrika, northern Sweden

symbolise this change and indicate the increasingly complex and diverse nature of Nordic heritage (Plate 9.3). However, this does not imply an end to heritage tourism. Instead, it shows its dynamic nature regarding supply and demand.

Discussion questions
(1) What are the major forms of heritage tourism in the Nordic countries?
(2) Does heritage tourism matter to regional economic development?
(3) What is the main supply of heritage tourism and how can it change in the future?

Essay questions
(1) Cultural tourism is a tricky concept – discuss the multiple relationships of tourism and culture.
(2) Scrutinise the web pages of selected municipalities and assess their cultural tourism supply with regards to geographical location and market appeal.

Key readings
Timothy and Boyd (2003) provide a useful overview of heritage tourism; meanwhile Richards (1999, 2001) collected papers representing the growing importance of cultural tourism in Europe. Butler and Hinch (2007) provided a state of the art update on indigenous peoples and tourism. Sami tourism is more explicitly addressed in a special issue of *Scandinavian Journal of Hospitality and Tourism* (2006: 1). The issue of authenticity is an important area of discussion in tourism with both *Tourism Management* volume 28 (4), and *Tourism Recreation Research* volume 32 (1) both hosting useful debates on the concept.

Web pages
Besides the web pages provided by the national tourism authorities, the following web pages offer interesting insights:

Heritage and museums
The National Heritage Board of Sweden	www.raa.se
The Heritage Agency of Denmark	www.kulturarv.dk
The National Board of Antiquities Finland	www.nba.fi
Norwegian Directorate for Cultural Heritage	www.riksantikvaren.no
The Archaeological Heritage Agency of Iceland	www.fornleifavernd.is
Åland's Museibyrån (only in Swedish)	www.museum.ax

Festivals
Finland Festivals	www.festivals.fi
Festivalinfo	www.festivalinfo.net

Sami
Sami Information Center	www.eng.samer.se

Chapter 10
Winter Tourism: Changing 'Snow Business'

Learning Objectives

After reading this chapter, you should be able to:

- Understand the concept of winter tourism in the Nordic countries.
- Understand the relationship of winter tourism to other tourism types such as adventure tourism, ecotourism and 'new tourism'.
- Appreciate the economic importance of winter tourism in peripheral areas.

Introduction

Winter tourism has a long history and tradition in Nordic countries. This chapter introduces the background and current patterns, activities and changes of winter tourism. The emergence of so-called 'new tourism' and its potential outcomes, such as the rise of adventure tourism, in Nordic winter tourism contexts are discussed. In addition, the crucial role of innovation strategies in winter tourism is considered.

The Development of Winter Tourism

Winter tourism and related recreation activities have a long history in the Nordic region. According to Hudson (2000: 8) documentation of skiing's earliest emergence as a form of recreation associated with tourism dates back to mid-19th century in Norway, and the world's oldest, longest and largest cross-country ski race, *Vasaloppet*, was started in 1922 in Sweden. In the tourism industry the growth in winter activities and tourist numbers were faced in the 1920s and 1930s when the first downhill skiing resorts, lifts and other facilities were developed in the Nordic countries, especially in Finland, Norway and Sweden (Nilsson, 2001).

The growth and modernisation of tourism, and especially the rise of nature-based winter tourism, brought new types of travellers onto the tourism scene. In Finland, for example, the Women's Exercise Association organised the first guided nature-based tourism tours to northern Finland and Lapland in the late 1920s and 1930s (see Kari, 1978; also see Chapter 3 on the concept of social tourism). The Association built the first cottage, a tourist hut in 1934, in the Pallas area in western Lapland to serve as a base camp for skiing schools for women. Later, the area became a National Park and these skiing schools and trips organised by the Association formed a basis for the notion of children's winter school holidays and family vacations in winter time, now one of the peak tourist seasons in the Finnish (and Nordic) tourism destinations.

A second phase of winter tourism infrastructure development started in the 1960s and early 1970s as a result of greater domestic and international tourism mobility in the Nordic countries and in northern Europe generally. In addition to the development of a basic level of infrastructure, such as skiing slopes and accommodation capacity, the first cable car was inaugurated in 1976 at Åre, Sweden. The growth of winter tourism was based on increasing free time, wealth and development of the transportation system in the Nordic countries. However, compared to the 1980s and 1990s this development phase was still rather moderate and based mainly on domestic demand in the winter season.

The fastest and most visible growth began in the mid-1980s when many of the present resorts and other winter tourism landscapes were further designed, constructed or expanded towards mass-scale resort structures. At the same time international demand and capital investments started to evolve. The major capacity building was done in the Finnish Lapland and Norwegian and Swedish mountain regions which provided natural advances to winter tourism development such as topography and a long, snow-covered season with relatively good accessibility. In addition, the regional policy at the time supported the economic development of peripheral areas and, compared to earlier decades, the regional policy-makers and developers recognised the potential role of tourism in regional development, which was an important policy aspect for expanding, not only the tourism infrastructure, but also supporting transportation networks.

Internationally, many traditional winter tourism destinations have recently faced stagnating markets (Flagestad & Hope, 2001) and other outdoor adventure activities have become increasingly popular in tourism markets (see Cater, 2006; Swarbrooke *et al.*, 2003). In addition there are evolving challenges to manage the negative consequences of ski development in mountain areas which has led to conflicts between tourism developers and environmental groups (Holden, 2000). The latter perspective has also been discussed in the Nordic context, but the current and increasing investments in winter tourism do not imply stagnation yet. For example, the Ruka tourist destination in northeastern Finland invested approximately 550,000 euros in new artificial snow-making techniques in 2006–2007 alone (see Case 10.1).

Cases and issues 10.1: The development process of Ruka tourist resort, Finland

PEKKA KAUPPILA

Ruka tourist resort is located in the municipality of Kuusamo in northeastern Finland near the border of Finland and Russia and is situated just 50 km below the Arctic Circle. Ruka is one of the most well-known and famous (ski) resorts in Finland. It has been estimated that about 1 million tourists visit the resort annually and most of them are domestic winter tourists.

According to previous studies (Kauppila, 1995, 2004), the development process of Ruka appears to follow Butler's (1980) resort life cycle model from the exploration stage to the consolidation stage. However, the consolidation stage of the resort implies national, rather than international, level consolidation (see Johnston, 2001; Keller, 1987). The sequential stage of the model, stagnation, is not evident in Ruka, and the contemporary stage resembles more rejuvenation. The resort is also moving on to the international level of the life cycle model.

The development stages of Ruka

Exploration (1880–1955). The exploration stage in Ruka began in the 1880s. At that time, the area was visited by occasional wanderers, scientists, writers, poets and artists. The number of tourists was small, and the tourists integrated into the local community. They even had their meals and spent their nights with the local inhabitants and employed local guides for their excursions. Travelling to Ruka and Kuusamo was laborious because of poor accessibility, and therefore trips took several days or weeks, even months.

In the exploration stage, tourism in Ruka was irregular. Trips to the Kuusamo area were touring in nature, with the Oulanka river valleys, lakes Kitka and Paanajärvi and Nuorunen fell as the main attractions. The first 'tourists' of Ruka were attracted by the unique nature, and they climbed up the slopes to admire the spectacular views of the area. Until the Second World War, Ruka was predominantly a summer destination.

The area provided no actual tourist services, and the control and financing of development was completely in local hands. In the early stage, tourism was not an important industry for the local population. The inhabitants of Ruka earned their living mainly in agriculture and forestry.

Entrepreneurship in Ruka is clearly connected to the stages of development. In the exploration stage, no enterprises operating exclusively in the tourist industry existed. Instead, tourists relied on the services provided by the centre of the municipality of Kuusamo.

Involvement (1955–1975). The starting point of all development in Ruka was a change in international politics; after the Second World War the first attractions in the Kuusamo area, lake Paanajärvi and Nuorunen fell, were located behind the new border with Russia. Through investments, Ruka was developed to the involvement stage. The first skiing slopes in the area were introduced in the mid-1950s. Ruka gradually became a winter resort, regularly visited by Finnish downhill skiing innovators, and ski lifts, accommodation and catering services emerged in the late 1950s. These were the first 'real' tourist facilities in the destination.

As accessibility improved, the length of stay of tourists shortened. Downhill skiing became the principal product, although views still attracted visitors in the summer. At this stage, the number of tourists already exceeded the number of inhabitants of the municipality of Kuusamo. The development of the area was mainly in local hands, but to some extent external involvement emerged. The development process can be characterised as catalytic, although to the population of Ruka, tourism was only a secondary industry.

Development (1975–1990). Changes in the ownership of tourist enterprises in the early 1970s were one indication of the transition into the development stage. The main tourist enterprises, Rukakeskus Oy (ski lift) and Rantasipi Rukahovi (hotel) companies, were bought by outside operators of the area and were now controlled at the national level. Tourism was mainly individualistic and domestic mass tourism in nature. In the end of the development stage, which can also be named the second home stage (see Chapter 8), great numbers of both private and commercial dwellings were built in the area. Improvements in accessibility and the introduction of air traffic made short lengths of stay possible. Ruka became a recurrently visited resort, particularly in the winter. In the summer, the resort was still mainly visited by tourists making a stop on a tour. In the 1980s at the latest, tourism in Ruka became institutionalised.

The appeal of the resort was increasingly based on attractions especially constructed for mass tourists, and these attractions provided diversity in the product range. In the diversification of the product range, services providing activities were significant. The choice in accommodation also increased, and a spa was built close to Ruka.

The control of the development process was also transferred into external hands. The development process can be characterised as integrated with the process dominated by the Rukakeskus Oy and Rantasipi Rukahovi companies. The local population was employed in the tourist industry, when tourism became the principal industry of the destination. The resort gained population, and self-sufficiency in jobs in the area rose above 100% with newcomers being to a great extent young people of working age.

After the mid-1970s, the rate of growth remained steady until the mid-1980s. In the early 1980s, some set-backs occurred although several indicators showed a pronounced growth in tourism in Ruka in the late 1980s, but in the early 1990s, the rate of growth seemed to slow down again.

Consolidation (1990–2000). All indicators demonstrate that during the early 1990s, the rate of growth of tourism in Ruka slowed down – at least a temporary consolidation stage was reached. However, this stage can be seen only a breather in preparation for new measures in development. In the case of Ruka, development was brought to a standstill because of the economic recession in Finland. Certain conflicts also arose between the tourist industry and conservationists over the need to expand the skiing slopes.

In the consolidation stage, Ruka had become a versatile tourist resort. Numerous services, such as shops, postal and bank services and personal services had been set up. A pronounced increase in the construction of second homes brought about an increase in the letting of these dwellings including for short-term rentals. The number of enterprises in accommodation and catering was no longer growing, nor was the ski lift capacity. Instead, growth in the construction and letting of second homes brought along further enterprises, such as managing and maintaining second homes. The enterprises in the area also became increasingly multi-functional, since they operated in several different lines of industry. However, because of this, the number of new enterprises did not increase as much as might have been expected.

Rejuvenation (2000–). In the 2000s, a new development period began in Ruka. Because of consolidation, the municipality of Kuusamo was the initiator of the tourism planning process regarding Ruka. As a result of this process, a new tourism strategy was created for the resort during the years 1998–2005. The goals of the strategy were year-round tourism, nature, culture and quality as attractions, international tourism and a versatile resort for residence, leisure and entrepreneurship.

At the beginning of 2000, a project was set up for increasing the number of international tourists to Ruka. The project, financed by the European Union, has been very successful. For example, during the winter season 2005–2006, four charter flights a week arrived at Kuusamo. These flights came from the UK and the Netherlands. The Christmas season seems to be more international year after year, too. For example, in 2005 more than 20 charter flights landed at Kuusamo, particularly from Russia, Ukraine and the UK. Regarding internationalisation the World Cup season in cross-country skiing, ski jumping and Nordic combining has been opened in Ruka since 2001.

Simultaneously with the internationalisation of the tourist structure, second home construction has proceeded very intensively in Ruka. This construction includes both private and commercial dwellings. Thus, it is possible that in the future tourists are to an increasing extent regular re-visitors, because of the great number of second homes.

As noted earlier, the operative tourism strategy emphasises international tourism. This can be interpreted so that both demand and supply of tourism are expected to internationalise. According to the strategy, there are attempts to attract international companies to Ruka, because the resources of development are limited in peripheral areas. At the time of writing, there is only one international hotel chain in Ruka.

From periphery to core

On the basis of these findings, the shape of the life cycle of the Ruka resort is a prolonged exploration stage and a pronounced growth in the development stage towards the end of the 1980s. In the 1990s, the development process of the resort steadied, but in the beginning of new millennium tourism in Ruka seems to be on a growth path again, which is a consequence of the internationalisation of the tourist structure. Basically, the shape of the life cycle of Ruka closely resembles Butler's (1980) model and the letter S.

Nowadays there are both natural and built attractions in Ruka. The product range and accommodation services in the area are very varied and the number of jobs and enterprises in the resort has increased since the mid-1990s. There are more jobs in the resort than there are employed persons. This implies that Ruka offers job opportunities for persons living outside the resort. Ruka is a full-service year-round resort with a diverse structure of economic life. In some ways it can even be argued that the resort resembles more the characteristics of core areas (population and economic growth) than the surrounding, declining periphery: Ruka has evolved as an enclave on the local level in the northern periphery.

Ruka website: http://www.ruka.fi/ (the seasonality of the destination is illustrated by the resorts summer (http://www.ruka.fi/summer_eng/) and winter (http://www.ruka.fi/winter_eng/) season web pages.

The present fourth development phase of Nordic winter tourism aims to turn, or have already turned, many of the major skiing resorts into all year round tourist destinations. In addition to summer season activities the winter season has also been expanded to the 'first snow season' in October–November and the Christmas season in November–early January. The growth is widely based on both internationalisation and globalisation of the former regional tourism industry (and the utilisation of developed techniques in snow-making and better accessibility by air). For example, Ruka Resort (see Case 10.1) often promotes the fact that it is one of the first ski resorts in Europe to open, an issue of potentially considerable advantage in the longer term given the affects of climate change on alpine resorts in other parts of Europe (see Chapter 11). Customers and markets are increasingly international, but also the industrial basis is changing. International hotel groups and investors are nowadays visibly present in Nordic destinations which have become economically much more interconnected, but also dependent on distant places, actors, knowledge, capital and decision-making.

Winter Tourism in Mountain Areas

There is a wide variety of winter tourism destinations in Nordic countries. Most of the winter tourism activities and large scale tourist destinations, such as Levi, Åre and Lillehammer, are located in northern Finland and the Swedish and Norwegian mountain areas (Plate 10.1) where tourism has been an important and increasing part of the regional economy for several decades. With the ongoing decline of traditional economies in the northern rural and peripheral regions, tourism is taking a more significant role (Heberlein *et al.*, 2002). In Finnish Lapland tourism is providing more employment opportunities than any other economy using the region's natural resources (Saarinen, 2003).

In Sweden the mountain region is the third most important tourist destination after larger cities and coastal areas. In the winter season the southern part of the mountain region is the most popular destination (Sahlberg, 2001). The markets in Sweden and also in Norway are still relatively regional ('Nordic'). In Sweden, for example, only 5% of tourists were from outside Nordic countries, and between 1995–1999 almost half of the Swedish adult population visited at least once the Swedish mountain region in their vacation (Heberlein *et al.*, 2002). Case 10.2 details the resort labour market in the Swedish mountain range. However, there are ongoing processes of change; for example, Åre resort's development is nowadays largely based on international markets. Similarly in Norway the country's biggest skiing and winter sports centre Lillehammer, the Olympic Winter Games host from

Plate 10.1 Olympic ski jump, Lillehammer, Norway

1994, offers both alpine and Nordic winter sport activities increasingly to foreign markets (see Case 4.3 on the aftermath of the Olympic Games in Lillehammer).

In Nordic winter tourism the main tourist activities are skiing (cross-country, down hill and snowboarding) and snowmobiling. In general (but depending on a specific location) the season for these activities starts in November and ends by early May. The improved techniques to produce snow artificially have expanded the winter tourism season at both ends. This has benefited the southern tourist destinations and early winter season, but for institutional and sociocultural reasons the skiing season practically stops at the end of April. Most of the domestic visitors in mountain resorts are from the southern parts of their countries whose areas are already turning to the summer season and recreational activities of the time. This creates a clear and perhaps too deep contrast and the assumption is that people are not interested in getting back to 'winter' while they are waiting for the summer season. However, this kind of social construction of tourist seasons has been relatively little studied in Nordic tourism.

Although the dominant winter tourism activities are still cross-country and downhill skiing, the winter tourism markets are under visible transformation in which new activities and customers with new kinds of needs are evolving. In addition to recently increased snowboarding and snowmobiling, other kinds of adventure tourism activities have become familiar and visible. Many of these new products are labelled under the terms new tourism, experience economy and adventure tourism.

Case 10.2 Mountain resort labour market: The case of the Swedish mountain range

LINDA LUNDMARK

Many mountain areas are struggling with demographical problems, such as low population densities and distorted population compositions, caused by low birth rates and negative net-migration. In terms of socioeconomic development several processes have caused the current situation. Increasing efficiency and effectiveness has resulted in layoffs in the once labour intensive primary sectors with significant restructuring of the labour market as a result. These economic changes alongside shifts between sectors of the labour market, are contributing to unemployment as well as out-migration. The Swedish mountain range is no exception (Figure 10.1). In the period 1980–2001 the population in the municipalities decreased by 15% or around 25,000 people, with the largest drop in the age cohort 25–34. The population development has been negative for all municipalities in the area. Some municipalities, like Dorotea and Storuman for example, have lost up to 12% of the population in 10 years. Least negative development is observed in Åre and Krokom municipalities with a 4% population loss (Lundmark, 2006). During the past 20 years there

232 Nordic Tourism

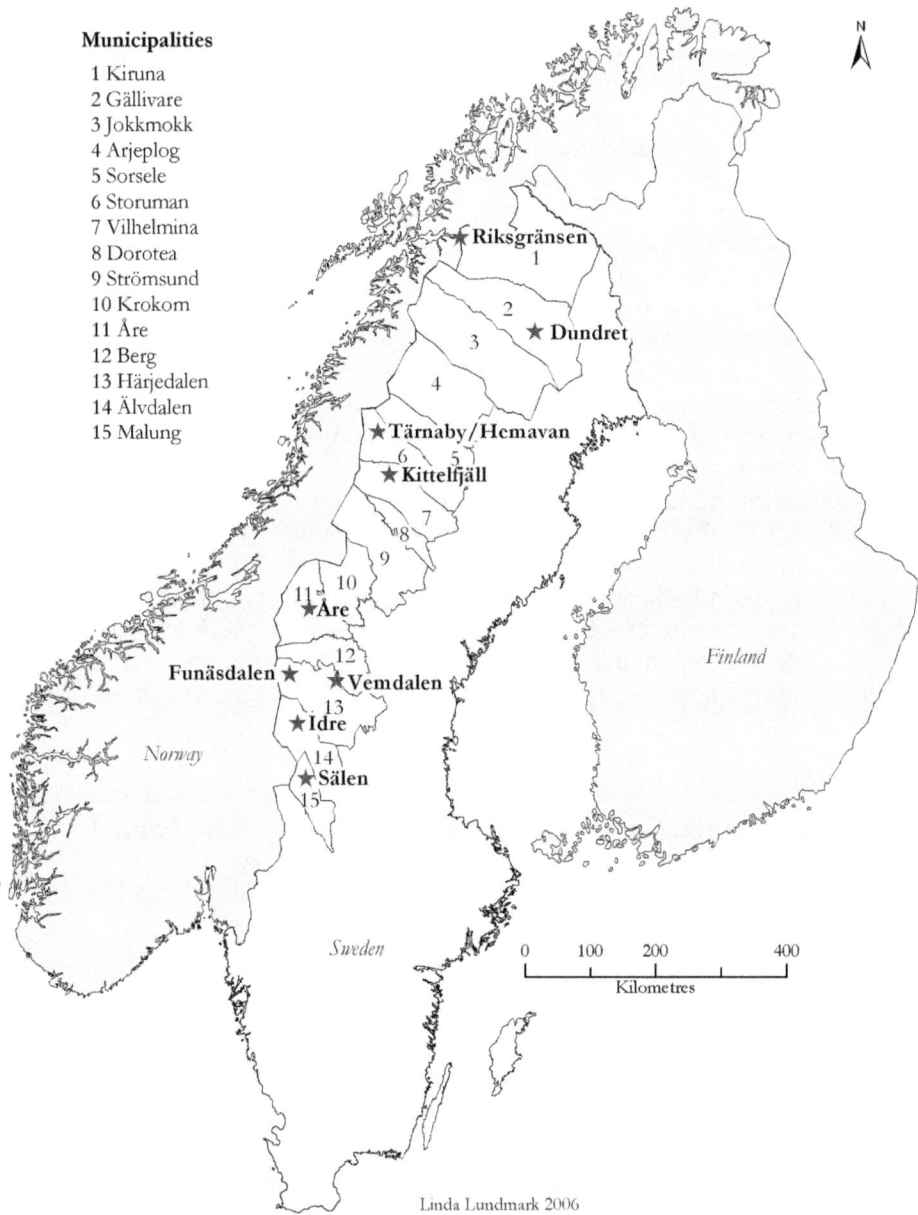

Figure 10.1 A selection of ski resorts in the Swedish mountain range

has also been a decline in the overall work force participation in the 15 municipalities. This is connected to the aging population increasingly outside the workforce.

In the context of a changing labour market and population structures, tourism is regarded as an opportunity to turn the negative development into a positive one. In terms of restructuring, tourism employment is also seen as being positive through its inherent place dependence: tourism must be consumed on site which should support local employment effects. In this way tourism could help counteract the negative effects of population change and economic change.

Mountain tourism has increased in Sweden during the past 15 years and has become more concentrated in the downhill ski areas (Fredman & Heberlein, 2003) in resorts like Sälen, Idre and Åre (Figure 10.1). Fifty of the most visited ski establishments in the mountain range accounts for around 6,500,000 visits (SLAO). The concentration on winter sports emphasises the importance of the winter season, and as the Swedish mountain region first and foremost is a leisure space and not a destination of business-related travel, the dependence of one season is accentuated. The ski establishments in the southern mountain areas in the Dalarna and Jämtland counties account for around 80% of the total downhill skiing turnover in the country (Fredman & Heberlein, 2003). The Sälen area accounts for one-third of the total turnover and the Åre-Edåsdalen area accounts for around one-fifth. Among the 10 largest in terms of turnover are also Idre-Grövelsjön-Fjätervålen, Vemdalen, Funäsdalsfjällen and the Tärnaby-Hemavan establishments. The largest resorts are all situated in the south of the mountain area, apart from Tärnaby-Hemavan. The turnover in the north is lower but some important establishments in terms of visits are Riksgränsen, ranked on 22nd place in 2004/2005 with 67,000 visits and Dundret with 49,000 ranked in 28th place. Kittelfjäll received 28,000 visits and is ranked 41st (SLAO).

There has been an increase in the share of those employed in the tourism sector from 5.9% in 1985 to around 8% during the 1990s and this figure has been stable during the early 2000s. Åre and Malung municipalities receive a large share of the tourism demand and hence tourism employment in the mountain range. This means that the importance of tourism is high in these municipalities. The geographical location has been important for the tourism business development and a concentration to more densely populated areas has taken place during the 1990s (Lundmark, 2005). This development is connected to the so called 'urbanisation' of the rural areas, where sparsely populated areas tend to become sparser while the larger settlements gain population.

Tourism employment is commonly understood as being highly seasonal with an emphasis on women and young employees. In the case of the municipalities and ski resorts in the Swedish mountain range two of these classifications are confirmed (Figure 10.2). The age distribution among employees in

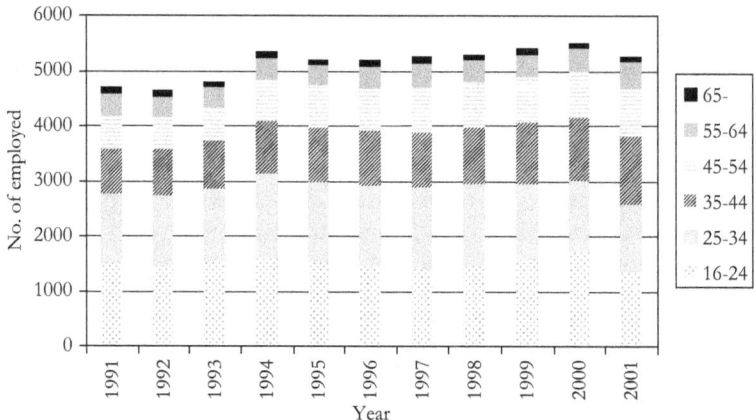

Figure 10.2 Age profile and number employed in tourism in the mountain municipalities

the tourism sector is concentrated to the younger segments of the population and there has even been an increase in the number of those employed in tourism in the same population strata that are decreasing in these municipalities. To be noted, though, is that there seems to be no imbalance in male and female tourism employment participation in number of employed, albeit the type of work performed by men and women might be of different character. The high dependence on the winter season indicates that there are many job opportunities during parts of the year, further putting the seasonality in focus.

According to statistics it is clear that many mountain residents are employed in tourism. However, not only local labour is employed and increasing seasonal labour migration due to tourism businesses is taking place. Approximately 25% of the labour force is recruited outside of the daily commuting distance. This means that a large part of the tourism labour force work part time and/or on a seasonal basis by means of long-distance commuting. The number of people whose official place of residence was in the Stockholm municipality with a workplace in the mountain area was 4% (around 200 employees) of the total tourism employment in 1985 and 7% (around 450 employees) in 1999. The origin of the long-distance commuters is mainly the larger cities along the coast, notably around 28% came from the Stockholm municipality in 1999 (Lundmark, 2005).

The southern part of the area is more likely to have seasonal migration and long-distance commuting than the northern parts. This is explained by the closeness to areas with higher population densities. Apart from this, the infrastructure connecting the south of Sweden to the southern mountain chain is better giving easier access for tourists as well.

In Åre and Malung, where the largest ski resorts are, the share of long-distance commuting to employment in tourism from other parts of the country is considerably higher than on average (Lundmark, 2006). This could be explained by simple demand and supply logics. The demand is higher than in the mountain range in general due to the popular ski resorts of Sälen and Åre. At the same time, the population characteristics do not differ from the other mountain municipalities. This means that the demand for tourism workers is too high for the local labour market to supply. In general the tourism industry does not absorb the local labour force as efficiently as expected. First, there is the connection to the characteristics of the work with high seasonal demands and flexible working hours (Lafferty & van Fossen, 2001; Montanari & Williams, 1995). Second, part of the explanation is found in the demography of the mountain area. It is significant that the accessible work force is not employable in the sense that they do not have the skills needed nor the human capital looked for in the service oriented tourist industries (Shaw & Williams, 1994; Stynes & Pigozzi, 1983). Young people are more mobile, both between sectors and geographically. This means that young people living in the cities will be able to take the seasonal type of work offered in tourism even if it includes long-distance commuting.

The motivation for the long-distance commuters to come to the mountain resorts is both income and life-style related. Seasonal tourism employment makes it possible to combine these two motives because life-style motivated seasonal migrants are able to combine leisure and labour market participation. In the case of the ski resorts in Åre and Malung this is an explanation for the high amount of long-distance commuters working in tourism.

Åre and Malung are examples of municipalities where tourism has stimulated migration of young people to the mountain range (Lundmark, 2006). Work in tourism has led to more in-migration, contributing to a positive development of the local service base, infrastructure and public facilities. Such positive outcomes can be perpetuated by further making the places attractive to investors, tourists and possible in-migrants. However, the positive influences of economic restructuring towards tourism in general in the mountain municipalities are small with regards to increased in-migration. The permanent population in the area continues to move out because tourism employment is not enough to absorb the local workforce and seasonal in-migration does not give a positive development in the basic services in general. Tourism employees from municipalities outside the area do not increase the tax revenues necessary to improve infrastructure and other public services, because they do not pay taxes in the area. A successful tourism development is dependent on having a functioning service base but this might not be possible with a small permanent population.

New Tourism in Winter Landscapes

New tourism and the experience economy: 'New wrappings for old gifts'?

During the last two decades structural changes related to tourism have been visibly present in the development of Nordic winter tourism. Traditionally, northern winter tourism has been based on cross-country and downhill skiing by domestic tourist segments. In many of the Finnish skiing resorts, for example, major capacity owners were trade unions, larger private companies and governmental organisations which offered accommodation to their members or employees. Starting from the late 1980s and early 1990s, however, new activities and segments started to evolve. These changes were largely based on wider movements in Western consumption systems which were also reflected in the tourism industry.

The growth of more individual and international tourism and the decrease in the relative importance of mass tourism has long been highlighted in the literature (see Poon, 1993). These estimated but less studied and especially empirically analysed changes are based on wider shifts in consumption and production in Western countries. In the literature this shift is often described as moving from Fordist production towards post-Fordist production and the related new ways of consuming (Urry, 1990). In tourism this transformation towards more individually oriented production, marketing and consumption is indicated by the term 'new tourism' (Poon, 1993, 2003). New tourism and new tourists are considered fundamentally different from the 'old' ones; they are seen to be more flexible, independent, experienced, quality and environment conscious, allocentrics and spontaneous (see Duffy, 2002; Fennell, 1999). Old tourism is also seen as more homogenous, stable and predictable than new tourism, which is less structured, active and hybrid (Poon, 1993). New tourism is regarded as being able to combine very different kinds of elements of nature, culture and services in the same product; a new tourism winter holiday, for example, can include an overnight dog sledge safari to the wilderness and after that downhill skiing over-night stays in a big resort environment with spa facilities – all this during the same holiday 'package' with optional activities individually selected. Similarly, a winter destination and experience can be combined with summer 'sea-sand-sun' resort experiences. For example, international tour operators link Rovaniemi, Arctic Circle, and its winter activities with the summer seaside resort of Phuket, Thailand, in the same 12 day package holiday (see Seiviaggi, 2006).

The rise of new tourism is largely based on the debated assumption of the crisis in modern tourism, specifically the mass tourism industry. According to Poon (1993, 2003) changes in technology, production and consumption modes, management strategies and socioeconomic contexts will evidently lead to the end of mass tourism and towards alternative forms of tourism. This may well be an overstatement and there is no clear empirical evidence yet that the core of the industry, mass tourism, is really declining either globally (see Honkanen, 2003) or in Nordic winter tourism destinations. However, it is clear that in the future some of the present forms and practices of mass and winter tourism will most probably change due to the rising energy (especially transport and aviation fuel) prices and global climate change, for example (see Gössling & Hall, 2006a).

It is also evident that mass tourism itself is currently affected by the wider changes in production and consumption modes (see Mowforth & Munt, 1998). In addition many tourism-related technologies in marketing and reservation systems have changed the industry and led to the development of new tourist practices in consumption. However, using the internet instead of a travel agent in holiday trip reservation, for example, does not necessary turn a person into an individual tourist (drifter, explorer or allocentric). In addition, based on existing evidence and knowledge it is still a matter of opinion whether the previous examples of the combination of dog sledge safaris to the cold wilderness with major ski resort activities or the combination of winter and summer resort experiences represent the rise of new tourism as suggested by Poon or whether they are new products and packages indicating structural and operational changes inside the ('old') mass tourism industry.

Critically thinking the new tourism does not necessarily represent a specific form of tourism which is totally separated from the old tourism, but rather a new style or dimension of production, with increasing flexibility, individuality and hybridity; features which are also present in the representations and new uses of the winter landscapes and their use in mass tourism (Saarinen, 2004). Some of the new uses are linked to the present growth of the so-called experience economy, with its increased tendency for individualisation in production and consumption (see Beck & Beck-Gernsheim, 2001; see also Chapter 2). For Pine and Gilmore (1999), the experience economy represents a logical development in production and the progression of economic value. For example, agricultural commodities are extracted, fungible market-based products, while the goods in the next phase are manufactured, tangible and user-oriented products. This is followed by the production of services, characterised by their delivered nature and intangible activities for a particular client. Pine and Gilmore argue that the production of experiences therefore represents the higher mode of production.

In tourism these staged experiences and memorable events engage individuals in a personal way as guests. On the other hand tourism has always been an 'experience economy'. Tourism ties space, time and human experiences together in systems in which individuals seek to perform a variety of activities. New products of the experience economy are good contemporary examples of this development and connection between spatiality, time and experiences, which nowadays often involve spectacles, exercise, deep contrasts and speed (see Ryan *et al.*, 2000; Thrift, 2001). But this kind of production and more effective segmentation of consumers does not, however, lead automatically to the end of mass tourism, but rather the appearance of new forms of tourism and tourists in the market.

Changing activities in Nordic winter

The northern winter landscapes have traditionally offered touristic opportunities for recreation activities based on nature. This basis has not changed, but during the last decade winter tourism has experienced both quantitative and qualitative changes. Snowmobile trekking has become one of the most central and visible forms of the new nature-based winter tourism activities, especially in Finland and Sweden.

New Nordic winter activities such as snowmobile trekking, ice sailing and dog or reindeer sledge safaris are pursued through enterprises that are related specifically to the new notions of an experience economy. Many of them rely on changes in production and consumption, international influences and new images and uses of nature. Dog sledge safaris, for example, are mainly based on representations of the Arctic and cold, snowy areas promoted by tourist operators focusing initially on French markets. French tourists most probably evaluate northern nature, its uses and attractions in the light of their historical relations with the Canadian Arctic where the dog has been traditionally used in winter season transportation. The local equivalent of the husky dog in Nordic contexts is the reindeer, the role of which is growing in the nature-based winter tourism market but is still marginal compared with that of the sledge dog or especially the modern snowmobile.

Traditional Nordic skiing is, however, still the most important form of nature-based winter tourism activities that directly use northern nature (Lapin Matkailumarkkinointi, 1999). Nordic skiing (and also majority of downhill skiing) is mainly based on domestic markets and the share of international demand has been significant in sledge dog and reindeer safaris, as well as snowmobile trekking. Generally, the degree of internationalisation of the tourist structure and the tourism in Nordic countries has risen in recent years (see Heberlein & Vuorio, 1999; Saarinen, 2003) and the growth of international demand has also created markets for new types of visitors such as adventure tourists.

Winter Tourism and Changing 'Snow Business' Products

Adventure tourism

In recent years outdoor adventure activities have become increasingly popular in tourism. Adventure tourism has become part of the new tourism modes. Some commentators estimate that adventure tourism may be growing 10–15% every year (see Cater, 2006) but there are no official or unofficial statistics that would indicate the numbers of adventure tourists regionally, nationally or internationally. The concept of adventure tourism is also subject to debate in the tourism research community (Gyimóthy & Mykletun, 2004) and its distinction from the general idea of activity tourism is perceived to be problematic (Swarbrooke *et al.*, 2003). For some adventure tourism is seen as an active recreational practice which requires deep involvement: 'being, doing and seeing' (Cloke & Perkins, 1998: 189) rather than 'touristic staying', and adventure tourism is often described in terms of its motivation. Based on a motivational approach tourist activities that may contain serious risk taking and threats to a participant's health or life are regarded as adventure tourism activities. However, this motivational ground of definition makes it difficult to conceptualise the different activities as adventurous or non-adventurous; subjectively the same activity, such as an overnight skiing trip to the wilderness may represent a very different level of risk taking and danger for different people.

It is evident that adventure tourism, activity tourism and different modes of so-called new tourism do overlap with each other. Therefore, many authors see that the concepts form a continuum. Swarbrooke *et al.* (2003), for example, analyse the adventure spectrum in activity tourism by using examples ranging from activity tourism that is not at all adventurous to activity tourism that is extremely adventurous. For example, a visit to Kemi Snow Castle, Finland, or a guided tour in Akureyri, Iceland, would usually represent the former while a skiing trip to Svalbard may refer to the latter.

In a well-cited paper Fennell and Eagles (1990) separated tourism, ecotourism and adventure tourism by analysing the degree of preparation, training and unknown risks and results. Adventure tourism includes a lack of certainty and safety and is motivated by self-learning and personal fulfilment, whereas tourism involves a lower level of personal preparation, a high degree of safety and group organisation undertaken for the tourist. According to Fennell and Eagles (1990) ecotourism is located somewhere between the two opposites covering both adventure and tour travel aspects in motivation and degree of risk, security and training.

For Millington *et al.* (2001: 67) adventure tourism is a leisure activity that takes place in an unusual, exotic, remote or wilderness destination. It is often associated with high levels of activity by the participant, most of it outdoors, and adventure tourists expect to experience various levels of risk. Many of the Nordic peripheries, especially in winter season offer possibilities for such tourist activities. For example, Gyimóthy and Mykletun (2004) studied extreme winter trekking in Svalbard which offers long self-served Arctic trekking trips by snowmobile or Telemark skis. In Svalbard adventure tourists were found to seek the presence of risk, insight seeking (competence development) and play. These three elements of adventure were closely interrelated, which led the authors to the conclusion that together they create the conditions of Arctic adventure tourism (see also Viken, 2006; and Case 6.1).

The history of travel and tourism is filled with examples of adventurous journeys by 'brave' travellers, often told in travel writing or, in more recent times, on the *Discovery* or *National Geographic* channels on television. The present idea of adventure tourism may therefore represent a modern continuum for such endeavours and the present adventure tourism is clearly an organised activity inside the tourism industry. Indeed, according to Swarbrooke *et al.* (2003) adventure tourism is situated at the cutting edge of shifts in tourism production and consumption. Although adventure tourism is still a relatively limited activity in the Nordic countries the images of 'wild and free' and 'harsh and dangerous' are utilised also in less risk requiring form, for the promotion and development of Nordic winter tourism.

Competing Santa Clauses, snow castles and ice hotels

In recent years one of the most international fields of winter tourism in Nordic countries has been based on the Christmas season. In addition to Santa Claus, Nordic winter tourism is characterised by a variety of new products and attractions, such as

IceHotel (Sweden), Snow Castle and ice breaker trips with ice dips in Kemi (Finland), dining on ice in Luleå (Sweden) and hot baths and swimming with nature-based tourism trips in Iceland, which all demonstrate the evolving flexibility and hybridity of winter tourism production and consumption.

The growing trend of Christmas tourism is relatively new. Until the 1980s Christmas time and Santa Claus actually represented rather negative elements in Nordic tourism markets by keeping people home or creating 'visiting friends and relatives' (VFR) mobility which was not regarded as providing wider economic opportunities for the tourism industry. Finland was one of the first Nordic countries to develop a Christmas tourism season. In the 1980s Finnish tourist authorities launched a new tourism strategy based on the idea of 'Santa Claus Land'. It included a new kind of marketing and development of the Santa Claus Village in the Rural Municipality of Rovaniemi. The location along the main road to North Cape and next to the Rovaniemi airport served development purposes and after a modest start the place evolved into a large-scale tourist attraction (Pretes, 1994). In the late 1990s an additional theme park, Santa Park, was created near the Village (Plate 10.2). It was designed by, and for, the British who form the major segment of Christmas tourists to Finnish Lapland; for example, during the season 2005 over 200 direct charter flights arrived from the UK to Finnish Lapland and most of the visitors made a day trip to the Rovaniemi area.

According to Pretes (1994) the history of the Santa Claus industry in northern Finland is the history of the commodification of Christmas. For him Santa Claus and the Village have become features of the post-modern tourism landscape,

Plate 10.2 Santa Park, Rovaniemi, Finland

a landscape marked by a spectacle where tourists consume the marker and thereby consume a nostalgic conception of Christmas that was left behind in the modernisation of Western societies. This Finnish example of utilising Santa Claus has been followed in other Nordic countries, but the early start has given a certain advantage to the Finnish interpretation of Santa Claus and location of his dwelling.

In contrast to Santa Claus, the use of snow and ice in tourism operations and international marketing has been more successful in Sweden than in the neighbouring countries. Originally the IceHotel in Jukkasjärvi in Swedish Lapland started off as a summer tourism destination in the 1970s, but by the end of the 1980s the activities were reorganised based on the utilisation of the dark and cold winter as an advantage (see Gertner & Kotler, 2004; Gibson, 2006). Thus, the unique elements of the arctic winter environment were regarded as an asset. In 1990 the ice and snow-based small art exhibition igloo was used for 'voluntary' guest accommodation. In the following years facilities made out of ice were built to serve as more exotic accommodation (see Kvist, 2005) and the place promotion has been done in cooperation with internationally known products such as Sweden's Absolut vodka (see IceHotel, 2006) one of the world's best known vodka brands. A similar utilisation of the cold environment and combination of cold, dark and 'warm hospitality' is used in the dining on ice concept in Luleå (see Nilsson & Ankre, 2006).

Summary and Discussion

Winter tourism activities have a long history in Nordic countries and are an important part of the broader Nordic image. In spite of the fact that larger winter tourist destinations are currently transforming or have already transformed to all-season destinations, the winter still represents the major season for most of these destinations. Winter tourism has a major role especially in Finland, Norway and Sweden and more specifically in their northern and/or mountainous areas which provide a topography and natural environment suitable for most of the winter season activities.

The core of Nordic winter tourism is based on cross-country and downhill skiing but new modes of tourism production and consumption are evolving. The most visible among new tourism activities have been snowboarding and snowmobiling. There are also possibilities to develop adventure tourism products based on the wilderness environments and activities. In addition to new tourist activities, Nordic winter tourism has faced new kinds of tourist attractions. In addition to Santa Claus and the Christmas season many of the attraction elements are based on the utilisation of the previous disadvantages: darkness, coldness and icy and snowy landscapes and the contrast they can create with the elements of the hospitality industry and tourists' home environments.

The rise and the future prospects of Nordic 'snow business' are based on ongoing innovation strategies that currently range from the snow and ice castles with tourism facilities, ski tubes that allow skiing also during the summer season (or warmer winter seasons), adventure trekking, ice fishing and car and tyre testing laboratories in extreme conditions. These innovation strategies cover multiple actors, scales and public–private partnerships. Research results indicate that tourism operators' capacity to innovate to a large extent depends on the wider social, economic, cultural and political environment of which they are part. Thus, the tourist destinations and cooperation with the different firms and actors inside the destination and with larger governmental and research bodies hold a central role in the innovation processes (Nordin & Svensson, 2005).

Innovation processes may also include adaptation strategies for a changing environment, such as climate conditions. Winter tourism and ice and snow-based tourism activities are considered very climate-sensitive human activities (Scott, 2006). In the European context the impacts of climate change and rising average temperatures may create temporal advantages for northern destinations in competition for winter tourists in the near future compared to skiing resorts in the Alps, for example. However, climate change in Nordic area most probably requires the utilisation of special adaptation strategies and innovation mechanisms which may go beyond artificial snow-making techniques, an issue that is examined in more detail in the next, and final, chapter.

Discussion questions

(1) What are the development phases of Nordic winter tourism?
(2) What are the reasons for the recent growth of adventure tourism?
(3) What is 'new' in new tourism?

Essay questions

(1) What is the relative importance of winter tourism activities for Nordic tourism development?
(2) What are the future challenges and prospects of Nordic winter tourism?

Key readings

Winter tourism has been studied especially in north America and central and northern Europe. A majority of research has focused on skiing destinations with special interest on the development and seasonality of winter tourist resorts (see Davidson, 1981; Fredman & Heberlein, 2003; Gill, 2000; Heberlein *et al.*, 2002). For those interested in the tourism area life cycle and its relationship to resort development see the two volume review in Butler (2006). Recently, global climate change and the vulnerability of winter tourism has been under focus (see Elsasser & Messerli, 2001; OECD, 2007; Scott, 2006). In spite of an intensive research tradition there are relatively few general volumes dealing specifically with winter tourism, such as Hudson's (2000) *Snow Business* with a strong marketing studies orientation.

All the major winter tourism destinations (e.g. Levi, Lillehammer, Åre) in Nordic countries have web pages. In addition the ice and cold-based products such as IceHotel (www.icehotel.com) and Kemi Ice Breaker (www.sampotours.com) have their web pages.

Chapter 11
The Future of Nordic Tourism: Regional and Environmental Change

Learning Objectives

After reading this chapter, you should be able to:

- Understand the changing concept of Nordicity and its role in region-building.
- Understand some of the implications of climate change for tourism in the Nordic countries.

Introduction

This chapter examines some of the potential issues facing the future of Nordic tourism. After a discussion of changing notions of Nordicity and its implications, the chapter examines the renewed interest in the north in the light of environmental change in the Arctic region. These changes then bring us to examine some of the implications of climate change for tourism in the Nordic countries.

Changing Notions of Nordicity

In a world in which issues of image and representation are of commercial as well as cultural importance, the concept of Nordicity is of considerable significance for tourism in terms of place competition. For centuries the north has constituted a mythological space, constantly defined and redefined, by centuries of writers, explorers and other image makers. As noted several times in this book, its contemporary portrayal is borne out of literary beginnings found in ancient

Greek writings and Viking sagas. The northern 'other' was also arguably defined in other European cultures, especially British, French and German, before being defined in Norden itself. Yet even here understanding of Nordic was related to new Romantic nationalist identities, especially in Finland and Norway, that were connected to the northern environment. Significantly, travel played an extremely important role in developing and popularising artistic and intelligensia understandings of Nordicity through both the travel writing genre and through tourism as a means of extending elite culture into popular culture. Canadian geographer Edmond Hamelin (1979) was however the first who moulded the concept scientifically in an attempt to find and delimit this space geographically, proposing an index covering both natural and cultural features since he felt that no single variable could capture what the north represented in various ways.

Tourism became especially important because of how it commodified and represented northern spaces, including some of the first national parks and areas of national landscape, to the point where it can be argued that the Nordic landscape is as much a social construction as it is a physical one. However, as the 20th century developed so representation and understanding of Nordicity also shifted, greater attention being given to indigenous culture of the Sami and the Inuit, while the Nordic countries are also far more diverse in terms of ethnicity and culture than ever before. A situation perhaps best represented by the picture of the Buddha at the end of Chapter 9. Such changes are not just the concern of students of cultural studies or literature, but have practical impact in terms of social and political relations as well as place representation and tourism. Indeed, for many people around the world place understanding of Nordicity is as much tied into such things as Abba, Ericsson, Ikea, Lego, Nokia and Volvo as they are to lakes, forests, sauna and snow, while new products and images are also changing the face of Nordland (see Case 11.1).

Cases and issues 11.1: Lordi and Finnish tourism

Even if we lose the contest, we have already won', Lordi says. 'Many Finns would rather have sent someone boring and acceptable than to be represented by freaks like us. (quoted in Bilefsky, 2006)

The hard rock heavy metal band Lordi gave Finland its first ever Eurovision song contest victory in 2006 with a record that won 292 points for their song 'Hard Rock Hallelujah'. Although the band was criticised by some Finnish religious leaders as being evil and unFinnish because of lyric content and horror genre inspired costumes (Bilefsky, 2006), they received substantial acclaim for their success. Most notably, being presented a bronze key flag award for exemplary Finnish work by the Finnish President Tarja Halonen at an open-air concert in Helsinki in front of 90,000 people to celebrate Lordi's success at the Eurovision song contest. In recognition of their Eurovision win the Suomen Posti Oyj (Finnish postal service) issued a Lordi postage stamp in 2007.

In tourism terms the band are also significant as the budget airline Ryanair used the win for route promotion and offered 10,000 tickets for payment only of taxes and service charges on its route to Tampere in Finland. The Finnish Tourist Board also sought to utilise the band's success to attract new visitors to Finland, particularly as Finland hosts many summer music festivals including the internationally renowned Sauna Open Air Metal Festival held annually in Tampere. The size of the festival is significant with the 2007 event attracting over 22,000. As a result of Lordi's victory Helsinki also gained the opportunity to host the 2007 Eurovision song contest.

In addition to their contribution to the Finnish image, Lordi are also significant at a regional tourism level. Given that the band was founded in Rovaniemi by Tomi Putaansuu (referred to as 'Mr Lordi') the city recognised the band's achievement by renaming a square in the city centre after the band. The band has also directly contributed to economic development through opening a restaurant, Lordi's Rocktaurant, in Rovaniemi in December 2006. The restaurant features Lordi memorabilia, background heavy rock music as well as a quality value for money lunchtime buffet.

Lordi: http://www.lordi.fi/
Sauna Open Air Metal Festival: http://www.sauna-open-air.fi/

Yet, rather than being a break with tradition's of Nordic representation, the success of Lordi arguably extends various dimensions of Nordic culture and their commodification. The background storylines of Lordi are to be found as much in fairytales of goblins and trolls as they are in Alice Cooper and Kiss. Moreover, they share an important connection with present-day interest in the creative and experience economy concepts as part of place competitiveness in which tourism is also heavily implicated (see Chapter 4) and to which the Nordic countries have made a strong policy commitment. Furthermore, and importantly for the relationship between creativity and economic development, the success of Lordi has also been due, the comments of certain Christian groups aside, to Nordic notions of a tolerant society. Such factors are important for tourism as while there are commonalities in political and cultural expressions of Nordicity they suggest that a key theme in understanding Nordic tourism is that of continuity and change.

Norden is usually seen as a forum for crossing 'mental bridges' (Makarychev, 2002: 44). Nordic political values are characterised by 'transparency, egalitarianism, and consensual democracy which together form a distinct protestant identity' (Tiílikainen, 2001: 2). This has meant attention not only to region-building in terms of the cultural project to create a Nordic identity, but also to a particular style of state intervention and public–private partnership with respect to areas such as tourism. The Nordic region is itself also being reshaped as new relationships with the Baltic States and northwest Russia are formed. For example, Estonia identified strongly with the Nordic community with the then Foreign Minister (and President as at the time of writing), Toomas Hendrik Ilves, delivering a speech entitled 'Estonia as a Nordic Country' to the Swedish Institute for International Affairs in December, 1999, noting that 'It is time that we recognize that we are dealing with three very different countries in the Baltic area, with completely different affinities. There is no Baltic identity with a common culture, language group, religious tradition.' Instead he argued that there was a need for a paradigm shift 'in the mental geography of the Nordic region' that would be inclusive of Estonia.

New northern regionalisms

At the same time that the Norden concept is being embraced by Estonia, so there are other new regionalisms developing that overlap strongly with notions of Nordicity. The Barents Euro-Arctic Council has been part of the development of a Barents region that is gradually being opened up via increased trade, particularly as a result of mineral development in the north. Even if the Barents concept is only promoted by a 'very limited group of entrepreneurial people with particular interests in developing contacts at the other side of the border' (Hønneland, 1998: 288–289), it still remains significant for cross-border mobility and tourism. For example, as noted in Chapter 4, the growth of Haparanda following the opening of an Ikea store was regarded by the city mayor as being related to the Barents region concept and opportunities for cross-border shopping. As well as the Barents region there is also increased emphasis on the Baltic as a new regional model (which includes all the Nordic countries), particularly as a result of growing interrelationships between urban and regional centres. Such developments have already led to new tourism projects founded on joint transportation projects, such as the 'Blue Road', a highway and a tourist route crossing Norway, Sweden, Finland and Russian Karelia (Makarychev, 2002). In addition, there is also attention being given to the construction of a Baltic identity or Balticness in order to improve the region's place competitiveness (see Chapter 4). According to Martinsone (2007: 11), coordinator of Project Balticness for the Council of Baltic Sea States (CBSS) the overall goal of which is 'to raise awareness of Balticness and the region's competitive identity' in response to the question, what is Balticness?

> Project Balticness intends to promote awareness of the region's unique comparative advantages and values through raising the public recognition of Balticness – creativity, professionalism, skilfulness, dynamics, stability, multiculturalism, and openness of the region and its people. We hope for it to represent a call for broader and deeper cooperation in the Baltic Sea Region ... The project *Balticness* is our contribution to the branding of the region. First of all, we have to brand it internally – raising our own sense of belonging to the region, acknowledging our common values, culture; facing common challenges and future ....
>
> The word *Baltic* might have different meanings in the region – the sea, the three Baltic States, the eastern coast of the sea and so on, but for the rest of the

world there is only the Baltic Sea Region associated with the Baltic Sea and nations surrounded.

As strange as it may sound to many readers the concept of Nordicity is also applied outside of the Nordic nations. For example, Hamelin's (1979) equation of Nordicity with northernness, defining 'the North' as the area that exists above the tree line that – depending on which northern country you are in – lies between the 55th and 60th parallels. Arguing that this conforms with widely held geographic indicators of Nordicity based on latitude, annual temperatures below zero, accessibility (including airline connections), population density, types of ice and levels of economic activity. Nevertheless, as Roth (2005: 43) commented, 'innumerable sources over the centuries indicate that it is not, and never has been, science that has articulated the idea of the North in the North American public imaginary. Rather, it is our own mythical notions of "The North" that circumscribe our views'. For example, Shields (1991) argues that there are three different 'Norths' in Canada: the administrative north defined by federal and provincial political boundaries; the imaginary north which is 'a frontier, a wilderness, and empty "space" which, seen from southern Canada is white, blank'; and the ideological 'True North Strong and Free' (a line from the Canadian national anthem) which is an 'empty page onto which can be projected images of the essence of "Canadian-ness" and also images to define one's urban existence against' (Shields, 1991: 165).

Medvedev (2000: 1, cited in Joenniemi & Sergounin, 2003) in comparing Canadian, Finnish and Russian understandings of the north also argued that 'The North is more often communicated than experienced, imagined rather than embodied'. According to Medvedev (2001: 91), 'Lacking in rationality, the north is rich in mythos and implied meanings. In many traditional mythologies the north is singled out among other parts of the world as essentially being the *outer* fringe.' The north is left as 'the last Frontier, the only part of the world that holds the fascination of emptiness, a white space in our mental maps' with the implications of this being substantial for tourism because of the symbolic value of northernness. 'Europe's southernmost and easternmost points are hardly known to the public at all; on the contrary, an entire tourist industry has been built around pilgrimages to the Nordkap. The north turns out to be marketable precisely because of its remoteness, relative obscurity and anonymity' (Medvedev, 2001: 91). The north is therefore, 'the emptiness we are filling with our imagination, narratives and texts; a blank sheet of paper on which words are written and erased; an empty snow field on which lonely figures emerge, pass, and disappear' (Medvedev, 2001: 92).

Interestingly, similar issues appear in Russian understandings of the north. According to Makarychev (2002: 46–47) on the negative side, the north 'is associated with remoteness and cultural backwardness':

- the north can be seen as synonymous with vast loosely organised spaces, which have to be somehow preserved or conserved;
- the north is connoted with social conservatism and traditionalism;
- the north is a depopulated area;

- the northern provinces are perceived as prone to [raw] material separatism and even isolationism, and in this capacity they might contribute to the disintegration of the federation;
- the peoples of the north are on their way to emphasising their self-identities, which is a challenge for the federal authorities.

Such comments are significant as they reinforce the notion that the north has been substantially defined by those outside of it, 'the "symbolic north" has been fabricated by non-natives who have talked about it, analysed it, made statements about it, settled it, ruled it, authorized certain views of it, managed it, photographed it, and ultimately produced it as an "exotic" commodity for Southern consumption' (Roth, 2005: 44–45) (see also Payne, 2006). Not least of which has been for the benefits of visitors (Grenier, 2007), through what Sörlin (1998, 1999) describes as the 'articulation of territory'. The result of which are symbolic and mental landscapes that are deeply embedded in the image and self-understanding of nations and regions. Although he also stresses that the articulation of territory is in itself an important part of the historical emergence and growth of nationalism and regionalism.

Nevertheless, ideas and fictions of Norden and Nordicity have shifted and changed over time and will continue to do so now and in the future. Yet they remain significant in general, as well as directly for tourism, as they provide important articulations of governance and partnership, representation and promotion, and identity whether for self or for tourist others. Moreover, tourism is embedded in the soft regionalism of Nordicity that blurs the distinction between 'insiders' and 'outsiders' and is part of the 'open geography' in which there are no strict dividing lines between regions:

> Here regions are understood as mobile social and cultural constructs that might 'encounter', 'clash', 'inject their own stories', etc. Importantly, the idea that regional identity is determined in geographical categories actually always involves a choice (i.e., 'what we wish to belong to?'), because the social world is defined not only by physical constraints but also intellectually and spiritually. As such, there can be no single mode of spatial representation or articulation of spaces. Geography cannot lock up regions in a 'steel cage', and geographical affiliations are subject to re-writing and re-interpretation. (Makarychev, 2002: 47)

Given the sometimes fluid nature of northernness and Nordicity it is interesting to note the extent to which the Arctic has now become a significant space for political partnership and competition, an emerging space of economic development, as well as an important symbol for environmental change.

Arctic regionalism and environmental change

The Fourth International Polar Year 2007–2008 has attracted substantial attention to the Arctic however probably of greater significance is the related issues of climate change and increased interest in the Arctic region's energy resources. Both the Arctic and Nordic Councils, as well as member governments, have expressed concerns about the potential environmental changes that may occur as a result of energy exploration and extraction in the Arctic, particularly with respect to the impact on indigenous peoples (e.g. Hansen, 2008).

The Arctic Council's (2004) policy document with respect to Arctic Climate Impact Assessment reported that, 'Climate change, together with other stressors such as ultraviolet radiation, presents a range of challenges for human health, culture and well-being of Arctic residents, including indigenous peoples and communities, as well as risks to Arctic species and ecosystems' and identified 10 key findings.

(1) The Arctic climate is now warming rapidly and much larger changes are projected.
(2) Arctic warming and its consequences have worldwide implications.
(3) Arctic vegetation zones are projected to shift, bringing wide-ranging impacts.
(4) Animal species' diversity, ranges, and distribution will change.
(5) Many coastal communities and facilities face increasing exposure to storms.
(6) Reduced sea ice is very likely to increase marine transport and access to resources.
(7) Thawing ground will disrupt transportation, buildings, and other infrastructure.
(8) Indigenous communities are facing major economic and cultural impacts.
(9) Elevated ultraviolet radiation levels will affect people, plants, and animals.
(10) Multiple influences interact to cause impacts to people and ecosystems.
(Arctic Council, 2004: 4)

The changing Arctic climate has become one of the focal points for global attention due to the impacts of climate change, through concern over icon species such as polar bears, as well as for the wider landscape and seascape. For example, in the northern summer of 2007 ice cover was reduced to 4.13 million km^2, the smallest ever extent in modern times. The long-term average minimum, based on data from 1979 to 2000, is 6.74 million km^2. In comparison, 2007 was lower by 2.61 million km^2, an area approximately equal to the size of Alaska and Texas combined, or the size of 10 United Kingdoms. The decline in sea ice means that there are some estimates that the northern polar waters could start to be ice-free in summer by 2013 (Amos, 2007). Such changes have also come on top of other significant symbolic events to gain international coverage including the Northwest Passage, the most direct shipping route from Europe to Asia, being fully clear of ice for the first time since records began (BBC, 2007); while the Ayles Ice Island off the coast of Ellesmere Island split in two (Shukman, 2007).

The ACIA scientific report concluded that tourism is 'becoming an increasingly important economic factor in many arctic regions. Impacts on these economic sectors in monetary terms are difficult to project and quantify since factors other than climate, including future regional economic development, play a major role'

(ACIA, 2005: 1003). It is noticeable that tourism was recognised as a likely beneficiary of climate change, at least in the short term, as a result of the potential expansion of marine routes especially during the summer navigation season (ACIA, 2004: 16). However, in the longer term significant concerns exist over the biotic resources of nature-based tourism, including for example, the availability of fish and game for fishing and hunting, as well as the impacts of natural resource change on indigenous culture. Anisimov *et al.* (2007: 655) conclude, 'Impacts on food accessibility and availability, and personal safety are leading to changes in resource and wildlife management and in livelihoods of individuals (e.g. hunting, travelling) (high confidence). The resilience shown historically by Arctic indigenous peoples is now being severely tested (high confidence)', while also noting that 'Increased eco-tourism may increase incentives for protection of environmental areas. Taking advantage of these potentially positive impacts will, however, require institutional flexibility and forms of economic support' (Anisimov *et al.*, 2007: 673). Indeed, one possible implication of evidence of climate change is that it may spark a 'see it while you can' response from tourists and the tourism industry, thereby, leading to even greater stress on the Arctic environment and on capacities to govern tourism.

Clean, Green Nordic Tourism?

Climate change presents a major challenge to Nordic tourism – and not just in the most northern regions. For example, the Baltic sea basin has been substantially affected by climate change. According to Pawlak and Leppänen:

> The warming trend for the entire globe was about 0.05°C/decade from 1861–2000, while the trend for the Baltic Sea basin has been somewhat larger, 0.08°C/decade. This warming trend has been reflected in a decrease in the number of very cold days during winter as well as a decrease in the duration of the ice cover and its thickness in many rivers and lakes ... In addition, the length of the frost-free season has increased and an increasing length of the growing season in the Baltic Sea basin has been observed during this period ... The projected decrease of ice cover by the end of the 21st century is dramatic, with the Bothnian Sea, large areas of the Gulf of Finland and the Gulf of Riga, and the outer parts of the southwestern archipelago of Finland becoming, on average, ice free. The length of the ice season would decrease by 1–2 months in the northern parts of the Baltic Sea and by 2–3 months in the central parts. (Pawlak & Leppänen, 2007: 3)

In addition, there would be a change in precipitation patterns both geographically and seasonally associated with future warming with the largest increases in the northern areas of the Baltic basin. Such changes may have impacts not only on the availability of snow for winter tourism but also on biodiversity. Also of significance is the potential of increased eutrophication marked by algal blooms in the Baltic Sea and Nordic lakes as a result of the combination of warmer water temperatures and nutrient inputs.

While the Nordic image is socially constructed its environmental image and identity, which is grounded in representations of nature, is based on the physical environment. Global environmental change, of which climate change is one very important component, therefore has significant implications not just for Nordic landscape and biodiversity but also as to how this affects branding and promotion. As the Nordic Council of Ministers Tourism Ad Hoc Working Group highlighted:

> The tourism industry in the Nordic region faces a range of challenges in the future. Its products are sold using images of a clean nature, high standards of living and a unique cultural heritage. At the same time, many of its markets are becoming increasingly aware of the pressures that the tourism industry can have on these values. The Nordic countries are also under pressure from the highest political levels in the international arena to develop tourism policy that takes into account the need to preserve the sustainability of these values, in other words a sustainable tourism development. (Nordic Council of Ministers Tourism Ad Hoc Working Group, 2003: 2)

All the Nordic countries are seeking to reduce greenhouse gas emissions under their commitments to the Kyoto Protocol but also under other international protocols and agreements (see Case 11.2). For example, the EU has an aim to reduce its greenhouse gas emissions by at least 20% by 2020. Significantly, it is expected that aviation would also be brought into the EU's Emissions Trading Scheme (ETS) within that time period, an important step given that the EU is responsible for about half of the CO_2 emissions generated by international air transport in developed countries (European Commission, 2005), while tourism is specifically recognised by the European Commission in its climate change strategy (European Commission, 2007).

Cases and issues 11.2: Tourism mobility and climate change: The case of Sweden

STEFAN GÖSSLING

Tourism, mobility and climate change

Tourism is 'the social science of mobility' (Hall, 2005a). Clearly, all tourism is based on movement, involving different means of transport. While some forms of transport are generally environmentally sustainable, such as travel by bike, others are not. Air travel in particular can hardly ever be sustainable as it causes comparably large amounts of greenhouse gas (GHG) emissions, which are released at climatically sensitive altitudes. As average travel distances have continuously increased in recent years, while reductions in fuel use through technological progress have remained moderate, the overall amount of fuel used for tourism and recreation is growing, along with emissions. At the same time, there is scientific consensus that emissions of carbon dioxide (CO_2) need to be reduced to avoid 'dangerous climate change' (Schellnhuber *et al.*, 2006), that is, a corresponding average maximum warming of 2°C by 2100 (*cf.* Meinshausen, 2006). This demands considerable reductions in GHG emissions in the short- and medium-term future and consequently poses a substantial problem: transport sustainability is one of the greatest challenges for tourism (*cf.* Gössling *et al.*, 2002; Hall & Higham, 2005).

Swedish mobility in the context of the Kyoto Protocol

Two parameters affect the overall environmental impact of transport in terms of their contribution to climate change: (1) the distances travelled over a given period of time; and (2) the modes of transport used. Distances are measured in passenger kilometres (pkm), while emissions are expressed in kilograms of carbon dioxide equivalents (kg CO_2-e). The term 'equivalent' indicates that the contribution to climate change of gases other than CO_2 is considered in the calculation, which is of particular relevance for aviation (*cf.* Sausen *et al.*, 2005). CO_2-e emissions vary considerably between 0.01 kg CO_2-e/pkm for mopeds, 0.019 kg CO_2-e/pkm for coaches, 0.026 kg CO_2-e/pkm for trains, 0.079 kg CO_2-e/pkm for cars and 0.325 kg CO_2-e/pkm for aircraft (values for European transport; *cf.* Gössling *et al.*, 2005), even though there can also be considerable variation within each transport mode, primarily depending on the type and age of the engines used as well as occupancy rates. Aviation is, from a climate change perspective, the most environmentally harmful means of transport, both in comparison to other transport modes and in terms of its share (60–95%) of the overall emissions caused by an air travel-based vacation, including transport, accommodation and activities (Gössling *et al.*, 2005).

Swedes fly on average some 3000 pkm per year, which is about six times the annual per capita distance flown on global average. Swedish air travel contributes to emissions of about 8 million tons of CO_2-e per year, corresponding to 25% of all emissions of Swedish transport and 10% of overall Swedish emissions (Åkerman, 2006, personal communication). Total Swedish emissions of GHGs were in the order of 70 million t CO_2-e in 2002 or about 8 t per capita (excluding air travel outside Sweden). While emissions from Swedish aviation are expected to double within the coming 20 years (Åkerman, 2005), there is a juxtaposed obligation to reduce emissions to avoid 'dangerous interference with the climate system'. Should current growth within air travel continue unabated, even in a scenario of moderate growth (3% per year) air travel alone would account for 2.3 t CO_2-e per capita per year by 2050. Note that this scenario assumes that aircraft will become 40% more fuel-efficient (Åkerman, 2006, personal communication). Several questions arise from this conclusion. First, which levels of leisure transport are, in the future, still acceptable in Sweden, and how will this affect travel behaviour? Second, how can sustainable levels of tourism mobility be achieved? And, finally, what consequences will this have for the tourism industry?

Swedish leisure mobility: A climate change perspective

The vast majority of scientists agree that climate change is a serious threat for the economic systems societies depend upon, not last for tourism (*cf.* Gössling & Hall, 2006a). For Sweden, emission reductions of −4% by 2012 have been voluntarily agreed upon in the context of the Kyoto Protocol (base year: 1990), with considerably larger emission reductions of −60% needed by 2050 (*cf.* Tyndall Centre for Climate Change, 2005). Assuming that reductions would be equally shared among all human consumption sectors (food, housing and transport), the transport sector, which is currently responsible for about one-third of Swedish greenhouse gas emissions, should account for no more than 2.7 t of CO_2-e per capita per year by 2012. This corresponds to a travel distance of about 8000 pkm by aircraft, 34,000 pkm by car or more than 100,000 pkm by train. Note, however, that this includes all forms of transport, not only leisure mobility. From these calculations it becomes clear that one single flight to a medium distance destination (for example, Stockholm–Tunisia) would correspond to an entire year's travel budget, while it would be sustainable to travel more than 270 pkm per day by train. Once again, this points at the importance of favouring public transport over other means of transport such as cars and aircraft. The trend, however, is otherwise. Vacations are characterised by more frequent, usually shorter trips to more distant destinations, increasingly involving air travel.

As technological progress is not going to solve the problem (*cf.* Gössling & Peeters, 2007), while individual behaviour continues to develop towards more energy-intense lifestyles, options to achieve sustainable mobility patterns are limited, including taxation, emission trading and, potentially, incentives rewarding pro-environmental choices. One of the reasons why air travel, the most problematic form of transport, is comparably cheap is that kerosene is not taxed internationally. This leads to huge differences in fuel costs for different means of transport. In Sweden, as of early 2006, the cost of one litre of kerosene was four Swedish Crowns (about €0.44), while the price of fuel for cars fluctuated around 11–12 Swedish Crowns (about €1.2–1.3). One important step in the direction of more sustainable transport would thus be to tax all transport modes evenly. In this context, it is also important to note that aviation is not part of the Kyoto Protocol, and emissions remain unaccounted for. Inclusion of aviation in the Kyoto Protocol is likely by 2013. By then, however, emissions from this sector will have grown by 25% in comparison to 2006 (Åkerman, 2006, personal communication). Should aviation become part of the emission trading system, another problem is that this only foresees CO_2, even though the more relevant aspect of aviation might be its contribution to climate change through other, non-carbon gasses. Potentially, the inclusion of aviation in the Kyoto Protocol as currently planned could thus paradoxically increase rather than decrease its contribution to climate change (Lee & Sausen, 2000), unless the emission trading system is adjusted for non-carbon emissions.

Taxation and emission trading would increase the costs of travel, and could, potentially, reduce its environmental impact. However, air travellers are known to not be price-elastic, this is, higher prices are not likely to influence travel behaviour, unless charges become substantially higher (*cf.* Brons et al., 2002). Per capita carbon contingents and individual emission trading have thus been suggested as measures to develop sustainable mobility patterns (e.g. UK Commission on Sustainability, 2006). In such a system, each citizen of a given society would have a fixed individual annual carbon budget, calculated in accordance with the Kyoto Protocol. Anyone exceeding this budget would have to buy emission rights from someone else using less than his/her own budget. Such a system would keep emissions within the margins of sustainability, while having the advantage of being just and 'visualising' the individual limits of consumption.

While these systems are built on regulations to reduce mobility, forcing people to change their travel behaviour through increasing costs, there are as yet few incentives to change travel behaviour. For instance, people using a bicycle to get to work do not receive compensation in Sweden, while anyone using a car is rewarded with a compensation of 1.6 Swedish Crowns (€0.18) per kilometre (in 2006). Similarly, within the current transport system, it is often cheaper to travel by aircraft than by train. Clearly, hidden and open subsidies thus have to be re-adjusted in order to create incentives for pro-environmental travel choices.

Changing travel patterns: Changing tourism?

Structural adjustments in mobility will obviously influence tourism in various ways. For instance, the Norwegian consultancies Aniara Markedsanalyser and Mimir AS recently recommended to develop Fly & Drive tourism (Enger & Jervan, 2005b), that is, to increase the number of international travellers flying to Norway in order to drive around with rented cars. Likewise, a recent Swedish study has attempted to show the great potential of low-fare air travel for otherwise often marginal destinations (Turismens utredningsinstitut, 2006). Should climate change be taken seriously, such forms of tourism would hardly have a future. Industry lobbyists have thus sought to demonstrate that controlling the growth of aviation will affect employment and the growth of national economies. However, it seems likely that the growth in aviation entails job losses in other traffic sectors (e.g. ferries, railways), while tourist travel abroad leads to leakages of financial resources. Friends of the Earth UK (2005), for instance, calculated that spending abroad by UK residents resulted in net losses for the UK economy of some €22 billion in 2004. While this only addresses the economic side of air travel, it is clear that current aviation-based patterns of tourism have evolved in less than 50 years, while the aviation-based mass tourism phenomenon has developed in just 15 years. New directions in tourism are thus needed in order to address climate change in the context of changing mobility lifestyles.

Under the so-called 20/20/20 strategy the EU has set the unilateral target to cut its greenhouse gas emissions by 20% by 2020 compared to 1990 levels. The European Council also agreed that developed countries should commit to collectively cutting their emissions by about 30% by 2020, compared to 1990 levels, as part of an international agreement, and by 60–80% by 2050. The Council therefore supported a 30% cut in the EU's emissions by 2020, provided that such an international agreement is concluded. With its action plan on energy policy for the period 2007–2009, the European Council supported the following goals:

- to improve energy efficiency to save 20% of the EU's energy consumption compared to forecasts for 2020;
- to raise the share of renewable energy to 20% of EU overall energy consumption by 2020;
- to raise the share of biofuels to at least 10% of total petrol and diesel consumption for transport in the EU by 2020 (European Commission, 2007).

Implementation of the policies will lead to new energy targets for the Nordic countries and industries such as tourism. Under the proposal Nordic member states' greenhouse gas emission limits by 2020 compared to 2005 will need to drop 20% for Denmark, 16% for Finland and 17% for Sweden (European Commission, 2008a), while the target share of energy from renewable sources will need to increase to 30% by 2020 for Denmark (from 17% in 2005), to 38% for Finland (from 28.5%), and to 49% for Sweden (from 39.8%) (European Commission, 2008b). In terms of the non-EU Nordic states Norway has set a target to be 'carbon neutral' by 2030, which is 20 years earlier than its previous target. In 2007 the Norwegian government stated that Norway would aim to cut net emissions of CO_2 to zero by 2050 by reducing emissions at home and investing abroad in environmental projects that will give Norway CO_2 reduction credits. Tourism related initiatives include investment in mass transport and reduction of emissions in the transport sector plus additional taxes on diesel and petrol (Office of the Prime Minister, 2008). Under Iceland's obligation to the Kyoto Protocol Iceland is committed in the period from 2008 to 2012 not to increase emissions more than 10% from 1990 levels. However, although Iceland has substantial amounts of renewable energy, longer term commitments are embroiled within broader debates over energy projects and smelter development, as well as the significance of transport to the Iceland economy. Nevertheless, the long term vision of the Icelandic government for the reduction of net emissions of greenhouse gases by 50–75% until the year 2050, using 1990 emissions figures as the baseline (Ministry for the Environment, 2007).

National goals clearly set significant challenges for the tourism industry, especially given its dependence on transport (see Case 11.2). However, there is considerable variance within the tourism industry with respect to making businesses and public agencies more sustainable. Case 11.3 looks at such issues in the Swedish context as an example of the broader issues facing the development of sustainable tourism businesses northern wide.

Cases and issues 11.3: Climate change responses of Swedish tourism actors: An analysis of actor websites

STEFAN GÖSSLING and C. MICHAEL HALL

Key actors in Swedish tourism include transport companies, tour operators, accommodation and public and private sector tourism organisations. In order to understand how these key actors in Swedish tourism currently consider emissions and climate change, a total of 19 websites were evaluated in September 2007. Websites were searched for information relating to climate change or other aspects of the environment, including annual reports, environmental and sustainability reports, environmental and sustainability policies and related documents. Whenever such documents were found, these were analysed based on criteria, related to the position taken with regard to climate change and action taken to address the problem. Based on the results of the evaluation, tourism actors were then ranked with regard to two parameters: (1) awareness versus ignorance of climate change as represented by the extent and nature information available on the website; and (2) responsibility versus irresponsibility with respect to climate change as indicated by offering opportunities to customers to reduce their travel impact, for example, by making carbon offsets available or by utilising accredited sustainable business programmes. The basic assumption for this approach is that there is now global awareness of climate change, as there has been extensive coverage of the topic by international and national media following the release of the 4th Assessment Report (AR4) by the International Panel on Climate Change (IPCC) in February to May 2007, as well as substantial media coverage of the Stern Review (Stern, 2006). Moreover, the Swedish government's initiative to curb emissions by 20% on the EU level by 2020 has received broad media coverage in Sweden.

Table 11.1 seeks to chart the various ways in which Swedish tourism actors represent their perspectives on climate change and environmentally sustainable tourism issues to their consumers. These are assessed on the basis of two continuums with respect to these issues (awareness/ignorance; responsibility/irresponsibility). If actors choose to ignore climate change in their public representations this may be interpreted as corresponding to ignorance. Consequently, actors not mentioning climate change or environmental issues at all in publicly available material are described as 'ignorant'. In contrast, actors can seek to actively acquire information on the issue and represent them to their customers. Where this results in public statements regarding climate change, this would mirror 'awareness'. 'Responsibility', the second parameter, would then reflect actors' behaviour and actions with regard to climate change, that is, whether steps are taken to mitigate the contribution to climate change within the actor's field of activity. 'Irresponsibility', on the other hand, would correspond to inaction, denial of climate change or its seriousness, or the spreading of false information

Table 11.1 Climate change and key actors in Swedish tourism

Company	Awareness	Responsibility
Transport		
Scandinavian Airlines	+/−	−
Malmö Aviation	−−	−−
Swedish Railways (SJ)	++	++
Tour operators		
My Travel	+	+/−
Apollo	+	+/−
Fritidsresor	++	+
Scandorama	−−	−−
Novasol	−−	−−
Sembo	−−	−−
Accommodation		
Choice Hotels	−−	−−
Scandic Hotels	++	++
Radisson Hotels	++	+
Rica Hotels	+	+
Svenska Turistföreningen	+	+
Svenska Möten	+/−	+/−
Organisations		
NUTEK	+	−
Visit Sweden	−	−
Svenska Hotell & Restaurang företagare	+	+/−
Swedish Civil Aviation Authority	++	−

Notes: − no information provided (awareness) or no action taken (responsibility).
++ detailed information provided on website (awareness) or strong action taken (responsibility).

running counter to accepted scientific findings. The ranking of the two parameters was expressed in between −− that is, double minus, to ++, that is, double plus in order to indicate the actors' relative position in the awareness-responsibility continuum. It is important to note that the assessment is not of their actual emissions and impacts on climate change, but on their environmental sustainability reporting with respect to climate change. Note that the assessment only considers information contained on the actors' respective websites in September, 2007, and will almost certainly be different to an analysis conducted while you are reading this chapter. The results of the analysis are indicated in Table 11.1 while the remainder of the case reviews information provided on websites and seeks to arrive at an assessment of actual emissions and contribution to climate change.

Scandinavian Airlines (SAS) (http://www.sasgroup.net/)

The airline is the largest in Scandinavia and makes a considerable contribution to overall emissions from Swedish tourism. SAS acknowledges its contribution to climate change, and publishes an annual sustainability report. The company also provides an emissions calculator on its website and offers climate compensation to its customers. While this is positive, there were also several weaknesses in SAS' environmental sustainability strategy, such as the fact that SAS maintained a relatively old fleet of aircraft (average emissions: 0.134 kg CO_2/rkm) (Scandinavian Airlines, 2007); and had a communication strategy that played down the role of aviation in climate change. Voluntary offsetting was available to customers. The 126-page sustainability report (Scandinavian Airlines, 2007) contained three pages on environmental sustainability, emphasising that technology will solve the problem. With emissions of 6.2 Mt CO_2 in 2006 (Scandinavian Airlines, 2007), SAS is likely to be the most important single contributor to tourism-related emissions in Sweden.

Malmö Aviation (http://www.malmoaviation.se/)

The website contains a short environmental policy (Malmö Aviation, 2007). Climate change was not mentioned anywhere on the website. The company focuses on volume growth, offering low fares and rewarding frequent flyers. Emissions are not stated by the company, but are estimated here to be in the order of 0.14 Mt CO_2 (1.2 million passengers at 0.150 kg CO_2/pkm times an estimated average flight distance of 400 km).

SJ (Swedish Railways) (http://www.sj.se/)

The company has taken a very proactive stance to climate change. It provides customers with an emissions calculator (Swedish Railways, 2007), comparing emissions of various means of transport. Although this could be interpreted as a sales argument, SJ has made a decision to only use electricity from renewable sources, while seeking to constantly reduce its energy use. The company's environmental strategy goes beyond energy throughput, and also includes issues such as recycling on board trains as well as of the trains themselves. The company has an environmental management system (ISO 14001), and seeks to constantly reduce its impact on the environment. SJ's overall contribution to tourism related climate change is also relatively small, due to the use of renewable energy.

MyTravel (http://www.mytravel.com/)

The company informed customers that every flight contributes to climate change and offers carbon compensation (MyTravel, 2007). Carbon compensation focused on tree planting, which although a relatively common approach is regarded as a debatable strategy to reduce emissions in the longer term (Gössling *et al.*, 2007). The company also added that it was seeking to increase the fuel-efficiency of its aircraft, but there was no further information provided on the website. MyTravel referred customers to the emissions calculator provided by the offset provider Climate Care. There was no information on emissions, but these were estimated to be in the order of 0.4 Mt CO_2 per annum (calculated on the basis of 15,000 passengers/week, emissions of 0.130 kg/CO_2 (high load factor) and an average one-way flight distance of 2000 km).

Apollo (http://www.apollo.se/)

The company informed consumers about climate change and offers climate compensation (Apollo, 2007). Offsetting projects focused on forestry. There is also general information that Apollo operates a new fleet of aircraft, and seeks to achieve a high load factor. The company has 400,000 customers per year. No information on emissions is made available, though is here estimated at 0.3 Mt CO_2 per annum (calculated on the basis of 400,000 passengers × 0.13 kg CO_2 and an average flight distance of 3000 km).

Fritidsresor (http://www.fritidsresor.se/)

The company made a very clear statement that flying cannot be positive for the environment (Fritidsresor, 2007). It informed customers about the advantage of charter flights, which offer higher load factors and direct flights. There was also information on a range of measures taken to reduce energy use. No information on the age of the fleet was provided. According to the website, the company can save 2% of fuel use through winglets, which corresponds to 200–300 t fuel/year. Based on this information, it is assumed that total emissions are in the order of 0.04 Mt CO_2 per year. The company also offered climate compensation through atmosfair, as well as Climate Care (which according to Gössling *et al.* (2007) follows a more dubious approach to offsetting). Fritidsresor also offered customers the opportunity to go by train to Italy to avoid flights.

Sembo (http://www.sembo.se)

No information on climate change or environmental issues was contained on the website. The company sends about 120,000 travellers around the world each year, mostly on self-organised trips. Emissions are therefore also difficult to assess, and are here assumed to be in the order of 0.03 Mt CO_2 per year (at the assumed world average of 250 kg CO_2 per trip).

Scandorama (http://www.scandorama.se)

No information on climate change or environmental issues was contained on the website. However, as the company mainly organises journeys by bus, it is assumed that its overall contribution to climate change is moderate.

Novasol (http://www.novasol.se)
No information on climate change or environmental issues was contained on the website. As the company's focus is the letting of summerhouses it is assumed that its overall contribution to climate change is limited.

Choice Hotels (http://www.choicehotels.no)
No information on climate change or environmental issues was contained on the website. At the time of the survey the company had about 150 hotels in Scandinavia. At an assumed occupancy rate of 65% and 100 rooms per hotel, emissions would amount to about 0.02 Mt CO_2 per year (at 60 kWh per guestnight).

Scandic Hotels (http://www.scandic-hotels.se/)
The chain has put down considerable effort into environmental management and is regarded as one of the leading sustainable tourism businesses in the Nordic region. As early as 1994, Scandic hotels acknowledged their responsibility and began to re-structure their hotels towards sustainability (Scandic Hotels, 2007). Today, all hotels are certified with the Nordic Swan, an environmental certification widely used in the Nordic countries, and they focus on energy-/water reduction, recycling, environmentally friendly constructions and refurbishment, ecological food and waste minimisation. The company causes an estimated 0.04 Mt of CO_2 emissions per year (own calculation based on Bohdanovicz & Martinac, 2007).

Radisson Hotels (http://www.radisson.com)
There was no information on climate change or environmental issues contained on the Swedish website of Radisson Hotels. However, the Rezidor group's annual report contained a detailed section on energy use, as well as the use of fresh water and waste management. The group sought to make meetings carbon neutral through cooperation with The Carbon Neutral Company. Overall, the group emitted 0.233 Mt CO_2 in 2005, according to their own assessment (Rezidor, 2007). Various individual hotels are also environmentally certified.

Rica Hotels (http://www.rica.no/)
The chain provides some information about environmental initiatives (Rica Hotels, 2007). This includes demands on suppliers (environmental policy, the use of environmentally certified products and the reduction of non-renewable package), recycling, reduced paper use, energy use, water use, and the encouragement of guests to use public transport. Information provided on the website is not detailed, though. The company has 90 hotels in Norway and Sweden. At an occupancy rate of 65% and 80 rooms per hotel, emissions would amount to 0.001 Mt CO_2 per year (at 60 kWh per guest night).

Svenska Turistföreningen (http://www.svenskaturistforeningen.se)
The organisation manages 315 youth hostels, eight mountain stations and 40 mountain huts. There is a recent focus on environmental issues in the annual

report, with the goal to increase the number of youth hostels that are environmentally certified. Generally, the focus is more on local environmental issues than global climate change. Energy use by the organisation for running youth hostels is assumed to be low.

NUTEK (http://www.nutek.se/)

The Swedish Agency for Economic and Regional Growth is the government agency responsible for enabling tourism development. The website contained no document on climate change and/or other environmental issues in the context of tourism. The agency published several documents with regard to 'environment-driven economic development', but no reference is made to tourism in these documents. One of its documents, *Facts on Swedish Tourism* (NUTEK, 2007) contains a small section indicating that regulation in tourism is necessary to meet environmental challenges. As an organisation, NUTEK accounts for rather small emissions, although its influence on development processes may potentially have a substantial impact on emissions in terms of the broader tourism and economic development system. Given the Swedish government's stated intention to reduce emissions and make Sweden more environmentally sustainable NUTEK can potentially have a significant role in encouraging low-emissions development. However, its role, plus that of other government agencies, will also be dependent on the political willingness of governments to actually implement their policies.

Visit Sweden (http://www.visitsweden.com/)

The organisation is responsible for the promotion of Sweden as a destination and has thus considerable influence on the targeting and development of markets. Climate change or other environmental issues are not mentioned on the website, but there are documents showing that Visit Sweden seeks to actively develop Asian and other long-distance markets, that is, those that will contribute to the major growth in emissions in the future (air travel, hotel accommodation).

Svenska Möten (http://www.konferensportalen.se/)

The website contained a very short 'environment' section, where it was stated that the company is environmentally certified (no name of the label is provided, nor any explanation of what this means in terms of action taken). Svenska Möten operates 100 conference locations, but given the information available it is impossible to estimate the amount of emissions caused by these.

Sveriges Hotell & Restaurang Företagare (SHR) (http://www.shr.se/)

The organisation asks its members to work 'environmentally consciously'. For members, there is an 'environmental guide' and an internet-based environmental programme for hotels and restaurants (SHR, 2007). The organisation also supports environmental certification, encouraging members, however, to first 'work towards their own environmental certification', as environmental labels are 'costly'.

Swedish Civil Aviation Authority (http://www.luftfartsstyrelsen.se/)

The website contained very detailed information on the impact of flying on the environment, but some of this information appeared to play down the importance of aviation as a contributor to climate change (Swedish Civil Aviation Authority, 2007). On the basis of the information provided it appears that there is no proactive work being done by the organisation to reduce the impact of aviation on the environment; rather, new fuels and the EU ETS are identified as solutions to lower emissions.

One of the key points to make in the website analysis is that there is substantial variability between tourism actors, whether public or private, with respect to their reporting on emissions and other aspects of environmental sustainability related to climate change. Very few organisations actually state what their contribution to emissions are. In many cases of course they may not have even sought to measure it. However, such a situation is problematic as, in some cases, the consumer is unable to find relevant environmental information or, if they wish to, seek to off-set or reduce their emissions. Moreover, it raises substantial questions about how possible will it be to actually reduce the emissions contribution of the tourism sector and achieve government policy unless such studies are conducted and then made publicly available. Finally, the lack of information raises yet further issues as to how 'green' Swedish tourism can actually be with respect to how it is represented in marketing and promotion unless there is actions and information of real substance to support the Swedish destination brand. While the tourism industry is clearly more environmentally aware than what it was in the 1990s, it is readily apparent that for all the discussion of sustainability and sustainable tourism development, there is still a considerable distance to go before individual tourism firms or organisations sustainable commitment can be fully evaluated.

Figure 11.1 is an attempt to relate emissions estimates to the representations of actors with respect to climate change by considering the relative emissions contribution of actors (size of bullets). It must be emphasised that these contributions are estimates only based on available information which, as indicated above, is often minimal. The figure indicates, encouragingly, that many tourism industry actors are aware and working with sustainability. However, only two could be identified who reported absolute reductions in emissions in their operations: Scandic Hotels and SJ. For most key actors in Swedish tourism, the business reality may instead be that there is a focus on volume growth with relative improvements in emission levels. Consequently, while their efficiency (and relative sustainability) may improve, overall emissions will still grow (as in the case of Scandinavian Airlines), thereby creating considerable challenges for the Swedish government to meet its emissions targets. It is also important to note that there is a range of actors who appear from their web site information as uninterested in the problem of sustainable tourism, supporting growth of highly energy-intense forms of tourism and/or not

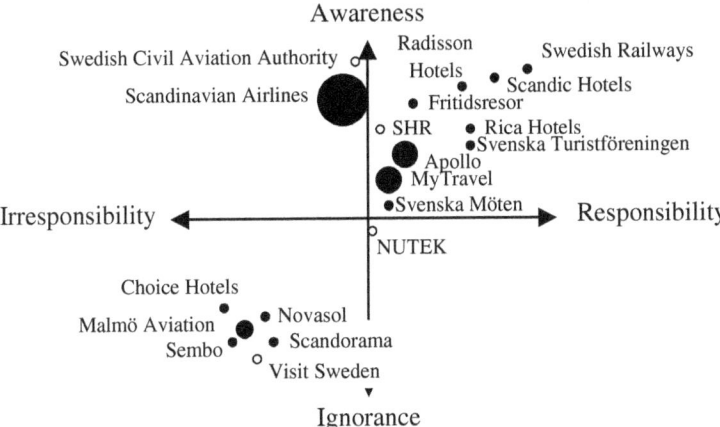

Figure 11.1 The awareness-responsibility spectrum of key Swedish tourism actors
Note: Bullet size is indicative of contribution to climate change. Bullets with white interior represent organisations without considerable emissions, but eventually large influence on tourism development.

acknowledging the emissions implications of the markets they seek to develop. Examples include Visit Sweden's focus on long-distance markets, or Turismen's utredningsinstitut (2006) report 'Low-fare air travel: a revolutionary opportunity for Swedish tourism' (authors' translation). This is not unique for Sweden: in Norway, fly and drive tourism is recommended as an innovation strategy (Enger & Jervan, 2006) (see Case 11.2), and Iceland Air is planning to annually fly 60,000 Chinese visitors to Iceland (Huijbens, personal communication, 2007). Of course the focus on growth represents 'business as usual', but the challenge for the Nordic tourism industry is whether such an approach is going to be possible in the longer-term.

Conclusions

This chapter has primarily focused on two main issues relating to the future of Nordic tourism. First, changing notions of Nordicity and regionalism. Second, environmental change, and climate change in particular. Both of these issues substantially affect the way in which Nordic tourism destinations are socially constructed by the tourist as well as having very real influence on the social and physical environment of Nordic communities and countries. Technological and economic change, or even political change, will also be significant for Nordic tourism. Indeed, it is perhaps ironic that at the same time attempts are being made to strengthen the environmental image of Sweden as a destination and reduce energy consumption and emissions so there are plans to develop space tourism out of Kiruna in the north of the country.

Yet the problem of tourism is that the decision to travel as well as the decision to accept visitors is the result of a vast range of forces many of which cannot be easily isolated, but which do provide the basis for seeking to articulate some sort of future vision for Nordic tourism. Undoubtedly, tourism does bring enormous economic benefits to the Nordic economies especially in the more peripheral regions, but that growth is being challenged by climate change concerns in the short term and by the increasing costs of energy for transport in the longer term. Arguably for a region who identity is grounded into being on the European, if not the world's, northern periphery the issues of distance, whether perceived or otherwise, and accessibility must become be a core issue for domestic and inbound tourism in the future (Hall, 2006d; Müller & Jansson, 2007a; Saarinen, 2005). Yet such a situation is potentially full or paradoxes and in exactly the same way that climate change may lead to winners as well as losers, for example, a warming of the climate and extension of the growing season is likely seen as positive by many Nordic communities.

In the case of the Nordic countries there is currently a substantial travel deficit in the sense that there are far more travellers leaving the Nordic countries than arriving. This then leads to a significant negative terms of tourism trade which represents a substantial loss from the Nordic economy (Table 11.2). If air travel became more restricted it is quite possible that more people would travel for holidays within the Nordic countries rather than fly to Spain or Portugal, particularly given the strength of the Nordic rail network. Thereby this may prove quite beneficial to the regional economy, as well as for the Nordic destinations that they visit, although long distance island destinations such as Iceland, Greenland and the Faroes may have greater challenges than mainland or near-shore destinations. Indeed, a broader issue is the extent to which global warming some destinations, such as the Caribbean or the Mediterranean may get too warm, therefore making the Nordic countries more attractive in summer as well as in winter (as even with global warming snow would be available in many

Table 11.2 Balance of tourism trade (tourism receipts and expenditure) 2005 (million euro)

Country	Credit	Debit	Net
Denmark	3977	5195	−1219
Finland	1757	2460	−704
Iceland	331	788	−458
Norway	2641	7841	−5200
Sweden	5957	8667	−2710
Total	14,663	24,951	−10,291

Note: Figures are subject to rounding.
Source: EuroStat 2007.

northern ski resorts for much of this century, although the season may be shortened) (Gössling & Hall, 2006a, 2006b). Therefore, perhaps a key insight is that the future of tourism in the Nordic region may be about to become more turbulent and uncertain.

The fact that there is no tourism without being able to get to the destination is one of the great truisms of tourism. Yet it is one that is often forgotten by business and industry. Destinations have rose and fell as a result of changes in access and the development of intervening opportunities for visitors and new competitors. The greatest challenge facing Nordic destinations in the immediate future is therefore to ensure that they remain accessible, visible, and yet desirable places to visit.

Self-review questions

(1) How has the concept of Nordicity changed in recent years?
(2) How might climate change affect tourism in the Arctic region?
(3) To what extent does Nordic tourism fulfil a green image?

Essay questions

(1) Conduct your own analysis of the sustainability and climate change reporting of a selection of tourism organisations and countries. Answer some or all of the following questions: What information is provided to the customer? What is not provided? Are customers able to carbon-offset? What marketing advantages may be gained from promoting 'green tourism credentials'?
(2) What are the greatest challenges facing Nordic tourism over the next 10 years?

Key readings

A 2003 special issue of *Geopolitics* (8 (1)) has a collection of papers on northern Europe presents a number of critical accounts of concepts of northernness and nordicity within the context of region building, that is, Aalto *et al.* (2003). On climate change see ACIA (2004, 2005), Anisimov *et al.* (2007) and Gössling and Hall (2008).

 Carbon Neutral Norway: www.CarbonNeutralNorway.no
 SusNordic Gateway: Governance for Sustainable Development in the Nordic Region: http://www.sum.uio.no/susnordic/

References

Aalto, P., Dalby, S. and Harle, V. (2003) The critical geopolitics of Northern Europe: Identity politics unlimited. *Geopolitics* 8 (1), 5–19.
ACIA (Arctic Climate Impact Assessment) (2004) *Impacts of a Warming Arctic: Arctic Climate Impact Assessment*. Cambridge: Cambridge University Press.
ACIA (2005) *Arctic Climate Impact Assessment*. Cambridge: Cambridge University Press.
Åkerman, J. (2005) Sustainable air transport – On track in 2050. *Transportation Research Part D* 10, 111–126.
Ålands landskapsregering/Näringsavdelningen (2004) *Ålands turismstrategi 2003–2010*. On WWW at http://www.ls.aland.fi/composer/uppload/modules/publikationer/turists.3.pdf. Accessed 1.4.08.
Aldskogius, H. (1968) *Studier i Siljansområdets fritidshusbebyggelse*. Uppsala: Geografiska institutionen.
Aldskogius, H. (1969) *Modelling the Evolution of Settlement Patterns: Two Studies of Vacation House Settlement*. Uppsala: Department of Geography.
Aldskogius, H. (1993) Festivals and meets: The place of music in 'Summer Sweden'. *Geografiska Annaler* 75B, 55–72.
Alexandersson, K. and Hirsch, D. (2007) Hotellåret 2006. *Restauratören* 10, 10–11.
Amager Strandpark (2007) Amager Strand. On WWW at http//www.amager-strand.dk. Accessed 2.7.07.
American Marketing Association (2008) AMA adopts new definition of marketing. On WWW at http://www.marketingpower.com/content21257.php. Accessed 1.4.08.
Amos, J. (2007) Arctic summers ice-free 'by 2013'. *BBC News*, 12 December. On WWW at http://news.bbc.co.uk/2/hi/science/nature/7139797.stm. Accessed 1.4.08.
Andersen, J.H. (2001) Danmarks Turistråds brandingstrategi i relation til lokalt niveau i dansk turisme. Master thesis, Aalborg University. On WWW at www.ihis.aau.dk/edu/snf/turisme. Accessed 1.4.08.
Anderson, L.T. (1995) *Guidelines for Preparing Urban Plans*. Chicago: Planners Press.
Andersson, T. (2007) The tourist in the experience economy. *Scandinavian Journal of Hospitality and Tourism* 7 (1), 46–58.
Andersson, T., Carlsen, J. and Getz, D. (2002) Family business goals in the tourism and hospitality sector: A cross-case analysis from Australia, Canada and Sweden. *Family Business Review* 15 (2), 89–106.
Anholt, S. (1998) Nation-brands of the twenty-first century. *Journal of Brand Management* 5 (6), 395–404.

Anisimov, O.A., Vaughan, D.G., Callaghan, T.V., Furgal, C., Marchant, H., Prowse, T.D., Vilhjálmsson, H. and Walsh, J.E. (2007) Polar regions (Arctic and Antarctic). In M.L. Parry, O.F. Canziani, J.P. Palutikof, P.J. van der Linden and C.E. Hanson (eds) *Climate Change 2007: Impacts, Adaptation and Vulnerability. Contribution of Working Group II to the Fourth Assessment Report of the Intergovernmental Panel on Climate Change* (pp. 653–685). Cambridge: Cambridge University Press.

Ankre, R. (2005) Visitor activities and attitudes in coastal areas: A case study of the Luleå archipelago, Sweden. *Etour Working Paper 2005: 1*. Östersund: Etour.

Anonson, N. (1954) Kollektiv trafik till fritidsområdena. *Plan* 6 (8), 11–12.

Apollo (2007) *Gör din resa lite grönare*. On WWW at http://www.apollo.se/apollo/GreenSeat.htm. Accessed 18.9.07.

Arctic Council (2004) *Arctic Climate Impact Assessment Policy Document*. Issued by the Fourth Arctic Council Ministerial Meeting, Reykjavík, 24 November 2004.

Arlov, T.B. (1996) *Svalbards Historie*. Oslo: Aschehough.

Arnesen, T. (2007) Involvement in outfield land use policy formation, who and why? In I. Balberg and H. Hofstad (eds) *Authority, Responsibility and Justice in Environmental Politics*. Papers from the 8th Nordic Environmental Social Science Research Conference June 18–20 2007 (pp. 123–139). NIBR Working Paper 2007: 112. Oslo: Norwegian Institute for Urban and Regional Research.

Arnesen, T., Ericsson, B. and Flygind, S. (2002) Fritidsboliger i Norge: Lokaliseringsmønster 1970–2002. *Utmark* 2002 (3). On WWW at www.utmark.org. Accessed 1.4.08.

Aronsson, L. (1993) *Mötet: En studie i Smögen av turisters, fritidsboendes och bofastas använding av tid och rum*. Karlstad: Högskolan i Karlstad.

Aronsson, L. (1994) Sustainable tourism systems: The example of sustainable rural tourism in Sweden. *Journal of Sustainable Tourism* 2 (1–2), 77–92.

ÅSUB (Ålands statistik- och utredningsbyrå) (2004a) *Inkvarteringsstatistik 2003*, Statistik 2004:4. Mariehamn: Ålands Statistik och Utredningsbyrå.

ÅSUB (2004b) *Turismens samhällsekonomiska betydelse för Åland 2003*. Rapport 2004:2, Mariehamn: Ålands Statistik och Utredningsbyrå.

ÅSUB (2005) *2005 Statistical Yearbook of Åland/Statistisk årsbok för Åland*. ÅSUB. On WWW at http://www.asub.aland.fi/informer/img/asub2/arsbok05.pdf. Accessed 1.4.08.

ÅSUB (2006) Statistikmeddelande Inkvartering 2006:12. Mariehamn: Ålands statistik – och utredningsbyrå.

Ashworth, G.J. and Tunbridge, J.E. (1999) Old cities, new pasts: Heritage planning in selected cities of Central Europe. *GeoJournal* 49, 105–116.

Ashworth, G.J. and Tunbridge, J. (2000) *The Tourist-Historic City: Retrospect and Prospect of Managing the Heritage City*. Oxford: Pergamon.

Ashworth, G.J. and Voogd, H. (1990) *Selling the City: Marketing Approaches in Public Sector Urban Planning*. London: Belhaven Press.

ÅTF (2004) *Verksamhetsberättelse 2003*. Mariehamn: Ålands Turistförbund.

Bagrow, L. (1985) *History of Cartography* (revised and enlarged by R.A. Skelton). Chicago: Precedent Publishing.

Baldersheim, H. and Ståhlberg, K. (2003) Guided democracy or multi-level governance? New trends in regulatory regimes in central-local relations in the Nordic countries. In P. Carmichael and A. Midwinter (eds) *Regulating Local Authorities: Emerging Patterns of Central Control* (pp. 74–90). London: Frank Cass.

Baum, T. and Twining-Ward, L. (1997) Utmaningar för turismen på Åland. *Radar* 1/1997, 5–12.

Bayliss, D. (2004) Denmark's creative potential: The role of culture within Danish urban development strategies. *International Journal of Cultural Policy* 10, 5–28.

Bayliss, D. (2007) The rise of the creative city: Culture and creativity in Copenhagen. *European Planning Studies* 15 (7), 889–903.

BBC (2007) Warming 'Opens Northwest Passage', *BBC News, Americas*, 14 September. On WWW at http://news.bbc.co.uk/2/hi/americas/6995999.stm. Accessed 1.4.08.

Beck, U. and Beck-Gernsheim, E. (2001) *Individualization*. London: Sage.

Berg, P. and Lofgren, O. (2000) Studying the birth of a transnational region. In P. Berg, A. Linde-Laursen and O. Löfgren (eds) *Invoking a Transnational Metropolis: The Making of the Øresund Region*. Lund: Studentlitteratur AB.

Berglund, C. (2005) *"Ja, vi elsker dette landet ... og": En studie kring norskägda fritidshus i svenska delen av Inre Skandinavien*. Arbetsrapport 2005: 15. Karlstad: CERUT.

Berry, S. and Ladkin, A. (1997) Sustainable tourism: A regional perspective. *Tourism Management* 18 (7), 433–440.

Bianchi, R. (2004) Tourism restructuring and the politics of sustainability: A critical view from the European periphery (The Canary Islands). *Journal of Sustainable Tourism* 12 (6), 495–529.

Bilefsky, D. (2006) Monster Band has Finland fretting over face it shows. *International Herald Tribune* (European edition). Monday 22 May. On WWW at http://www.iht.com/articles/2006/04/17/news/finn.php. Accessed 1.4.08.

Boden, M. and Miles, I. (2000) *Services and the Knowledge-Based Economy*. London: Continuum.

Bœrenholdt, J.O. and Haldrup, M. (2006) Mobile networks and place making in cultural tourism: Staging Viking ships and rock music in Roskilde. *European Urban and Regional Studies* 13, 209–224.

Bohdanowicz, P. and Martinac, I. (2007) Determinants and benchmarking of resource consumption in hotels – case study of Hilton International and Scandic in Europe. *Energy and Buildings* 39, 82–95.

Bohlin, M. (1982) *Fritidsboende i den regionala ekonomin*. Uppsala: Kulturgeografiska institutionen.

Boo, E. (1990) *Ecotourism: The Potentials and Pitfalls* (Vol. 1). Lancaster: WWF & U.S. Agency for International Development.

Bosmina (2006) Med Bosmina till Sandskär. *Bosmina*. On WWW at http://www.bosmina.bd.se. Accessed 24.11.06.

Bramwell, B. (ed.) (2003) *Coastal Mass Tourism: Diversification and Sustainable Development in Southern Europe*. Clevedon: Channel View Publications.

Bristow, G. (2005) Everyone's a 'Winner': Problematising the discourse of regional competitiveness. *Journal of Economic Geography* 5, 285–304.

Brons, M., Pels, E., Nijkamp, P. and Rietveld, P. (2002) Price elasticities of demand for passenger air travel: A meta-analysis. *Journal of Air Transport Management* 8 (3), 165–175.

Bülow U. (2006) Branding af Danmark – på den globale bane. *Børsen* 18.01.2006.

Bunce, M. (1994) *The Countryside Ideal: Anglo-American Images of Landscape*. London: Routledge.

Burns, P. (1999) Paradoxes in planning: Tourism elitism or brutalism? *Annals of Tourism Research* 26 (2), 329–348.

Burtenshaw, D., Bateman, M. and Ashworth, G.J. (1991) *The European City. A Western Perspective*. London: Wiley.

Butler, R.W. (1980) The concept of a tourist area cycle of evolution: Implications for management of resources. *The Canadian Geographer* 24 (1), 5–12.

Butler, R.W. (1999) Sustainable tourism: A state-of-the-art review. *Tourism Geographies* 1 (1), 7–25.

Butler, R.W. (ed.) (2006) *The Tourism Area Life Cycle* (2 vols). Clevedon: Channel View Publications.

Butler, R.W. and Boyd, S. (eds) (2000) *Tourism and National Parks: Issues and Implications*. Chichester: Wiley & Sons.

Butler, R.W. and Hall, C.M. (1998) Image and reimaging of rural areas. In R. Butler, C.M. Hall and J. Jenkins (eds) *Tourism and Recreation in Rural Areas* (pp. 251–258). Chichester: John Wiley.

Butler, R.W., Hall, C.M. and Jenkins, J.M. (eds) (1998) *Tourism and Recreation in Rural Areas*. Chichester: John Wiley.

Butler, R.W. and Hinch, T. (eds) (1996) *Tourism and Indigenous Peoples*. London: International Thomson Business Press.

Butler, R.W. and Hinch, T. (eds) (2007) *Tourism and Indigenous Peoples: Issues and Implications* (2nd edn). Amsterdam: Butterworth-Heinemann.

Campbell, L.M. (1999) Ecotourism in rural developing communities. *Annals of Tourism Research* 26 (3), 534–553.

Capek, K. (1995 [1936]) *En Reise til Norden (Cesta na Sever)*. Oslo: Ex Libris.

Cater, C. and Cater, E. (2007) *Marine Ecotourism: Between the Devil and the Deep Blue Sea*. Wallingford: CABI.

Cater, E. (1993) Ecotourism in the Third World: Problems for sustainable development. *Tourism Management* 14 (2), 85–90.

Cater, E. (1994) Introduction. In E. Cater and G. Lowman (eds) *Ecotourism: A Sustainable Option?* (pp. 3–17). New York: John Wiley & Sons.

Cater, E. (2003) Between the devil and the deep blue sea: Dilemmas for marine ecotourism. In B. Garrod and J.C. Wilson (eds) *Marine Ecotourism: Issues and Experiences* (pp. 37–47). Clevedon: Channel View Publications.

Cater, E. (2006) Adventure tourism: Will to power. In A. Church and T. Coles (eds) *Tourism, Power and Space* (pp. 63–82). London: Routledge.

Chalip, L. and McGuirty, J. (2004) Bundling sport events with the host destination. *Journal of Sport & Tourism* 9 (3), 267–282.

Chatterton, P. and Hollands, R. (2002) Theorising urban playscapes: Producing, regulating and consuming nightlife city spaces. *Urban Studies* 39 (1), 95–116.

Chatterton, P., Hollands, P. and Aubrey, M. (2002) *Youth Nightlife and Culture in Bristol*. Discussion Paper, Newcastle-upon-Tyne: University of Newcastle.

Chen, J.S. (1998) Travel motivation of heritage tourists. *Tourism Analysis* 2, 213–215.

Clawson, M. and Knetsch, J. (1966) *The Economics of Outdoor Recreation*. Baltimore: John Hopkins University Press.

Clement, K. (2004) Structural fund programmes as instruments for sustainable regional development a review of Nordic effectiveness. *Innovation: The European Journal of Social Sciences* 17 (1), 43–60.

Cloke, P. (1992) The countryside: Development, conservation and an increasingly marketable commodity. In P. Cloke (ed.) *Policy and Change in Thatcher's Britain* (pp. 61–88). Oxford: Pergamon Press.

Cloke, P. (ed.) (2003) *Country Visions*. Harlow: Pearson.

Cloke, P. and Perkins, H. (1998) "Cracking the canyon with the awesome foursome": Representation of adventure tourism in New Zealand. *Environment and Planning D: Society and Space* 16 (2), 185–218.

Cloke, P. and Little, J. (eds) (1997) *Contested Countryside Cultures*. London: Routledge.

Cohen, E. (1974) Who is a tourist? A conceptual clarification. *Sociological Review* 22, 527–555.

Coles, T. and Hall, C.M. (eds) (2008) *Tourism and International Business*. London: Routledge.

Coles, T. and Timothy, D. (eds) (2004) *Tourism and Diaspora*. London: Routledge.

Coles, T., Duval, D. and Hall, C.M. (2004) Tourism, mobility and global communities: New approaches to theorising tourism and tourist spaces. In W. Theobold (ed.) *Global Tourism* (3rd edn) (pp. 463–481). Oxford: Heinemann.

Coles, T., Hall, C.M. and Duval, D. (2005) Mobilising tourism: A post-disciplinary critique. *Tourism Recreation Research* 30 (2), 31–41.

Coles, T., Hall, C.M. and Duval, D. (2006) Tourism and post-disciplinary inquiry. *Current Issues in Tourism* 9 (4–5), 293–319.

Cooper, C. and Hall, C.M. (2008) *Contemporary Tourism: An International Approach*. Oxford: Butterworth-Heinemann.

Coppock, J.T. (ed.) (1977) *Second Homes: Curse or Blessing?* Oxford: Pergamon.

Council of Europe (1988) Rural tourism in Europe. *European Campaign for the Countryside* (Study No. 2). Strasbourg: Council of Europe.

Countryside Commission (1995) *Sustaining Rural Tourism*. Cheltenham: Countryside Commission.

Crouch, D. (1999) Introduction: Encounters in leisure/tourism. In D. Crouch (ed.) *Leisure/tourism Geographies: Practices and Geographical Knowledge* (pp. 1–16). London: Routledge.

Cruise Baltic (2006) Cruise Baltic: Northern Europe. On WWW at http://www.cruisebaltic.com. Accessed 11.10.06.
Cruise Lines International Association (CLIA) (2006) *CLIA 2006 Cruise Market Profile, Report of Findings*. Fort Lauderdale: CLIA.
Danmarks Turistråd (2000a) *Branding Danmark. Danmarks nye ansigt i verden*. København: Danmarks Turistråd.
Danmarks Turistråd, Turismens Fællesråd, Turismens Udviklingscenter (2000b) *Den tyske udfordring. Analyseresultater og anbefalinger*. København: Danmarks Turistråd.
Dansk Turismeudvikling I/S and Strange, M. (1997) *Nordjyllands Amts Turistpolitiske Handlingsprogram – Evaluering*. Dansk Turismeudvikling I/S and the County of Northern Jutland.
Danter, K.J., Griest, D.L, Mullins, G.W. and Norland, E. (2000) Organizational change as a component of ecosystem management. *Society and Natural Resources* 13, 537–547.
Davidson, R. (1981) Ski development in Scotland. *Scottish Geographical Magazine* 97 (2), 110–123.
Douglas, N. and Douglas, N. (2004) *The Cruise Experience. Global and Regional Issues in Cruising*. French's Forest: Pearson Education Australia.
Dowling, R.K. (2006) The cruising industry. In R.K. Dowling (ed.) *Cruise Ship Tourism* (pp. 1–17). Wallingford: CABI.
Drucker, P. (1973) *The Practice of Management*. London: Pan Books.
Duffy, R. (2002) *A Trip Too Far: Ecotourism, Politics and Exploitation*. London: Earthscan.
Duval, D. (2007) *Tourism and Transport*. Clevedon: Channel View Publications.
Duval-Smith, A. (2007) Blizzards, reindeer, darkness: New Klondike is hottest place in Europe. Shoppers from four countries flock to buy flatpacks as global warming ignites the Arctic economy. *The Observer*, Sunday 7 January 2007. On WWW at http://www.guardian.co.uk/environment/2007/jan/07/retail.theobserver. Accessed 1.4.08.
Dwyer, L. and Forsyth, P. (1996) Economic impacts of cruise tourism in Australia. *Journal of Tourism Studies* 7 (2), 36–43.
Dwyer, L., Forsyth, P. and Rao, P. (2000a) The price competitiveness of travel and tourism: A comparison of 19 Destinations. *Tourism Management* 21 (1), 9–22.
Dwyer, L., Forsyth, P. and Rao, P. (2000b) Sectoral analysis of destination price competitiveness: An international comparison. *Tourism Analysis* 5, 1–12.
Ek, R. (2005) Regional experiencescapes as geo-economic ammunition. In T. O'Dell (ed.) *Experiencescapes: Tourism, Culture and Economy* (pp. 69–89). Copenhagen: Copenhagen Business School Press.
Eligh, J., Welford, R. and Ytterhus, B. (2002) The production of sustainable tourism: Concepts and examples from Norway. *Sustainable Development* 10 (4), 223–234.
Elsasser, H. and Messerli, P. (2001) The vulnerability of the snow industry in the Swiss Alps. *Mountain Research and Development* 21 (4), 335–339.
Enberg, J. (2005) IKEAs etablering i Haparanda: Planerade och förväntade effekter för regionen Haparanda-Torneå. Unpublished thesis, Huddinge: Södertörn University College, School of Life Sciences.
Enger, A. and Jervan, B. (2006) *Fly & drive markedet for Norge. Kartlegging av barrierer og utviklingsmuligheter for økt utleie av biler til utenlandske turister, med vekt på sammenligninger av forholdene i Sverige og Finland*. Larvik/Jar, Mimir AS and Aniara Markedsanalyser for reiselivet.
Erämaakomitean mietintö (1988) Komiteamietintö 39. Helsinki.
Erämaalaki (1991) Suomen säädöskokoelma 1991 no 62, 129–131.
Ericsson, B., Arnesen, T. and Overvåg, K. (2005) *Fra hyttefolk til sekundærbosatte*. ØF-rapport 04/2005. Østlandsforskning: Lillehammer.
European Commission (1997) *Marketing Quality Rural Tourism*. Brussels: EU LEADER programme.
European Commission (2000) *Towards Quality Rural Tourism*. Brussels: Enterprise Directorate-General, Tourism Unit.

European Commission (2005) *Communication from the Commission to the Council, the European Parliament, the European Economic and Social Committee and the Committee of the Regions of 27 September 2005: Reducing the Climate Change Impact of Aviation.* COM (2005) 459. Brussels: Commission of the European Communities. Brussels: Commission of the European Communities.

European Commission (2007) *Communication from the Commission, of 10 January 2007, entitled: Limiting Global Climate Change to 2 Degrees Celsius – The Way Ahead for 2020 and Beyond.* COM(2007) 2. Brussels: Commission of the European Communities.

European Commission (2008a) *Proposal for a Decision of the European Parliament and of the Council on the Effort of Member States to Reduce their Greenhouse Gas Emissions to Meet the Community's Greenhouse Gas Emission Reduction commitments up to 2020.* COM (2008) 17 final, 2008/0014 (COD). Brussels: Commission of the European Communities.

European Commission (2008b) *Proposal for a Directive of the European Parliament and of the Council on the Promotion of the Use of Energy from Renewable Sources.* COM (2008) 19 final, 2008/0016 (COD). Brussels: Commission of the European Communities.

Eurostat (2007) *Inbound and Outbound Tourism in Europe*, 52/2007. Luxembourg: Eurostat.

Fagerlund, N. (1997) The special status of the Åland Islands in the European Union. In L. Hannikainen and F. Horn (eds) *Autonomy and Demilitarisation in International Law: The Åland Islands in Changing European Context* (pp. 189–256). The Hague: Kluwer Law International.

Fawcett, C. and Cormack, P. (2001) Guarding authenticity at literary tourism sites. *Annals of Tourism Research* 28, 686–704.

fDi (2006) European Cities of the Future 2006/7, *fDi*, 6 February. On WWW at http://www.fdimagazine.com/news/fullstory.php/aid/1543/EUROPEAN_CITIES_OF_THE_FUTURE_2006_07.html. Accessed 1.4.08.

Fennell, D.A. (1999) *Ecotourism: An Introduction.* London: Routledge.

Fennell, D.A. (2006) *Tourism Ethics.* Clevedon: Channel View Publications.

Fennell, D.A. and Eagles, P.F.J. (1990) Ecotourism in Costa Rica: A conceptual framework. *Journal of Park and Recreation Administration* 8 (1), 23–34.

Finnish Forest and Park Service (2008) Metsähallituksen kansallispuistojen käyntimäärät 2001–2007. On WWW at http://www.metsa.fi/page.asp?section=2847. Accessed 17.04.08.

Finnveden, B. (1969) Den dubbla bosättningen och sommarmigrationen: Exempel från Hallandskustens fritidsbebyggelse. *Svensk Geografisk Årsbok,* 36, 58–84.

Fischer, J. (2007) *Union of International Associations (UIA), International Meeting Statistics for the Year 2006.* Press Release. Brussels: UIA.

Fitje, A. (1999) Fjordene. In J.K.S. Jacobsen and A. Viken (eds) *Turisme: Stedet i en Bevegelig Verden* (pp. 176–182). Oslo: Universitetsforlaget.

Flagestad, A. and Hope, C.A. (2001) Strategic success in winter sports destinations: A sustainable value creation perspective. *Tourism Management* 22 (4), 445–461.

Fleischer, A. and Pizam, A. (1997) Rural tourism in Israel. *Tourism Management* 18 (6), 367–372.

Flognfeldt Jr, T. (2004) Second homes as a part of a new rural lifestyle in Norway. In C.M. Hall and D.K. Müller (eds) *Tourism, Mobility and Second Homes: Between Elite Landscape and Common Ground* (pp. 233–243). Clevedon: Channel View Publications.

Foss Hansen, H. (2003) *Evaluering i staten.* Frederiksberg: Samfundslitteratur.

Frändberg, L. and Vilhelmson, B. (2003) Personal mobility – a corporeal dimension of transnationalisation. The case of long-distance travel from Sweden. *Environment and Planning A* 35, 1751–1768.

Franklin, A. (2003) *Tourism: An Introduction.* London: Sage.

Fredman, P. and Heberlein, T. (2003) Changes in skiing and snowmobiling in Swedish mountains. *Annals of Tourism Research* 30 (2), 485–488.

Freimund, W.A. and Cole, D.N. (2001) Use density, visitor experience, and limiting recreational use in wilderness: Progress to date and research needs. In *Visitor Use Density and Wilderness Experience: Proceedings.* USDA Forest Service. RMRS-P-20.

Fridgen, J.D. (1984) Environmental psychology and tourism. *Annals of Tourism Research* 11, 19–40.
Friends of the Earth (2005) *Why Airport Expansion is Bad for Regional Economies*. On WWW at http://www.foe.co.uk/resource/briefings/regional_tourism_deficit.pdf. Accessed 18.8.05.
Fritidsresor (2007) *Reko resande/miljö*. On WWW at http://www.fritidsresor.se/tui.aspx?id=11760. Accessed 18.9.07.
Frost, W. and Hall, C.M. (eds) (2009) *Tourism and National Parks*. London: Routledge.
Gallent, N. and Tewdwr-Jones, M. (2000) *Rural Second Homes in Europe: Examining Housing Supply and Planning Control*. Aldershot: Ashgate.
Gallent, N., Mace, A. and Tewdwr-Jones, M. (2003) Dispelling a myth? Second homes in rural Wales. *Area* 35 (3), 271–284.
Gallent, N., Mace, A. and Tewdwr-Jones, M. (2005) *Second Homes: European Perspectives and UK Policies*. Aldershot: Ashgate.
Garcia, B. (2004) Urban regeneration, arts programming and major events. *International Journal of Cultural Policy* 10 (1), 103–118.
Garrod, B. and Wilson, J.C. (eds) (2003) *Marine Ecotourism: Issues and Experiences*. Clevedon: Channel View Publications.
Gertner, D. and Kotler, P. (2004) How can a place correct a negative image? *Place Branding* 1 (1), 50–57.
Gibson, L. (2006) Learning destinations – the complexity of tourism development. PhD dissertation, Faculty of Social and Life Sciences, Sociology. Karlstad University Studies 2006: 41.
Gill, A.M. (2000) From growth machine to growth management: The dynamics of resort development in Whistler, British Columbia. *Environment and Planning A* 32, 1083–1100.
GoÅ (Government of Åland) (2004) *Ålands turismstrategi*. On WWW at http://www.ls.aland.fi/naringsavd/allmanna/naringslivsutveckling.pbs. Accessed 23.1.06.
GoÅ (Government of Åland) (2006) *Åland in Brief*. On WWW at http://www.aland.fi/alandinbrief/. Accessed 30.1.06.
Godde, P., Price, M. and Zimmermann, F. (2000) Tourism development in mountain regions: Moving forward into the new millennium. In P. Godde, M. Price and F. Zimmermann (eds) *Tourism Development in Mountain Regions* (pp. 1–25). New York: CABI Publishing.
Goodwin, H. (1996) In pursuit of ecotourism. *Biodiversity and Conservation* 5 (3), 277–291.
Gössling, S. and Hall, C.M. (eds) (2006a) *Tourism and Global Environmental Change*. London: Routledge.
Gössling, S. and Hall, C.M. (2006b) An introduction to tourism and global environmental change. In S. Gössling and C.M. Hall (eds) *Tourism and Global Environmental Change* (pp. 1–33). London: Routledge.
Gössling, S. and Hall, C.M. (2008) Swedish tourism and climate change mitigation: An emerging conflict? *Scandinavian Journal of Hospitality and Tourism* 8 (2), 141–158.
Gössling, S. and Hultman, J. (eds) (2006) *Ecotourism in Scandinavia*. Wallingford: CAB International.
Gössling, S. and Mattsson, S. (2002) Farm tourism in Sweden: Structure, growth and characteristics. *Scandinavian Journal of Hospitality and Tourism* 2 (1), 17–30.
Gössling, S. and Peeters, P. (2007) 'It does not harm the environment!' – An analysis of discourses on tourism, air travel and the environment. *Journal of Sustainable Tourism* 15 (4), 402–417.
Gössling, S., Borgström-Hansson, C., Hörstmeier, O. and Saggel, S. (2002) Ecological footprint analysis as a tool to assess tourism sustainability. *Ecological Economics* 43 (2–3), 199–211.
Gössling, S., Broderick, J., Upham, P., Peeters, P., Strasdas, W., Ceron, J-P. and Dubois, G. (2007) Voluntary carbon offsetting schemes for aviation: Efficiency and credibility. *Journal of Sustainable Tourism* 15 (3), 223–248.
Gössling, S., Haglund, L., Källgren, H., Revahl, M. and Hultman, J. (2008) Swedish Air travellers and voluntary carbon offsets: Towards the co-creation of environmental value? *Current Issues in Tourism*, in press.

Gössling, S., Peeters, P.M., Ceron, J-P., Dubois, G., Patterson, T. and Richardson, R.B. (2005) The eco-efficiency of tourism. *Ecological Economics* 54, 417–434.

Grenier, A. (2004) *The Nature of Nature Tourism*. Acta Universitatis Lapponiesis, University of Lapland.

Grenier, A.A. (2007) The diversity of polar tourism: Some challenges facing the industry in Rovaniemi. *Polar Geography* 30 (1/2), 55–72.

Grumbine, E.R. (1994) What is ecosystem management? *Conservation Biology* 8 (1), 27–38.

Guðmundsson, R. (2005) *Norðurland: ferðamenn og gistinætur 2001–2004*. Reykjavík, Ísland: Rannsóknir og ráðgjöf ferðaþjónustunnar.

Guldbrandsen, C. (1997) A holistic approach to the evaluation of publicly assisted tourism projects in Greenland. In L. Lyck (ed.) *Socio-economic Developments in Greenland and in Other Small Nordic Jurisdictions*. Revised conference papers of a meeting held in January, 1996, sponsored by Nordic Arctic Research Forum. Frederiksberg: Forlaget Samfundslitteratur.

Gummesson, E. (1994) Making relationship marketing operational. *International Journal of Service Industry Management* 5 (5), 5–20.

Gunn, C.A. with Var, T. (2002) *Tourism Planning: Basics, Concepts, Cases* (4th edn). New York: Routledge.

Gunnarsdóttir, G. (2005) History and horses: The potential of destination marketing in a rural community. Unpublished MBA thesis, University of Guelph.

Gyimóthy, S. and Mykletun, R. (2004) Play in adventure tourism: The case of Arctic trekking. *Annals of Tourism Research* 31 (4), 855–878.

Hägerstrand, T. (1954) Somamrflyttningen från sydsvenska städer. *Plan* 6 (8), 3–9.

Håkansson, H. and Waluszewski, A. (2005) Developing a new understanding of markets: Reinterpreting the 4Ps. *Journal of Business & Industrial Marketing* 20 (3), 110–117.

Haldrup, M. (2004) Laid-back mobilities: Second-home holidays in time and space. *Tourism Geographies* 6, 434–454.

Haldrup, M. and Larsen, J. (2003) The family gaze. *Tourist Studies* 3, 23–46.

Haldrup, M. and Larsen, J. (2006) Material cultures of tourism. *Leisure Studies* 25 (3), 275–289.

Halewood, C. and Hannam, K. (2001) Viking heritage tourism: Authenticity and commodification. *Annals of Tourism Research* 28 (3), 565–580.

Halkier, H. and Therkelsen, A. (2004) *Umbrella Place Branding. A Study of Friendly Exoticism and Exotic Friendliness in Coordinated National Tourism and Investment Promotion*. SPIRIT Discussion Paper 26, Aalborg University.

Hall, C.M. (1992) *Wasteland to World Heritage: Preserving Australia's Wilderness*. Melbourne: Melbourne University Press.

Hall, C.M. (1998) Historical antecedents of sustainable development and ecotourism: New labels on old bottles? In C.M. Hall and A.A. Lew (eds) *Sustainable Tourism: Geographical Perspective* (pp. 13–24). Longman: New York.

Hall, C.M. (2002) Tourism in capital cities. *Tourism: An International Interdisciplinary Journal* 50 (3), 235–248.

Hall, C.M. (2003) *Tourism and Temporary Mobility: Circulation, Diaspora, Migration, Nomadism, Sojourning, Travel, Transport and Home*. Paper presented at International Academy for the Study of Tourism (IAST) Conference, 30 June–5 July 2003, Savonlinna, Finland.

Hall, C.M. (2005a) *Tourism: Rethinking the Social Science of Mobility*. Harlow: Prentice-Hall.

Hall, C.M. (2005b) Reconsidering the geography of tourism and contemporary mobility. *Geographical Research* 43 (2), 125–139.

Hall, C.M. (2006a) Policy, planning and governance in ecotourism. In S. Gössling and J. Hultman (eds) *Ecotourism in Scandinavia* (pp. 193–206). Wallingford: CAB International.

Hall, C.M. (2006b) Space-time accessibility and the TALC: The role of geographies of spatial interaction and mobility in contributing to an improved understanding of tourism. In R.W. Butler (ed.) *The Tourism Area Life Cycle, Vol. 2, Conceptual and Theoretical Issues* (pp. 83–100). Clevedon: Channel View Publications.

Hall, C.M. (2007a) National tourism competitiveness indices: Or, why don't we just be done with it and say you want to deregulate? *Proceedings, Nordic Symposium in Tourism &*

Hospitality Research, Lund University, Helsingborg, Sweden, September 2007. Helsingborg: Lund University.

Hall, C.M. (2007b) *Introduction to Tourism in Australia* (5th edn). Melbourne: Pearson.

Hall, C.M. (2007c) Tourism and regional competitiveness. In J. Tribe and D. Airey (eds) *Advances in Tourism Research, Tourism Research, New Directions, Challenges and Applications* (pp. 217–230). Oxford: Elsevier.

Hall, C.M. (2007d) North-south perspectives on tourism, regional development and peripheral areas. In D.K. Müller and B. Jansson (eds) *Tourism in Peripheries: Perspectives from the Far North and South* (pp. 19–37). Wallingford: CAB International.

Hall, C.M. (2008) *Tourism Planning* (2nd edn). Harlow: Prentice-Hall.

Hall, C.M and Boyd, S. (eds) (2005) *Nature-based Tourism in Peripheral Areas: Development or Disaster?* Clevedon: Channel View Publications.

Hall, C.M. and Härkönen, T. (eds) (2006) *Lake Tourism: An Integrated Approach to Lacustrine Tourism Systems.* Clevedon: Channel View Publications.

Hall, C.M. and Higham, J. (eds) (2005) *Tourism, Recreation and Climate Change.* Clevedon: Channel View Publications.

Hall, C.M. and Jenkins, J.M. (1995) *Tourism and Public Policy.* London: Routledge.

Hall, C.M. and Jenkins, J.M. (1998) The policy dimensions of rural tourism and recreation. In R.W. Butler, C.M. Hall and J.M. Jenkins (eds) *Tourism and Recreation if Rural Areas* (pp. 19–42). Chichester: John Wiley.

Hall, C.M. and Lew, A.A. (1998) The geography of sustainable tourism: Lessons and prospects. In C.M. Hall and A.A. Lew (eds) *Sustainable Tourism: A Geographical Perspective* (pp. 199–203). New York: Longman.

Hall, C.M. and McArthur, S. (1998) *Integrated Heritage Management: Principles and Practice.* London: Stationary Office.

Hall, C.M. and Müller, D.K. (eds) (2004a) *Tourism, Mobility and Second Homes: Between Elite Landscape and Common Ground.* Clevedon: Channel View Publications.

Hall, C.M. and Müller, D.K. (2004b) Introduction: Second homes, curse or blessing? Revisited. In C.M. Hall and D.K. Müller (eds) *Tourism, Mobility and Second Homes: Between Elite Landscape and Common Ground* (pp. 3–14). Clevedon: Channel View Publications.

Hall, C.M. and Page, S.J. (2006) *The Geography of Tourism: Environment, Place and Space* (3rd edn). London: Routledge.

Hall, C.M. and Williams, A. (2008) *Tourism and Innovation.* London: Routledge.

Hall, P. (1992) *Urban and Regional Planning* (3rd edn). London: Routledge.

Hallikainen, V. (1998) The Finnish wilderness experience. *Metsäntutkimuslaitoksen tiedonantoja* 711, 1–288.

Halme, M. (2001) Learning for sustainable development in tourism networks. *Business Strategy and the Environment* 10 (2), 100–114.

Hamelin, L-E. (1979) *Canadian Nordicity: It's Your North, Too.* Montreal: Harvest House.

Hanell, T., Aalbu, H. and Neubauer, J. (2002) *Regional Development in the Nordic Countries.* Nordregio Report. Stockholm: Nordregio.

Hannam, K. (2003) *Viking Themed Festivals.* Paper presented at the Journeys of Expression II IFEA 2003 Conference, 7–11 March 2003, Center for Tourism and Cultural Change, International Festivals Events Association (IFEA), Vienna, Austria.

Hansen, C. and Jensen, S. (1996) The impact of tourism on employment in Denmark: Different definitions, different results. *Tourism Economics* 2 (4), 283–302.

Hansen, H.K. and Winther, L. (2007) The spaces of urban economic geographies: Industrial transformation in the outer city of Copenhagen. *Geografisk Tidsskrift Danish Journal of Geography* 107 (2), 45–58.

Hansen, J. (2008) Tough Fight for Arctic Resources. Arctic Council 1 February. On WWW at http://arctic-council.org/article/2008/2/tough_fight_for_arctic_resources. Accessed 1.4.08.

Hansen, J.C. (1969) Fritidsbebyggelsen i Eidanger. *Ad Novas – Norwegian Geographical Studies* 8, 36–47.

Hansson, U. and Medin, S. (1954) Halmstads, Jönköpings, Kalmars och Växjös sommarortsfält. *Svensk Geografisk Årsbok* 30, 179–185.

Hardy, A., Beeton, R.J.S. and Pearson, L. (2002) Sustainable tourism: An overview of the concept and its position in relation to conceptualisations of tourism. *Journal of Sustainable Tourism* 10 (6), 475–496.
Harrison-Hill, T. and Chalip, L. (2005) Marketing sport tourism: Creating synergy between sport and destination. *Sport in Society* 8 (2), 302–320.
Harvey, D. (1988) Voodoo cities. *New Statesman and Society*, 30 September, 33–35.
Harvey, D. (1989) From managerialism to entrepreurialism: The transformation in urban governance in late capitalism. *Geografiska Annaler* 71B, 3–17.
Hauge, J. and Jensen, Ø. (2002) *Fjord Norge AS: Vurderinger og framtidsperspektiv*. Rapport RF-2002/323. Stavanger: Rogaland Research Institute.
Haukeland, J.V. and Lindberg, K. (2001) *Sustainable Tourism Management in Nature Areas*. TØI report 550/2001. Oslo: Institute of Transport Economics.
Haukeland, J.V., Ludviksen, J. and Ryntveit, G.O. (1994) *Marketing of Norwegian Tourist Industry in International Distribution Channels*. TØI report 253/1994. Oslo: Institute of Transport Economics.
Heberlein, T.A., Fredman, P. and Vuorio, T. (2002) Current tourism patterns in the Swedish Mountain Region. *Mountain Research and Development* 22 (2), 142–149.
Heberlein, T.A. and Vuorio, T. (1999) People and tourism in 1/3 of Sweden: Current status and recent trends. *ETOUR Working Paper* 1999:29.
Hecock, R.D. (1993) Second Homes in the Norwegian mountains: Cultural and institutional contexts for continuing development. *Tourism Recreation Research* 18, 45–50.
Hem, L.E. and Iversen, N.M. (2004) How to develop a destination brand logo: A qualitative and quantitative approach. *Scandinavian Journal of Hospitality and Tourism* 4 (2), 83–106.
Hendee, J., Stankey, S. and Lucas, P. (1990) *Wilderness Management*. Golden: Fulcrum.
Herbert, D.T. (ed.) (1997) *Heritage, Tourism and Society*. London: Pinter.
Hewison, R. (1991) Commerce and Culture. In J. Corner and S. Harvey (eds) *Enterprise and Heritage: Crosscurrents of National Culture* (pp. 162–177). London: Routledge.
Hilding-Rydevik, T., Lähteenmäki-Smith, K. and Storbjörk, S. (2005) *Implementing Sustainable Development in the Regional Development Context – A Nordic Overview*. Nordregio Report 2005: 5. Stockholm: Nordregio.
Hill, P. (1999) Tangibles, intangibles and services: A new taxonomy for the classification of output. *Canadian Journal of Economics* 32 (2), 426–446.
Hiltunen, M.J. (2007) Environmental impacts of rural second home tourism. Case Lake District in Finland. *Scandinavian Journal of Hospitality and Tourism* 7 (3), 243–265.
Hirsch, D. (2007) Rezidor tappar till Choice. *Restauratören*. 2007-11-05.
Hjalager, A. (2005) Innovation in tourism from a welfare state perspective. *Scandinavian Journal of Hospitality and Tourism* 5 (1), 46–62.
Hjalager, A-M. (1996) Agricultural diversification into tourism: Evidence of a European Community development programme. *Tourism Management* 17 (2), 103–111.
Holden, A. (2000) Winter tourism and the environment in conflict: The case Cairngorm, Scotland. *International Journal of Tourism Research* 2 (4), 247–260.
Holden, A. (2003) In need of new environmental ethics for tourism. *Annals of Tourism Research* 30 (1), 94–108.
Holloway, C. (2002) *The Business of Tourism*. Harlow: Prentice Hall.
Honkanen, A. (2003) Menneisyyden tulevaisuus: Postmodernit matkailuteoriat ja vapa-ajan matkailun muutokset eräissä Euroopan unionin jäsenvaltioissa vuosina 1985 ja 1997. Matkailualan verkostoyliopisto, *Keskustelua ja raportteja* No. 5.
Hønneland, G. (1998) Identity formation in the Barents Euro-Arctic Region. *Cooperation and Conflict: Nordic Journal of International Studies* 33 (2), 277–297.
Hopstock, C. and Madsen, T.S. (1969) *Baroniet i Rosendal*. Oslo: Universitetsforlaget.
Hubbard, P. and Lilley, K. (2000) Setting the past: Heritage-tourism and place identity in Stratford-upon-Avon. *Geography* 85 (3), 221–232.
Hudson, S. (2000) *Snow Business*. London: Cassell.
Hughes, H.L. (2002) Culture and tourism: A framework for further analysis. *Managing Leisure* 7, 164–175.

Huuskonen, M. (2006) Chinatown rises in Kouvola: Space for 120 Chinese companies in disused dairy. *Helsingin Sanomat*, International Edition – Business & Finance, 24 May.
ICCA, The International Congress & Convention Association (2007) ICCA publishes country and city rankings 2006: U.S.A. and Vienna once again top country and city respectively, Press Release. Amsterdam, The Netherlands: ICCA. On WWW at http://www.iccaworld.com/npps/story.cfm?ID=1305. Accessed 1.4.08.
IceHotel (2006) On WWW at http://icehotel.com/. Accessed 25.5.06.
Ilves, T.H. (1999) Estonia as a Nordic country. Speech by Toomas Hendrik Ilves, Minister of Foreign Affairs, to the Swedish Institute for International Affairs, 14 December 1999. On WWW at http://www.vm.ee/eng/nato/1210.html. Accessed 1.4.08.
Ingold, T. (2000) *The Perception of the Environment: Essays on Livelihood, Dwelling and Skill*. London: Routledge.
Ioannides, D., Nielsen, P.A. and Billing, P. (2006) Transboundary collaboration in tourism: The case of the Bothnian Arc. *Tourism Geographies* 8 (2), 122–142.
Jaakson, R. (1986) Second-home domestic tourism. *Annals of Tourism Research* 13, 367–391.
Jacobsen, J.K.S. (1994) *Arctic Tourism and Global Tourism Trends*. Thunder Bay, Ontario: Lakehead University Centre for Northern Studies.
Jacobsen, J.K.S. (1997) The making of an attraction: The case of North Cape. *Annals of Tourism Research* 24, 341–356.
Jacobsen, J.K.S. (2004) Roaming romantics: Solitude-seeking and self-centredness in scenic sightseeing. *Scandinavian Journal of Hospitality and Tourism* 4, 5–23.
Jacobsen, J.K.S., Dybedal, P. and Skalpe, O. (1996) *Strategies in Norwegian Tourism Industry. An Evaluation of Public Financial Support to Tourism Organisations*. TØI report 329/1996. Oslo: Institute of Transport Economics.
Jansson, B. and Müller, D.K. (2003) *Fritidsboende i Kvarken*. Umeå: Kvarkenrådet.
Jansson, J. and Power, D. (2006) *Image of the City: Urban Branding as Constructed Capabilities in Nordic City Regions*. Oslo: Nordic Innovation Centre.
Jensen, Ø. (1998) Kjøper-selger-relasjoner innenfor internasjonal reiseliv snæring. En studie av samaspillet mellom bedriftene ut fra ønske om utvikling av varige konkurransefordeler. PhD theses (1998: 4), Aarhus School of Business, Faculty of Business.
Jensen, O.B. (2005) Branding the Contemporary City – Urban branding as regional growth agenda? Plenary paper for Regional Studies Association Conference 'Regional Growth Agendas', Aalborg, 28–31 May 2005.
Joenniemi, P. and Sergounin, A.A. (2003) *Russia and the European Union's Northern Dimension: Encounter or Clash of Civilisations?* Nizhny Novgorod: Nizhny Novgorod Linguistic University Press.
Jóhannesson, G. Þ., Skaptadóttir, U.D. and Benediktsson, K. (2003) Coping with social capital? The cultural economy of tourism in The North. *Sociologia Ruralis* 43 (1), 3–16.
Johnston, C.S. (2001) Shoring the foundations of the destination life cycle model, part 1: Ontological and epistemological considerations. *Tourism Geographies* 3 (1), 2–28.
Johnston, C.S. (2006) The ontological foundation of the TALC. In R.W. Butler (ed.) *The Tourism Area Life Cycle Vol. 2: Conceptual and Theoretical Issues* (pp. 7–28). Clevedon: Channel View Publications.
Jokinen, A. (2002) Free-time habitation and layers of ecological history at a southern Finnish lake. *Landscape and Urban Planning* 61 (4), 99–112.
Jones, K.W. (1957) *To the Polar Sunrise*. London: Museum Press.
Karppi, I. (2001) *Competitiveness in the Nordic Economies*, Nordregio WP2001:2. Stockholm: Nordregio.
Kaltenborn, B.P. (1996) Tourism in Svalbard: Planned management or the art of stumbling through? In M. Price (ed.) *People and Tourism in Fragile Environments* (pp. 89–108). Chichester: John Wiley & Sons.
Kaltenborn, B.P. (1997a) Nature of place attachment: A study among recreation homeowners in southern Norway. *Leisure Sciences* 19, 175–189.

Kaltenborn, B.P. (1997b) Recreation homes in natural settings: Factors affecting place attachment. *Norsk Geografisk Tidsskrift* 51, 187–198.
Kaltenborn, B.P. (1998) The alternate home: Motives of recreation home use. *Norsk Geografisk Tidsskrift* 52, 121–134.
Kaltenborn, B.P. (1999) Setting preferences of Arctic tourists: A study of some assumptions in the recreation opportunity spectrum framework from the Svalbard Archipelago. *Norwegian Journal of Geography* 53, 45–55.
Kaltenborn, B.P. (2000) Arctic-Alpine environments and tourism: Can sustainability be planned? Lessons learned on Svalbard. *Mountain Research and Development* 20 (1), 28–31.
Kaltenborn, B.P. (2002) Bo i naturen: Meningen med hyttelivet. *Utmark* 2002 (3). On WWW at www.utmark.org. Accessed 1.4.08.
Kaltenborn, B.P., Haaland, H. and Martinac, I. (2001) The public right of access – some challenges to sustainable tourism development in Scandinavia. *Journal of Sustainable Tourism* 9 (5), 417–433.
Kari, K. (1978) *Haltin valloitus*. Helsinki: Kisakalliosäätiö.
Kauppalehti (2005) Ahvenenmaa – Liite avoimesta ja yrittäjähenkisestä saarimaakunnasta. *Kauppalehti* 28.11.2005.
Kauppila, P. (1995) Matkailukeskuksen elinkaari – esimerkkinä Kuusamon Ruka. *Nordia Geographical Publications* 24 (1).
Kauppila, P. (2004) Matkailukeskusten kehitysprosessi ja rooli aluekehityksessä paikallistasolla: esimerkkeinä Levi, Ruka, Saariselkä ja Ylläs. *Nordia Geographical Publications* 33 (1).
Kautto, M., Fritzell, J., Hvinden, B., Kvist, J. and Uusitalo, H. (eds) (2001) *Nordic Welfare States: In the European Context*. London: Routledge.
Kautto, M., Heikkilä, M., Hvinden, B., Marklund, S. and Ploug, N. (eds) (1999) *Nordic Social Policy: Changing Welfare States*. London: Routledge.
Kearsley, G.W., Coughlan, D.P., Higham, J.E.S., Higham, E.C. and Thyne, M.A. (1998) *Impacts of Tourist Use on the New Zealand Backcountry*. Research Paper 1, Dunedin: Centre for Tourism, University of Otago.
Keen, D. and Hall, C.M. (2004) Second homes in New Zealand. In C.M. Hall and D. Müller (eds) *Tourism, Mobility and Second Homes: Between Elite Landscape and Common Ground* (pp. 174–195). Clevedon: Channel View Publications.
Keller, C.P. (1987) Stages of peripheral tourism development – Canada's Northwest Territories. *Tourism Management* 8 (1), 20–32.
Kildal, N. and Kuhnle, S. (eds) (2005) *Normative Foundations of the Welfare State: The Nordic Experience*. London: Routledge.
Kirschenblatt-Gimblett, B. (1998) *Destination Culture. Tourism, Museums and Heritage*. Berkeley: University of California Press.
KNBF (2004) KNBF-nytt november 2004. Kongelig Norsk Båtforbund. On WWW at http://knbf.no/archivess/12/knbfnytt_novemb.html. Accessed 17.1.07.
Komppula, R. (2004) Success and growth in rural tourism micro-businesses in Finland: Financial or life-style objectives? In R. Thomas (ed.) *Small Firms in Tourism: International Perspectives* (pp. 115–138). Amsterdam: Elsevier.
Komppula, R. (2005) Pursuing customer value in tourism – A rural tourism case-study. *Journal of Hospitality & Tourism* 3 (2), 83–104.
Konttinen, J-P. (2005) *Tourism Satellite Account and Regional Economic Effects*. Ministry of Trade and Industry, KTM Rahoitetut tutkimukset 4/2005. Helsinki: Ministry of Trade and Industry.
Kotler, P. and Gertner, D. (2002) Country as brand, product, and beyond: A place marketing and brand management perspective. *Brand Management* 9 (4–5), 249–261.
Kotler, P. and Levy, S.J. (1969) Broadening the concept of marketing. *Journal of Marketing* 33, 10–15.
Kotler, P., Asplund, C., Rein, I. and Haider, D. (1999) *Marketing Places Europe*. London: Pearson Education.
Kotler, P., Haider, D.H. and Rein, I. (1993) *Marketing Places: Attracting Investment, Industry, and Tourism to Cities, States, and Nations*. New York: Free Press.

Krippendorf, J. (1982) Towards new tourism policies. *Tourism Management* 3 (1), 135–148.
Kvist, A-K.J. (2005) Needs and expectations of inbound tourist visiting a peripheral area: A case study in Northern Sweden. Licentiate thesis, Luleå University of Technology, Department of Business Administration and Social Sciences, 2005:07.
Kvistgaard, P. (2005) *Bæredygtige kompetencer i Midt – og Nordjysk turisme*, published electronically by the regional tourism development agency, Midt-Nord Turisme. On WWW at http://baeredygtig-turisme.dk/Evaluering_dec05.pdf. Accessed 1.4.08.
Lafferty, G. and van Fossen, A. (2001) Integrating the tourism industry: Problems and strategies. *Tourism Management* 22, 11–19.
Lane, B. (1994) What is rural tourism? *Journal of Sustainable Tourism* 2 (1), 7–21.
Langdalen, E. (1980) Second homes in Norway: A controversial planning problem. *Norsk Geografisk Tidsskrift* 34, 139–144.
Lapin Matkailumarkkinointi (1999) *Lapin matkailu 1998*. Rovaniemi: Lapin matkailumarkkinointi Oy.
Larsen, J. (2001) Tourism mobilities and the travel glance: Experiences of being on the move. *Scandinavian Journal of Hospitality and Tourism* 1, 80–98.
Larsen, S. and Mossberg, L. (2007) Editorial: The diversity of tourist experiences. *Scandinavian Journal of Hospitality and Tourism* 7 (1), 1–6.
Latour, B. (1999) On recalling ANT. In J. Law and J. Hassard (eds) *Actor Network Theory and After* (pp. 15–25). Oxford: Blackwell Publishers.
Lee, D.S. and Sausen, R. (2000) New directions: Assessing the real impact of CO_2 emissions trading by the aviation industry. *Atmospheric Environment* 34, 5337–5338.
Lee, H.F. (2001) Sustainable tourism destinations: The importance of cleaner production. *Journal of Cleaner Production* 9 (4), 313–323.
Lehtinen, A. (1991) Northern natures: A study of the forest question emerging within the timber-line conflict in Finland. *Fennia* 169, 57–169.
Lehtonen, L. (2006) Luonto-opastus on opastusviestintää. *Puistoväki* 3/2006.
Leiper, N. (1990) Tourist attraction systems. *Annals of Tourism Research* 17, 367–384.
Lennon, J.J. (eds) (2003) *Tourism Statistics: International Perspectives and Current Issues*. London: Continuum.
Lesjø, J.H. (2003) OL-saken: en prosess-sosiologisk studie av planlegging, politikk og organisering. Research Report no 100/2003. Lillehammer University College.
Lew, A.A., Hall, C.M. and Williams, A.M. (eds) (2004) *A Companion to Tourism*. Oxford: Blackwell.
Liiketaloudellinen tutkimuskeskus (2003) *Maaseutumatkailun Kuluttajatutkimus 2003*. Tampere: Tampereen yliopiston Liiketaloudellinen tutkimuskeskus.
Lindblom, A. (2001) Dismantling the social democratic welfare model? Has the Swedish welfare state lost its defining characteristics? *Scandinavian Political Studies* 24 (3), 171–193.
Lindgreen, A., Palmer, R. and Vanhamme, J. (2004) Contemporary marketing practice: Theoretical propositions and practical implications. *Marketing Intelligence & Planning* 22 (6), 673–692.
Lindström, R. and Salminen, D. (2004) *Kartläggning av de åländska stuguthyrarnas upplevda verksamhetsproblem och utbildningsbehov*. Åland Polytechnic. Degree thesis, 20/2004.
Lindvall, J. and Rothstein, B. (2006) Sweden: The fall of the strong state. *Scandinavian Political Studies* 29 (1), 47–63.
Linkoaho, R. (1962) Sommerhaussiedlung und Sommeraufenthalt der Stadtbevölkerung von Tampere. *Fennia* 87 (4).
Ljungdahl, S. (1938) Sommar-Stockholm. *Ymer* 58, 218–242.
Löfgren, A. (2000) A thousand years of loneliness? Globalization from the perspective of a city in a European periphery. *Geoforum* 31 (4), 501–511.
Löfgren, O. (1999) *On Holiday: A History of Vacationing*. Berkeley: University of California Press.
Lovelock, C. and Gummesson, E. (2004) In search of a new paradigm and fresh perspectives. *Journal of Service Research* 7 (1), 20–41.
Lowenthal, D. (1985) *The Past is a Foreign Country*. Cambridge: Cambridge University Press.

Lowenthal, D. (1997) *The Heritage Crusade and the Spoils of History*. London: Viking.
Lundgren, J.O. (2003) Spatial and evolutionary characteristics of Baltic Sea cruising: A historic-geographical overview. In R.K. Dowling (ed.) *Cruise Ship Tourism* (pp. 131–144). Wallingford: CABI.
Lundmark, L. (2005) Economic restructuring in a depopulating region? Tourism as alternative to traditional sectors in the Swedish mountain range. *Scandinavian Journal of Hospitality and Tourism* 5 (1), 23–45.
Lundmark, L. (2006) Mobility, migration and seasonal tourism employment. *Scandinavian Journal of Hospitality and Tourism* 6 (1), 54–69.
Lynch, K. (1960) *The Image of the City*. Cambridge, MA: MIT Press.
Lyngnes, S. and Viken, A. (1998) Samisk kultur og turisme på Nordkalotten. *BI Research Report* 8/1998. Sandvika: Institutt for kompetandeledelse.
MacCannell, D. (1976) *The Tourist: A New Theory of the Leisure Class*. New York: Schocken.
Makarychev, A.S. (2002) *Ideas, Images, and Their Producers: The Case of Region-Making in Russia's North West Federal District*. Copenhagen Peace Research Institute Working Papers, Copenhagen: Copenhagen Peace Research Institute.
Malecki, E.J. (2004) Jockeying for position: What it means and why it matters to regional development policy when places compete. *Regional Studies* 38 (9), 1101–1120.
Malmö Aviation (2007) *Miljöarbete*. On WWW at http://www.malmoaviation.se/o.o.i.s/1182. Accessed 18.9.07.
Marcusson, C.H. (2004) *The Baltic Sea Cruise Market 2003–2004 – Opportunities for Bothnian Arc 2005*. Nexø: Center for Regional and Tourism Research.
Marjavaara, R. (2007a) Route to destruction? Second home tourism in small island communities. *Island Studies Journal* 2 (1), 27–46.
Marjavaara, R. (2007b) The displacement myth: Second home tourism as scapegoat for rural decline. *Tourism Geographies* 9 (3), 296–317.
Marjavaara, R. and Müller, D.K. (2007) The development of second homes assessed property values in Sweden 1991–2001. *Scandinavian Journal of Hospitality and Tourism* 7 (3), 202–222.
Markkola, S. (2003a) Selvitys maaseudun matkailukapasiteetista vuonna 2002. *Maaseutu-Matkailu* 2003 (Spring), 6–9.
Markkola, S. (2003b) Maaseutumatkailun majoitustilojen käyttöasteet 2002. *Maaseutu-Matkailu* 2003 (Spring), 10–13.
Markusen, A. (1999) Fuzzy concepts, scanty evidence, policy distance: The case for rigour and policy relevance in critical regional studies. *Regional Studies* 33 (9), 869–884.
Marsden, T. (1999) Rural futures: The consumption countryside and its regulation *Sociologia Ruralis* 39 (4), 501–526.
Martin, V. and Inglis, M. (eds) (1983) *Wilderness: A Way Ahead*. Middleton: Findhorn Press and Lorina Press.
Martisone, S. (2007) Balticness – Belonging and identity in the Baltic Sea Region. *Baltinfo: The Official Journal of the Council of the Baltic Sea States* 88 (Sept/Oct/November), 10–11.
Maskell, P. and Törnqvist, G. (1999) *Building a Cross-border Learning Region: Emergence of the North European Øresund Region*. Copenhagen: Handelshøjskolens Forlag.
Mason, P. (2007) Visitor management in protected areas of the periphery: Experiences from both ends of the world. In D.K. Müller and B. Jansson (eds) *Tourism in Peripheries: Perspectives from the Far North and South* (pp. 154–174). Wallingford: CABI.
McIntosh, A.J. and Prentice, R.C. (1999) Affirming authenticity: Consuming cultural heritage. *Annals of Tourism Research* 26, 589–612.
McIntyre, N., Williams, D. and McHugh, K. (eds) (2006) *Multiple Dwelling and Tourism: Negotiating Place, Home and Identity*. Wallingford: CABI.
Medvedev, S. (2000) Glenn Gould, Russia, Finland and the North. Paper presented at ISA Congress in Los Angeles, 14–18 March 2000.
Medvedev, S. (2001) [the_blank_space] Glenn Gould, Russia, Finland and the North. *International Politics* 38 (1), 91–102.

Meethan, K. (2001) *Tourism in a Global Society: Place, Culture and Consumption*. New York: Palgrave.
Meinshausen, M. (2006) What does a 2°C target mean for greenhouse gas concentrations? A brief analysis based on multi-gas emission pathways and several climate sensitivity uncertainty estimates. In J. Schellnhuber, W. Cramer, N. Nakicenovic, T. Wigley and G. Yohe (eds) (2006) *Avoiding Dangerous Climate Change* (pp. 265–280). Cambridge: Cambridge University Press.
Mels, T. (1999) *Wild Landscapes: The Cultural Nature of Swedish National Parks*. Lund: Lund University Press.
Meyer, R. (1999a) Activity involvement, equipment, and geographic connection to recreation area: The case of boaters in southeastern Norway. *Norsk Geografisk Tidsskrift* 53, 17–27.
Meyer, R. (1999b) Encounter norms in a frontcountry boating area: A case study from the Nøtterøy/Tjøme Skerries in southeastern Norway. *Norsk Geografisk Tidsskrift* 53, 29–44.
Miller, M. (1993) The rise of coastal and marine tourism. *Ocean and Coastal Management* 21, 183–199.
Millington, K., Locke, T. and Locke, A. (2001) Adventure travel. *Travel and Tourism Analyst* 4, 65–97.
Ministry for the Environment (2007) *Iceland's Climate Change Strategy*. Reykjavik: Ministry for the Environment.
Ministry of Agriculture and Forestry (1996) *Forest Strategy of Lapland. Publications of Ministry of Agriculture and Forestry* 2/1996 (in Finnish).
Ministry of Trade and Industry (2006) *Matkailun aluetaloudelliset vaikutukset – matkailun alueellinen tilinpito* [Regional Economic Effects of Tourism – Regional Tourism Satellite Account]. Helsinki: Ministry of Trade and Industry.
Ministry of Trade and Industry (Faroe Islands) (2001) *Tourism Industry Policy (draft document)*. Tinganes: Ministry of Trade and Industry.
Ministry of Trade and Industry (Finland) (2006) *Annual Report 2005*. Helsinki: Ministry of Trade and Industry.
Molin, T., Müller, D.K., Paju, M. and Pettersson, R. (2007) *Kulturarvet och entreprenören*. Stockholm: Riksantikvarieämbetet.
Montanari, A. and Williams, A.M. (1995) *European Tourism. Regions, Spaces and Restructuring* (pp. 1–15). Chichester: John Wiley & Sons.
Morales-Moreno, I. (2004) Postsovereign governance in a globalizing and fragmenting world: The case of Mexico. *Review of Policy Research* 21 (1), 107–117.
Morgan, D.L. (1988) *Focus Groups as Qualitative Research*. Thousand Oaks: Sage.
Morgan, N. and Prichard, A. (2002) Contextualising destination branding. In N. Morgan, A. Pritchard and N. Pride (eds) *Destination Branding: Creating the Unique Destination Proposition* (pp. 11–41). Oxford: Butterworth-Heinemann.
Mossberg, L. (2007) A marketing approach to the tourist experience. *Scandinavian Journal of Hospitality and Tourism* 7 (1), 59–74.
Mowforth, M. and Munt, I. (1998) *Tourism and Sustainability: A New Tourism in the Third World*. London: Routledge.
Müller, D.K. (1999) *German Second Home Owners in the Swedish Countryside: On the Internationalization of the Leisure Space*. Umeå: Department of Social and Economic Geography.
Müller, D.K. (2002) German second home development in Sweden. In C.M. Hall and A.M. Williams (eds) *Tourism and Migration: New Realtionships between Production and Consumption* (pp. 169–186). Dordrecht: Kluwer.
Müller, D.K. (2004a) Mobility, tourism and second homes. In A. Lew, C.M. Hall and A. Williams (eds) *Companion to Tourism* (pp. 387–398). Oxford: Blackwell.
Müller, D.K. (2004b) Second homes in Sweden: Patterns and issues. In C.M. Hall and D.K. Müller (eds) *Tourism, Mobility and Second Homes: Between Elite Landscape and Common Ground* (pp. 244–258). Clevedon: Channel View Publications.

Müller, D.K. (2006) Unplanned development of literary tourism in two municipalities in rural Sweden. *Scandinavian Journal of Hospitality and Tourism* 6, 214–228.

Müller, D.K. and Hall, C.M. (2003) Second homes and regional population distribution: On administrative practices and failures in Sweden. *Espace Population Societes* 2003-2: 251–261.

Müller, D.K., Hall, C.M. and Keen, D. (2004) Second home tourism: Impact, planning and management. In C.M. Hall and D.K. Müller (eds) *Tourism, Mobility and Second Homes: Between Elite Landscape and Common Ground* (pp. 15–32). Clevedon: Channel View Publications.

Müller, D.K. and Jansson, B. (eds) (2007a) *Tourism in Peripheries: Perspectives from the Far North and South.* Wallingford: CABI International.

Müller, D.K. and Jansson, B. (2007b) The difficult business of making pleasure peripheries prosperous: Perspectives on space, place and environment. In D.K. Müller and B. Jansson (eds) *Tourism in Peripheries: Perspectives from the Far North and South* (pp. 3–18). Wallingford: CAB International.

Müller, D.K. and Pettersson, R. (2001) Access to Sami tourism in northern Sweden. *Scandinavian Journal of Hospitality and Tourism* 1, 5–18.

Müller, D. and Pettersson, R. (2005) What and where is the indigenous at an indigenous festival? Observations from the winter festival in Jokkmokk, Sweden. In C. Ryan and M. Aiken (eds) *Indigenous Tourism: The Commodification and Management of Culture* (pp. 201–218). Oxford: Elsevier.

Müller, D.K. and Pettersson, R. (2006) Sami heritage at the Winter Festival in Jokkmokk, Sweden. *Scandinavian Journal of Hospitality and Tourism* 6, 54–69.

Murdoch, J. and Pratt, A.C. (1997) From the power of topography to the topography of power: A discourse on strange ruralities. In P. Cloke and J. Little (eds) *Contested Countryside Cultures: Otherness, Marginalisation and Rurality* (pp. 51–69). London: Routledge.

Murray, I. and Haraldsdóttir, L. (2005) Culinary tourism where no crops are grown. In *Second international conference on culinary tourism, April 30–May 3* [CD-ROM]. Portland: International Culinary Tourism Association.

MyTravel (2007) *Climate Care.* On WWW at http://www.mytravel.com/AniteNextPage. asp?p=SPECIFICPROMOTION_63584&s=654834531. Accessed 18.9.07.

National Land Survey of Finland (2006) www.maanmittauslaitos.fi.

Nash, R. (1982) *Wilderness and the American Mind* (3rd edn). London: Yale University Press.

National Land Survey of Finland (2006) On WWW at www.maanmittauslaitos.fi. Accessed

Nature's Best (2006) A selection of best nature tours in Sweden. On WWW at http://www.naturensbasta.se. Accessed 24.11.06.

Negri, F. (1929 [1700]) *Viaggio Settentrionale di Francesco Negri.* Milan: Edizioni "Alpes".

Neubauer, J., Dubois, A., Hanell, T., Lähteenmäki-Smith, K., Pettersson, K., Roto, J. and Steineke, J.M. (2007) *Regional Development in the Nordic Countries 2007,* Nordregio Report 2007:1. Stockholm: Nordregio.

Niedomysl, T. (2004) Evaluating the effects of place-marketing campaigns on interregional migration in Sweden. *Environment and Planning A* 36, 1991–2009.

Nilsson, P.Å. (2001) Tourist destination development: The Åre Valley. *Scandinavian Journal of Hospitality and Tourism* 1, 54–67.

Nilsson, P.Å. (2002) Staying on farms: An ideological background. *Annals of Tourism Research* 29 (1), 7–24.

Nilsson, P.Å. (2005) *The Norwegian Coastal Express (Hurtigruten/Pikareitti): An Alternative for the Gulf of Bothnia?* Nexø: Center for Regional and Tourism Research.

Nilsson, P.Å., Marcusson, C.H., Pedersen, J. and Pedersen Munkgaard, K. (2005) *Cruise Tourism in the Baltic and Bothnian Sea: A Pilot Study on Maritime Tourism.* Nexø: Center for Regional and Tourism Research.

Nilsson, P-Å. and Ankre, R. (2006) The Luleå Archipelago, Sweden. In G. Baldacchino (ed.) *Extreme Tourism* (pp. 145–157). Amsterdam: Elsevier.

Nordic Council of Ministers Tourism Ad Hoc Working Group (2003) *A Road Map for Sustainable Tourism in the Nordic Countries Proposal for a Common Nordic Action Plan for Sustainable Tourism Development, Final Draft*. Nordic Council of Ministers Tourism Ad Hoc Working Group.

Nordic Statistical Yearbook (2007) Copenhagen: Nordic Council of Ministers.

Nordin, Å. (2007) *Renskötseln är mitt liv: Analys av den samiska renskötselns ekonomiska tillpassning*. Umeå: Vaartoe/Centre for Sami Research.

Nordin, S. and Svensson, B. (2005) *The Significance of Governance in Innovative Tourism Destinations*. ETOUR Working Paper 2005:2.

Nordin, U. (1993a) *Fritidshusbebyggelse för skärgårdsbor? Studier av fritidsboendets betydelse för sysselsättningen i Blidö församling, Norrtälje kommun 1945–1987*. Stockholm: Kulturgeografiska institutionen.

Nordin, U. (1993b) Second homes. In H. Aldskogius (ed.) *National Atlas of Sweden: Cultural Life, Recreation and Tourism* (pp. 72–79). Stockholm: SNA.

Nortek (2007) *Tourism and the Travel and Tourist Industry in Sweden, 2007 Edition*. Stockholm: Nortek.

NUTEK (2007) *Fakta om svensk turism och turistnäring. 2007 ars upplaga*. Stockholm: Nutek.

Nya Åland (2005) Åländsk turism börjar uppfattas som krisbransch. *Nya Åland*, 2005-04-08.

Nygård, S. (2003) *Villinge – villasamhället i skärgården* [Villinge – villa community in archipelago]. Helsingfors stadsmuseum, Serie: Memoria 16. Savonlinna: Painolinna.

O'Dell, T. (ed.) (2005) *Experiencescapes: Tourism, Culture and Economy*. Copenhagen: Copenhagen Business School Press.

OECD (1993) *What Future for Our Countryside? A Rural Development Policy*. Paris: OECD.

OECD (2006) *Territorial Review: Stockholm, Sweden*. Paris: OECD Publications.

OECD (2007) *Climate Change in the European Alps: Adapting Winter Tourism and Natural Hazards Management*. OECD: Paris.

Office of the Prime Minister (2008) *Broad Agreement to Boost National Climate Efforts*. Office of the Prime Minister Press Release no. 005-08, 18 January.

Olsen, K. (2006) Making differences in a changing world: The Norwegian Sami in the tourist industry. *Scandinavian Journal of Hospitality and Tourism* 6, 37–55.

Ooi, C.S. (2004) Poetics and politics of destination branding: Denmark. *Scandinavian Journal of Hospitality and Tourism* 4 (2), 107–128.

Orams, M. (1999) *Marine Tourism*. London: Routledge.

Ostertag, J. (2007) The definition and compilation of European city tourism statistics. Diploma thesis, Dreistetten.

Page, S.J. and Hall, C.M. (2003) *Managing Urban Tourism*. Harlow: Prentice-Hall.

Pálsson, G. (1995) *The Textual Life of Savants: Ethnography, Iceland, and the Linguistic Turn*. Chur: Harwood Academic Publishers.

Palttila, Y. and Niemi, E. (2003) Maaseutu EU-Ohjelmakauden 2000–2006 alussa – maaseutuindikaattorit. *Tilastokeskus, Katsauksia* 2003/2.

Pawlak, J.F. and Leppänen. J. (2007) *Climate Change in the Baltic Sea Area, Draft Helcom Thematic Assessment in 2007. Based on the BALTEX Assessment of Climate Change for the Baltic Sea Basin (BACC)*. Helsinki: Helsinki Commission, Baltic marine Environment Protection Commission.

Payne, C. (2006) Lessons with Leah: Re-reading the Photographic Archive of Nation in the National Fim Board of Canada's Still Photography Division. *Visual Studies* 21 (1), 4–22.

Periäinen, K. (2006) The summer cottage: A dream in the Finnish forest. In N. McIntyre, D. Williams and K. McHugh (eds) *Multiple Dwelling and Tourism. Negotiating Place, Home and Identity*. Cambridge, MA: CABI Publishing.

Pettersson, R. (2003) Indigenous cultural events – The development of a Sami winter festival in Northern Sweden. *Tourism* 51 (3), 319–332.

Pettersson, R. (2004) Sami tourism in Northern Sweden: Supply, demand and interaction. PhD thesis, ETOUR V 2004:14. Mid-Sweden University.

Pettersson, R. and Viken, A. (2007) Sami perspectives on indigenous tourism in Northern Europe: Commerce or cultural development? In R. Butler and T. Hinch (eds) (2007) *Tourism and Indigenous Peoples: Issues and Implications* (pp. 176–187). Amsterdam: Butterworth-Heinemann.

Pietarinen, J. (1987) Ihminen ja metsä: Neljä perusasennetta. *Silva Fennica* 21 (4), 323–331.

Pietarinen, J. (1996) The principle attitudes of humanity towards nature. In H.O. Oruka (ed.) *Philosophy, Humanity and Ecology: Philosophy of Nature and Environmental Ethics* (pp. 290–294). Darby: Diane Publishing.

Pigram, J.J. and Jenkins, J. (1999) *Outdoor Recreation Management*. London: Routledge.

Pihl Atmer, A.K. (1998) *Livet som leves där måste smaka vildmark: Sportstugor och friluftsliv 1900–1945*. Stockholm: Stockholmia.

Pine, J. and Gilmore, J. (1999) *The Experience Economy: Work is Theatre & Every Business a Stage: Goods & Services are no Longer Enough*. Boston: Harvard Business School Press.

Pitkänen, K. and Kokki R. (2005) *Mennäänkö mökille? Näkökulmia pääkaupunkiseutulaisten vapaa-ajan asumiseen Järvi-Suomessa* [Let's go to Mökki! Perspectives on Second Home Tourism of Helsinki metropolitans in the Finnish Lakeland]. University of Joensuu, Savonlinna Institute for Regional Development and Research, Publications 11.

Pitkänen, K. (2008) Second home landscape: The meaning(s) of landscape for second home tourism in Finnish lakeland. *Tourism Geographies* 10 (2), 169–192.

Poon, A. (1993) *Tourism, Technology and Competitive Strategies*. Wallingford: CAB International.

Poon, A. (2003) Competitive strategies for a 'new tourism'. In C. Cooper (ed.) *Classic Reviews in Tourism* (pp. 130–142). Clevedon: Channel View Publications.

Poria, Y., Butler, R. and Airey, D. (2003) The core of heritage tourism. *Annals of Tourism Research* 30, 238–254.

Power, D. and Hallencreutz, D. (2005) Cultural industry cluster building in Sweden. In P. Oinas and A. Lagendijk (eds) *Proximity, Distance and Diversity: Issues on Economic Interaction and Local Development* (pp. 25–45). London: Ashgate.

Power, D. (2003) The Nordic 'cultural industries': A cross-national assessment of the place of the cultural industries in Denmark, Finland, Norway and Sweden. *Geografiska Annaler* 85B, 167–180.

Power, T.M. (1996) *Lost Landscapes and Failed Economies*. Washington DC: Island Press.

Prahalad, C.K. and Ramaswamy, V. (2004) Co-creating experiences: The next practice in value creation. *Journal of Interactive Marketing* 18 (3), 5–14.

Prentice, R.C. (1994) Heritage: A key sector of the "new" tourism. *Progress in Tourism, Recreation and Hospitality Management* 5, 309–324.

Pretes, M. (1994) Postmodern tourism: The Santa Claus industry. *Annals of Tourism Research* 22 (1), 1–15.

PRF (2006) *Passagerarrederierna – en viktig del i rese- och turismnäringen*. Stockholm: Passagerarrederiernas förening PRF.

Price, M.F. (ed.) (1996) *People and Tourism in Fragile Environments*. Chichester: Wiley.

Pride, W.M. and Ferrell, O.C. (2003) *Marketing: Concepts and Strategies* (12th edn) Boston: Houghton Mifflin.

Prokkola, E. (2007) Cross-border regionalization and tourism development at the Swedish-Finnish border "Destination Arctic Circle". *Scandinavian Journal of Hospitality and Tourism* 7 (2), 120–138.

Puijk, R. (2000) Local implications of tourism: A case study from Western Norway. *Current Issues in Tourism* 3 (1), 51–80.

Rainisto, S.K. (2003) *Success Factors of Place Marketing: A Study of Place Marketing Practices in Northern Europe and the United States*. Helsinki University of Technology, Institute of Strategy and International Business, Doctoral Dissertations 2003/4, Espoo 2003. Helsinki: Helsinki University of Technology.

Rasmusson, E. (2004) *Tylösand i våran hjärtan*. Varberg: Utsikten.

Rezidor (2007) *Annual and Sustainability Report 2006. Now Watch Creative Hospitality Create Shareholder Value*. Brussels: The Rezidor Hotel Group.

Rhodes, R. (1996) The new governance: Governing without government. *Political Studies* 44, 652–667.
Rhodes, R.A.W. (1997) *Understanding Governance: Policy Networks, Governance, Reflexivity and Accountability*. Buckingham: Open University Press.
Rica Hotels (2007) *Miljø – et løft inn i fremtiden*. On WWW at http://www.rica.no/files/documents/miljo/Rica_miljo.pdf. Accessed 18.9.07.
Richards, G. (1996a) Introduction: Culture and tourism in Europe. In G. Richards (ed.) *Cultural Tourism in Europe* (pp. 3–18). Wallingford: CAB International.
Richards, G. (1996b) Production and consumption of European cultural tourism. *Annals of Tourism Research* 23, 261–283.
Richards, G. (2001) The development of cultural tourism in Europe. In G. Richards (ed.) *Cultural Attractions and European Tourism* (pp. 3–29). Wallingford: CAB International.
Rideng, A. and Dybedal, P. (2004) *Foreign Visitor Survey 2004*. TØI report 753/2004. Oslo: Institute of Transport Economics.
Riseth, J.Å. (2005) Nature protection and the colonial legacy- Sámi reindeer management versus urban recreation: The case of Junkerdal-Balvatn, Northern Norway. In T. Peil and M. Jones (eds) *Landscape, Law and Justice. Proceedings of a conference organised by the Centre for Advanced Study at the Norwegian Academy of Science and Letters, Oslo, 15–19 June 2003*. The Institute for Comparative Research in Human Culture, Serie B: Skrifter vol. CXVIII (pp. 173–186). Oslo: Novus forlag.
Roberts, L. and Hall, D. (2001) *Rural Tourism and Recreation: Principles to Practice*. Wallingford: CABI.
Robinson, M. and Smith, M. (2006) Politics, power and play: The shifting contexts of cultural tourism. In M. Smith and M. Robinson (eds) *Cultural Tourism in a Changing World: Politics, Participation and (Re)presentation* (pp. 1–17). Clevedon: Channel View Publications.
Roper, A., Jensen, O. and Jegervatn, R. (2005) The dynamics of the Norwegian package tour industry. *Scandinavian Journal of Hospitality and Tourism* 5 (3), 193–211.
Rossi, P.H., Freeman, H.E. and Lipsey, M.W. (1999) *Evaluation: A Systematic Approach*. Thousand Oaks CA: Sage Publications.
Rostgaard and Lehto, J. (2001) Health and social care systems: How different is the Nordic model. In M. Kautto, J. Fritzell, B. Hvinden, J. Kvist and H. Uusitalo (eds) *Nordic Welfare States: In the European Context* (pp. 137–166). London: Routledge.
Roth, L. (2005) *Something New in the Air: The Story of First Peoples Television*. Montreal: McGill-Queen's University Press.
Rural Policy and Rural Tourism Groups (RPRTG) (2000) *Strategy and Development Programme for Rural Tourism*. Helsinki: Ministry of Agriculture and Forestry.
Rural Tourism International (2006) On WWW at http://www.ruraltourisminternational.org/. Accessed 1.2.06.
Ryan, C. and Aicken, M. (eds) (2005) *Indigenous Tourism: The Commodification and Management of Culture*. Amsterdam: Elsevier.
Ryan, C., Hughes, K. and Chirgwin, S. (2000) The gaze, spectacle and ecotourism. *Annals of Tourism Research* 27 (1) 148–163.
Saarinen, J. (1998) Cultural influence on response to wilderness encounters: A case study from Finland. *International Journal of Wilderness* 4 (1), 28–32.
Saarinen, J. (1999) Representations of indigeneity: Sami culture in the discourses of tourism. In P.M. Sant and J.N. Brown (eds) *Indigeneity: Constructions and Re/Presentations* (pp. 231–249). New York: Nova Science Publishers.
Saarinen, J. (2001) *The Transformation of a Tourist Destination: Theory and Case Studies on the Production of Local Geographies in Tourism in Finnish Lapland*. Oulu: Oulu University Press.
Saarinen, J. (2003) The regional economics of tourism in Northern Finland: The socio-economic implications of recent tourism development and future possibilities for regional development. *Scandinavian Journal of Hospitality and Tourism* 3 (2), 91–113.

Saarinen, J. (2004) Tourism and touristic representations of nature. In A. Lew, C.M. Hall and A. Williams (eds) *Geography: A Companion to Tourism* (pp. 438–449). Oxford: Blackwell.

Saarinen, J. (2005) Tourism in the northern wilderness: Wilderness discourses and the development of nature-based tourism in Northern Finland. In C.M. Hall and S. Boyd (eds) *Nature-based Tourism in Peripheral Areas: Development Or Disaster?* (pp. 36–49). Clevedon: Channel View Publications.

Saarinen, J. (2006) Traditions of sustainability in tourism studies. *Annals of Tourism Research* 33 (4), 1121–1140.

Saarinen, J. and Hall, C.M. (eds) (2004) *Nature-Based Tourism Research in Finland: Local Contexts, Global Issues*. Finnish Forest Research Institute, Research Papers 916, Rovaniemi: Rovaniemi Research Station.

Saarinen, J. and Vaara, M. (2002) Mökki kansallispuiston laidalla: Loma-asukkaiden näkemyksiä Pyhäntunturin kansallispuiston käytöstä ja kehittämisestä. Research Reports of the Finnish Forest Research Institute, 845.

Saastamoinen, O. (1982) Economics of the multiple-use forestry in the Saariselkä fell area. *Communicationes Instituti Forestalis Fenniae* 104, 1–102.

Saethorsdottir, A.D. (2004) Adapting to change: Maintaining a wilderness experience in a popular tourist destination. *Tourism Today* 4, 52–65.

Sahlberg, B. (2001) Möten, människor och marknader inom upplevelsesamhället: turism och resande i tid och rum. ETOUR popularveteskapliga serien 11.

Sandell, K. and Sörlin, S. (2000) *Friluftshistoria: från 'härdande friluftslif' till ekoturism och miljöpedagogik*. Stockholm: Carlssons.

Sausen, R., Isaksen, I., Grewe, V., Hauglustaine, D., Lee, D.S., Myhre, G., Köhler, M.O., Pitari, G., Schumann, U., Stordal, F. and Zerefos, C. (2005) Aviation radiative forcing in 2000: An update on IPCC (1999). *Meteorologische Zeitschrift* 14 (4), 555–561.

Scandic Hotels (2007) *Vad gör Scandic speciellt*. On WWW at http://www.scandic-hotels.se/corporateinfo/900_CorporateInformation.jsp. Accessed 18.9.07.

Scandinavian Airlines (2007) *SAS koncernens Årsredovisning & Hållbarhetsredovisning 2006*. On WWW at http://www.sasgroup.net/SASGROUP_SUSTAINABILITY/CMSForeignContent/2006se.pdf. Accessed 18.9.07.

Schellnhuber, J., Cramer, W., Nakicenovic, N., Wigley, T. and Yohe, G. (eds) (2006) *Avoiding Dangerous Climate Change*. Cambridge: Cambridge University Press.

Scheyvens, R. (1999) Ecotourism and the empowerment of local communities. *Tourism Management* 20 (2), 245–249.

Schneider, V., Fink, S. and Tenbücken, M. (2005) Buying out the state: A comparative perspective on the privatisation of infrastructures. *Comparative Political Studies* 38 (6), 704–727.

Schouten, F. (2007) Cultural tourism: Between authenticity and globalization. In G. Richards (ed.) *Cultural Tourism: Global and Local Perspectives* (pp. 25–38). New York: Haworth Press.

Scott, D. (2006) US ski industry adaptation to climate change. In S. Gössling and C.M. Hall (eds) *Tourism and Global Environmental Change. Ecological, Social, Economic and Political Interrelationships* (pp. 262–285). London: Routledge.

Second Home Barometer (2004) *Nieminen M. Kesämökkibarometri 2003*. Statistics Finland & Ministry of the Interior/Island Committee.

Seiviaggi (2006) Lapponia, Scandinavia, Baltico Islanda, Svalbard, 11/2006–4/2007.

Sharpley, R. (2000) Tourism and sustainable development: Exploring the theoretical divide. *Journal of Sustainable Tourism* 8 (1), 1–19.

Sharpley, R. (2004) Tourism and the countryside. In A. Lew, C.M. Hall and A. Williams (eds) *A Companion to Tourism* (pp. 374–386). Oxford: Blackwell.

Sharpley, R. and Sharpley, J. (1997) *Rural Tourism: An Introduction*. London: Thompson.

Shaw, G. and Williams, A.M. (1994) *Critical Issues in Tourism: A Geographical Perspective*. Oxford: Blackwell.

Shields, R. (1991) *Places on the Margin: Alternative Geographies of Modernity*. London: Routledge.

Short, J.R. (1991) *Imagined Country: Society, Culture and Environment*. London: Routledge.
Shukman, D. (2007) Arctic ice island breaks in half. *BBC News*, Science/Nature 1 October. On WWW at http://news.bbc.co.uk/2/hi/science/nature/7022192.stm. Accessed 1.4.08.
Sievänen, T., Neuvonen, M. and Pouta, E. (2006) Finnish boaters and their outdoor activities. In C.M. Hall and T. Härkönen (eds) *Lake Tourism: An Integrated Approach to Lacustrine Tourism Systems* (pp. 149–165). Clevedon: Channel View Publications.
Sievänen, T., Pouta, E. and Neuvonen, M. (2007) Recreational home users – potential clients for countryside tourism? *Scandinavian Journal of Hospitality and Tourism* 7 (3), 223–242.
Skaptadóttir, U.D. and Jóhannesson, G.T. (2004) The role of municipalities in innovation: Innovations in three sectors of society in two municipalities in Iceland. In N. Aarsaether (ed.) *Innovations in the Nordic Periphery* (pp. 63–88). Stockholm: Nordregio.
Skärgårdarnas Riksförbund (2002) Vitbok för boende i skärgården. Runmarö.
Skjæveland, A. and Tøsdal, C.B. (1999) Land som merkevare – Hollywood versus Lofoten. *Magma* 2 (6), 52–62.
SLAO (Svenska Liftanläggningars Organisation). On WWW at http://www.slao.se. Accessed 11.1.06.
Sletvold, O. (1999) Kysten. In J.K.S. Jacobsen and A. Viken (eds) *Turisme: Stedet i en Bevegelig Verden* (pp. 183–190). Oslo: Universitetsforlaget.
Sletvold, O. (2006) The Norwegian Coastal Express: Moving towards cruise tourism? In R. Dowling (ed.) *Cruise Tourism: Issues, Impacts, Cases*. Oxford: CABI Publishing.
Smith, S.L.J. (1998) Tourism as an industry: Debates and concepts. In D. Ioannides and K.G. Debbage (eds) *The Economic Geography of the Tourist Industry: A Supply-side Analysis* (pp. 31–52). London: Routledge.
Smith, S.L.J. (2004) The measurement of global tourism: Old debates, new consensus, and continuing challenges. In A. Lew, C.M. Hall and A. Williams (eds) *A Companion to Tourism* (pp. 25–35). Oxford: Blackwell.
Sörlin, S. (1998) Monument and memory: Landscape imagery and the articulation of territory. *World Views: Environment, Culture, Religion* 2 (3), 269–279.
Sörlin, S. (1999) The articulation of territory: Landscape and the constitution of regional and national identity. *Norsk Geografisk Tidsskrift Norwegian Journal of Geography* 53 (2–3), 103–112.
SSB (2005) *Svalbard statistikk 2005*. Statistisk sentralbyrå. Statistics Norway. On WWW at www.ssb.no. Accessed 1.4.08.
Standl, H. (2004) Recent devevlopment in tourism in the Baltic states. *Greifswälder Beiträge zur Regional-, Freizeit- und Tourismusforschung* 15. Greifswald: Forum für Regional-, Freizeit- und Tourismusforschung and der Universität Greifswald.
Statistics Denmark (2005) StatBank Denmark. On WWW at www.statistikbanken.dk. Accessed
Statistics Denmark (2006) *Fritidssejlerturisme 2006*. Copenhagen.
Statistics Finland (2000) *Transport and Tourism* 2000: 11.
Statistics Finland (2005a) *Free-time Residences 2004*. Official Statistics of Finland. Housing 2005:7. Helsinki: Tilastokeskus.
Statistics Finland (2005b) *Finnish Travel 2004*. Official Statistics of Finland. Transport and Tourism 2005:9. Helsinki: Tilastokeskus.
Statistics Finland (2006) *Tourism Statistics 2006*. Official Statistics of Finland. Transport and Tourism 2006. Helsinki: Tilastokeskus.
Statistics Finland (2007) *Free-time Residences 2005*. Official Statistics of Finland. Housing 2007. Helsinki: Tilastokeskus.
Statistics Iceland (2005) Statistics. On WWW at www.statice.is. Accessed
Statistics Norway (2006) Statistikkbanken. On WWW at www.ssb.no. Accessed
Statistics Norway (2006a) Monthly Bulleting of Statistics, 31.01.2006.
Statistics Norway (2006b) *Tourism Satellite Accounts, Final Figures for 2004 and Preliminary Figures for 2005: Total Tourism Consumption NOK 90 billion*. Oslo: Statistics Norway. On WWW at http://www.ssb.no/turismesat_en/main.html. Accessed 1.4.08.
Statistics Sweden (2003) Statistikdatabasen. On WWW at www.scb.se. Accessed

Statistisk Sentralbyrå (SSB) (2004) *Ferieundersøkelsen 2004*. On WWW at http://www.ssb. no/emner/00/02/20/ferie/tab-2005-06-28-02.html; http://www.ssb.no/emner/00/02/ 20/ferie/tab-2005-06-28-07.html. Accessed 1.4.08.
Stern, N. (2006) *The Economics of Climate Change: The Stern Review*. London: Her Majesty's Treasury.
Støre, J.G, Singsaas, H., Brunstad, B., Ibenholt, K. and Røtnes, R. (2003) *Norge 2015 – en reise verdt? Scenarier for turisme-Norge*. Oslo: Kagge forlag.
Strapp, J.D. (1988) The resort cycle and second homes. *Annals of Tourism Research* 15, 504–516.
Stratton, J. (1990) *Writing Sites*. New York: Harvester Wheatsheaf.
Stynes, B.W. and Pigozzi, B.W. (1983) A tool for investigating tourism-related seasonal employment. *Journal of Travel Research* 21 (2), 19–24.
Sund, T. (1949) Sommer Bergen: Avisens adressforandringer som vitnesbyrd om Bergenserbes eandopphold sommeren 1974. *Norsk Geografisk Tidsskrift* 12 (2), 92–103.
Sund, T. (1981) Sommer Bergen: Avisens adressforandringar som vitnesbyrd om Bergensernes landopphold sommeren 1974. *Norsk Geografisk Tidsskrift* 12, 32–103.
Svalastog, S. (1981) Hytteplanering og planleggningsideologi. *Plan og arbeid* 4, 254–262.
Svalbard Reiselivsråd (2004) *Reiselivsutviklingen I Longyearbyen – Evalueringsrapport*. Svalbard reiselivsråd.
Svensson, G. (2002) Efficient consumer response – its origin and evolution in the history of marketing. *Management Decision* 40 (5), 508–519.
Svensson, H. (1954) En studie över sommarortsfältet för malmö stad. *Svensk Geografisk Årsbok* 30, 168–178.
Sveriges Hotell & Restaurang Företagare (SHR) (2007) *Miljö*. On WWW at http://www.shr. se/templates/Page.aspx?id=1590. Accessed 18.9.07.
Swarbrooke, J. and Horner, S. (2001) *Business Travel and Tourism*. Oxford: Butterworth-Heinemann.
Swarbrooke, J., Beard, C., Leckie, S. and Pomfret, G. (2003) *Adventure Tourism: The New Frontier*. London: Butterworth-Heinemann.
Swartz, L. (2007) Stockholms hotellpriser hotar framtida kongresser. *Restauratören* 2007-11-30.
Swedish Civil Aviation Authority (2007) *Flygets miljöpaverkan*. On WWW at http://www. luftfartsstyrelsen.se/templates/LS_InfoSida_70_30____38061.aspx. Accessed 18.9.07.
Swedish Railways (2007) *Miljö*. On WWW at http://www.sj.se/sj/jsp/polopoly.jsp?d= 260&l=sv. Accessed 18.9.07.
Swedish Tourism Authority (2003) *Tourism in Sweden 2003*. Stockholm.
Teigland, J. (2000) *Nordmenns friluftsliv og naturopplevelser. Et faktagrunnlag fra en panelstudie av langtidsendringer 1986–1999*. VF-rapport 07/2000. Sogndal: Vestlandsforskning.
Therkelsen A. (2003) Imagining places. Image formation of tourists and its consequences for destination promotion. *Scandinavian Journal of Hospitality and Tourism* 3 (2), 134–151.
Þjóðhagsstofnun (2000) *Yfirlit um þróun og umfang ferðaþjónustunnar. Greinargerð unnin fyrir samgönguráðuneytið*. Reykjavík, Ísland: Samgönguráðuneytið.
Thomson, C. and Thomson, J.S. (2006) Arctic cruise ship tourism. In G. Baldacciono (ed.) *Extreme Tourism: Lessons from the World's Cold Water Islands* (pp. 223–231). Amsterdam: Elsevier.
Thoreau, H.D. (1955) *Walden; Or, Life in the Woods*. Toronto: Dover Publishers.
Thrift, N. (2001) Still life in nearly present time: The object of nature. In P. Macnaghten and J. Urry (eds) *Bodies in Nature* (pp. 34–57). London: Sage.
Tiilikainen, T. (2001) *The Political Implications of the EU Enlargement to the Baltic States*. European University Institute Working Paper RSC N 2001/21.
Timothy, D.J. (2002) Tourism and the growth of urban ethnic islands. In C.M. Hall and A.M. Williams (eds) *Tourism and Migration: New Relationships Between Production and Consumption* (pp. 135–151). Dordrecht: Kluwer.
Timothy, D.J. and Boyd, S.W. (2003) *Heritage Tourism*. Harlow: Pearson Education.

Toivanen, E. (2006) Keynote speech: Nordic encounters. *Scandinavian Journal of Hospitality and Tourism* 6 (4), 349–354.
Tress, G. (2002) Development of second-home tourism in Denmark. *Scandinavian Journal of Hospitality and Tourism* 2, 109–122.
Turismens utredningsinstitut (2006) *Lågkostnadsflyget en möjlighetsrevolution för den svenska rese- och besöksnäringen* [Low-fare air travel: A revolutionary opportunity for Swedish tourism]. On WWW at http://www.shr.se/upload/dokument/Pressmeddelanden/L% E5gkostnadsflyget%20en%20 m%F6jlighetsrevolution.pdf. Accessed 11.5.06.
Turun Sanomat (2005) Ahvenanmaalla yöpymiset ovat vähentyneet lähes 10 prosenttia. *Turun Sanomat*. On WWW at www.turunsanomat.fi/verkkolehti/?ts=1,4:2:0:0,4:2:1:1. Accessed 10.7.05.
Tuulentie, S. (2006) The dialectic of identities in the field of tourism: The discourses of the indigenous Sami in defining their own and the tourists' identities. *Scandinavian Journal of Hospitality and Tourism* 6, 25–36.
Tuulentie, S. (2007) Settled tourists: Second homes as a part of tourist life stories. *Scandinavian Journal of Hospitality and Tourism* 7 (3), 281–300.
Tyndall Centre for Climate Change (2005) *Decarbonizing the UK. Energy for a Climate Conscious Future.* On WWW at http://www.tyndall.ac.uk/media/news/tyndall_decarbonising_the_uk.pdf. Accessed 11.5.06.
Tysfjord turistsenter (2006) Orca Tysfjord. On WWW at http:// www.tysfjord-turistsenter.no. Accessed 24.11.06.
UK Commission on Sustainability (2006) *I Will If You Will. Towards Sustainable Consumption.* On WWW at http://www.sd-commission.org.uk. Accessed 11.5.06.
UNESCO (2007) *World Heritage.* On WWW at whc.unesco.org. Accessed 12.12.07.
United Nations (1994) *Recommendations on Tourism Statistics.* New York: United Nations.
United Nations and United Nations World Tourism Organization (UN and UNWTO) (2007) *International Recommendations on Tourism Statistics (IRTS) Provisional Draft.* United Nations and United Nations World Tourism Organization: New York/Madrid.
United States Public Law (1964) Public law 88–577. 88th Congress. S. 4, 3 September 1964.
Urry, J. (1990) *The Tourist Gaze: Leisure and Travel in Contemporary Societies.* London: Sage Publications.
Urry, J. (1995) *Consuming Places.* London: Routledge.
Urry, J. (2000) *Sociology Beyond Societies.* London: Routledge.
Vahtikari, T. (2004) The (self-)perception of historic city: Case study of the Finnish World Heritage City Old Rauma. Paper presented at The Seventh International Conference on Urban History, Athens-Piraeus, October 2004. Session: Urban Images and Representations in Europe and beyond during the 20th Century.
Vail, D. and Heldt, T. (2000) Institutional factors influencing the size and structure of tourism: Comparing Dalarna (Sweden) and Maine (USA). *Current Issues in Tourism* 3 (4), 283–324.
Vail, D. and Heldt, T. (2004) Governing snowmobilers in multiple-use landscapes: Swedish and Maine (USA) cases. *Ecological Economics* 48 (4), 469–483.
Västsvenska Turistrådet (2003) Kortnytt nr 2 – Juni 2003. Göteborg.
Veneily (2007) Basfarleder för båttrafik. On WWW at http://www.veneily.fi/venevaylat.php. Accessed 17.1.07.
Vernon, J., Essex, S., Pinder, D. and Curry, K. (2005) Collaborative policymaking – local sustainable projects. *Annals of Tourism Research* 32 (2), 325–345.
Viken (2006) Ecotourism in Norway: Nonexistence or co-existence? In S. Gössling and J. Hultman (eds) *Ecotourism in Scandinavia: Lessons in Theory and Practice* (pp. 38–52). Wallingford: CABI.
Viken, A. and Müller, D.K. (2006) Introduction: Tourism and the Sami. *Scandinavian Journal of Hospitality and Tourism* 6, 1–6.
Viken, A. (1995) Tourism experiences in the Arctic – The Svalbard case. In C.M. Hall and M.E. Johnston (eds) *Polar Tourism – Tourism in the Arctic and Antarctic Regions* (pp. 73–84). Chichester: John Wiley & Sons.

Viken, A. (1997) Sameland tilpasset turistblikket. In J.K. Steen Jacobsen and A. Viken (eds) *Turisme: Fenomen og Naering* (pp. 174–180). Oslo: Universitetsförlaget.
Viken, A. (2002) Reiselivets organisering. In J.K.S. Jacobsen and A. Viken (eds) *Turisme. Fenomen og næring* (pp. 207–228). Oslo: Gyldendal Akademisk.
Viken, A. (2004) *Turisme. Miljø og Utvikling*. Oslo: Gyldendal Akademisk.
Viken, A. (2006) Svalbard, Norway. In G. Baldacchino (ed.) *Extreme Tourism* (pp. 129–142). Amsterdam: Elsevier.
Viking Line (2006) Viking Line. On WWW at http://www.vikingline.se. Accessed 8.10.06.
Vistad, O.I. (2002) Visitors and managers: Differing evaluations concerning recreational impacts and preferences for management actions? In A. Arnberger, C. Brandenburg and A. Muhar (eds) *Monitoring and Management of Visitor Flows in Recreational and Protected Areas. Conference Proceedings*. Vienna: Institute for Landscape Architecture and Landscape Management, University Vienna.
von Linné, C. (1969/1889) *Lapinmatka 1732*. Hämeenlinna: Karisto.
Vuori, O. (1966) *Kesähuvilanomistus Suomessa* [Ownership of Summer Villas in Finland]. Turku: Annales Universitatis Turkuensis, Series C, Tom. 3. (Deutsche Zusammenfassung).
Vuorio, T. (2003) *Information on Recreation and Tourism in Spatial Planning in the Swedish Mountains – Methods and Need for Knowledge*. Blekinge Institute of Technology Licentiate Dissertation Series 2003:03, ETOUR scientific book series V2003:12, Sweden.
Vuoristo, K-V. (2002) Regional and structural patterns of tourism in Finland. *Fennia* 180 (1–2), 251–259.
Wang, N. (1999) Rethinking authenticity in tourism experience. *Annals of Tourism Research* 26, 349–370.
Wang, N. (2000) *Tourism and Modernity*. Amsterdam: Pergamon.
Watson, A.E. and Kajala, L. (1995) Intergroup conflict in wilderness: Balancing opportunities for experience with preservation responsibility. In A-L. Sippola, P. Alaraudanjoki, B. Forbes and V. Hallikainen (eds) *Northern Wilderness Areas: Ecology, Sustainability, Values*. Arctic Centre Publications 7, pp. 251–270. Rovaniemi: University of Lapland.
Wawn, A. (2001) *The Viking Revival*. BBC History. On WWW at http://www.bbc.co.uk/history/ancient/vikings/revival_01.shtml. Accessed 1.4.08.
Westholm, E. (1999) From state intervention to partnerships. In E. Westholm, M. Moseley and N. Stenlås (eds) *Local Partnerships and Rural Development in Europe: A Literature Review of Practice and Theory* (pp. 137–156). Falun: Dalarna Research Institute.
Westin, L. and Paju, M. (2003) Bidrar det svenska kulturarvet till regionernas attraktivitet? *PLAN* 4 (2003), 17–21.
Wiik, P. (1958) Om stadsbefolkningens sommarbosättning i Sydösterbotten. *Terra* 70, 185–198.
Williams, A.M. and Hall, C.M. (2002) Tourism, migration, circulation and mobility: The contingencies of time and place. In C.M. Hall and A.M. Williams (eds) *Tourism and Migration: New Relationships between Production and Consumption* (pp. 1–52). Dordrecht: Kluwer.
Williams, D. (2001) Sustainability and public access to nature: Contesting the right to roam. *Journal of Sustainable Tourism* 9 (5), 361–371.
Williams, F. (2001) Quality rural tourism, niche markets and imagery: Quality products and regional identity. In L. Roberts and D. Hall (eds) *Rural Tourism and Recreation: Principles to Practice* (pp. 214–215). Wallingford: CABI Publishing.
Wilson, J.C. (2003) Planning policy issues for marine ecotourism. In B. Garrod and J.C. Wilson (eds) *Marine Ecotourism: Issues and Experiences* (pp. 48–65). Clevedon: Channel View Publications.
Wöber, K. (1997) Introducing a harmonization procedure for European city tourism statistics. In J.A. Mazanec (ed.) *International City Tourism: Analysis and Strategy* (pp. 26–38). London: Pinter.
World Economic Forum (WEF) (2001) *The Global Competitiveness Report 2001–2*. Geneva: World Economic Forum.
World Economic Forum (WEF) (2007) *The Travel and Tourism Competitiveness Report 2007: Furthering the Process of Economic Development*. Geneva: World Economic Forum.

World Economic Forum (2008) *The Travel and Tourism Competitiveness Report 2008: Balancing Economic Development and Environmental Sustainability*. Geneva: World Economic Forum.

World Tourism Organization (WTO) (1991) *Resolutions of International Conference on Travel and Tourism, Ottawa, Canada*. Madrid: WTO.

World Travel and Tourism Council (WTTC) (1997) Rejoinder: from the World Travel and Tourism Council. *Tourism Economics* 3 (3), 282–288.

WTO (1996) Rural tourism to the rescue of Europe's countryside. *WTO News* 3, 6–7.

WWF (2004) *Cruise Tourism on Svalbard: A Risky Business?* Oslo: WWF International Arctic Programme.

Ympäristöministeriö (2002) Ohjelma luonnon virkistyskäytön ja luontomatkailun kehittämiseksi. *Suomen ympäristö* 535, 1–48.

Zeppel, H. (1998) Land and culture: Sustainable tourism and indigenous peoples. In C.M. Hall and A.A. Lew (eds) *Sustainable Tourism: A Geographical Perspective* (pp. 60–74). Harlow: Longman.

Zillinger, M. (2006) The importance of guidebooks for the choice of tourist sites: A study of German tourists in Sweden. *Scandinavian Journal of Hospitality and Tourism* 6, 229–247.

Index

Adventure tourism 238-9
Arctic 24, 30-3, 60-1, 131-33, 156, 162-3, 216-9, 239-41, 248-50
ACIA (Arctic Climate Impact Assessment) 248-50
Åland Islands 2, 4, 111-4, 153, 154-5, 167, 175, 190-1, 204
– Mariehamn 2
Arctic Council 24, 60-3, 248-50
Arctic trekking 239
Attractions 30, 36, 212, 213, 216, 223, 230, 240, 244
Authenticity 38, 40, 120, 200, 223
Aviapolis 101

Balance of Tourist Trade 264
Baltic identity 246-7
Baltic Sea 246
Baltic states 245
Barents Euro-Arctic Council 60, 61, 63, 246
Barents region 246
Blue Road 246
Boating 146, 154, 166-7
Bothnian Sea 169, 250
Branding 35, 37, 38, 39, 40-4, 47, 57, 88-9, 91, 92, 99, 251
Buddha 223
Business city 100-1

Christmas 38, 228, 229, 239-40
Cities of the future 89
Climate change 220, 229, 237, 242, 249-263
Coastal tourism 133, 155-6. *See also* Marine tourism
Competitiveness 17-21, 68, 69, 88-96
Coordination 34, 47, 48, 54-6, 165
Council of the Baltic Sea States 62, 246

Creative city 92, 95
Cross-border tourism 7, 63, 92, 95, 100, 246
Cruise tourism 37, 47, 111, 128, 132, 133, 156-63
Cultural tourism 97-9, 197-223
Culture city 99

Danish Labour Market Holiday Foundation 57
Denmark 2, 3, 4, 18, 19, 23, 41-4, 46, 55, 61, 65, 66, 71-5, 92, 94, 97, 110, 166, 167, 169, 175, 178, 211, 255, 264
– Copenhagen 2, 35-7, 84, 89, 92-4, 100, 101, 155
– Faroe Islands 2, 4, 41, 61, 63, 68, 157, 165, 211, 264
– Greenland 2, 4, 61, 75, 156, 211, 257
– Helsingør 97
– Roskilde 97

Ecotourism 122, 131, 134, 139, 143, 154, 163-6, 239
Entrepreneurial state 56
Environmental change 58, 59, 60, 126, 248-50. *See also* Climate change
Estonia 62, 64, 157, 159, 245-6
– Tallinn 157
European Union 5, 53, 60, 62-64, 72, 100, 202, 228, 255, 256
Evaluation 36, 70-3
Experience economy 35, 40, 231, 236-7, 245

Farm tourism 117-21
Faroe Islands 2, 4, 61, 66, 157, 165, 211, 264
– Tórshavn 2
Father Christmas 38. *See also* Santa Claus
Ferry tourism 100, 113, 155-9, 190

Festivals 38, 92, 117, 169, 199, 213-20, 245
Finland 2, 4, 12, 14-15, 18, 19, 23, 38, 53, 55, 56, 61, 62, 64-7, 68, 85, 97, 99, 100, 110, 111, 121-5, 148-50, 157, 159, 166, 167, 168, 175, 178, 182, 185-8, 189, 190, 202, 211, 215, 220, 221, 225, 226-9, 241, 244-5, 246, 255, 264
– Hanko 164
– Helsinki 2, 80, 82, 86, 89, 97, 98, 145, 157, 244
– Kemi 231, 233, 239, 240
– Kouvola 99
– Kuusamo 226
– Lapland 218, 225, 240-1
– Oulu 80, 82, 215
– Pallas 218
– Rauma 87, 97
– Rovaniemi 38, 221, 232, 240, 245
– Ruka 145, 218, 225, 226-9
– Tampere 56-7, 245
– Tornio 100
– Åland Islands 2, 4, 111-4, 153, 154-5, 167, 175, 190-1, 204
Fishing tourism 163-4
Folk Life Museums 211

Governance 52, 55, 58-68, 195, 248
Government roles in tourism 54-58
Greenland 2, 4, 23, 38, 61, 62, 63, 75, 211, 264
– Nuuk 2

Hans Christian Andersen bicentenary 36-37
Heritage 21, 37-8, 46, 64-6, 87, 90, 97-8, 121, 132, 154, 162, 168, 172, 175, 198-202, 204-22. *See also* World Heritage
Historic city 86, 96-8
Hotels 101-3, 257, 260
Hurtigruten 156-7

Iceland 2, 4, 5, 18, 19, 38, 39-40, 53, 61, 62, 63, 65, 66, 67-8, 110, 117-20, 139-42, 157, 161, 173, 174, 178, 204, 213-6, 248, 256, 257
– Akureyi 231
– Reykjavik 2, 39
– Skutustadir 38
– Þingeyri 206-209
Ice hotels 239-40
Ikea 95, 100
Indigenous tourism 78, 216-21, 237, 244, 249, 250
Inuit 237, 242
Interpretation 202-3

Kalmar Union 4

Labour market 230-5
Local government 47, 56, 60
Lordi 244-5

Marine tourism 153-71
Marketing 3, 25-51, 63, 64-6, 69, 75, 84, 87-95, 104, 120, 140, 155, 198, 240-1, 262
Marketing management 44-6
Meetings 100-1
Mobility 16-17
Mountain areas 104, 131, 140-1, 146-8, 181, 183-4, 225, 230-5
Museums 56, 74, 83, 92, 96, 99, 199, 206, 211-2

National identity 205, 245, 246, 248
National parks and protected areas 11, 49, 56, 69, 76-80, 131, 138, 143-50, 166, 167, 202, 207, 225, 244
Nature-based tourism 59, 122, 130-52, 250
New tourism 236-8
Night life city 99
Nordic Council 4, 60, 62, 249
Nordic Council of Ministers 60, 62
Nordic Council of Ministers Tourism Ad Hoc Working Group 251
Nordic idea 1-4. *See also* Nordicity
Nordicity 243-4, 245-8. *See also* Nordic idea
Nordic Passport Union 5
Norway 2, 4, 5, 12, 13, 18, 19, 24, 38, 46-9, 53, 55, 63, 64, 65, 66, 67, 76-80, 92, 97, 104, 110, 117, 118, 121, 127, 154-5, 156-7, 161, 163-4, 169, 178, 183-5, 194, 204-8, 211, 212, 220, 224, 247, 255, 264
– Bergen 97
– Drøbak 38
– Lillehammer 104-6, 212, 230
– North Cape 30-33
– Oslo 2, 86, 98
– Rosendal 204-8
– Røros 97
– Stavangar 92
– Svalbard 128-129, 162, 231
– Voss 220
Norwegian Coastal Voyage 156-7

Ocean tourism, *see* Marine tourism
Øresund region 60, 65, 92-5

Peripheral regions 3, 22, 53, 63, 67, 111-4, 126, 127, 138, 143, 186, 191, 203, 216, 230, 264
Place competition and reimaging 85, 86, 87-89

Place marketing 87-95
Planning 26, 53-4, 55-6, 58, 67-8, 69-70, 80, 90, 92, 104-5, 141-3, 146, 167-9, 186, 191, 228
Policy 20, 53, 58, 60, 63-8, 76-7, 79, 88, 91, 95, 114, 123, 126, 142, 159, 165, 225, 249, 251
Promotion 35, 38, 40, 45, 46, 56-7, 59, 60, 63, 88, 89, 92, 95, 220
Protected areas. *See* National parks and protected areas
Public interest 58

Regional development 48, 53, 63, 67-8, 71, 95-6, 123, 143, 172, 201-2, 225
Regionalism 245-9
Regulation 27, 56, 67, 133, 134, 186, 254, 261
Reindeer 217, 218, 220-2, 232, 238
Resorts 116, 126, 131, 168-9, 182, 183, 224-9, 242
Rural tourism 109-29, 186
Russia (Russian Federation) 4, 61, 62, 64, 66, 100, 156, 162, 165, 220

Safaris 136, 236, 237, 238
Sami 4, 64, 78, 135, 216-22, 244
Santa Claus 38, 221, 239-41. *See also* Father Christmas
Santa Claus Village 240
Scandinavia 3-4
Second homes 6, 7, 16, 17, 105, 114, 149, 150, 169, 172-96
Service industry. *See* Services
Services 11, 12, 13, 22, 26-29
Shopping city 96, 99-100
Skiing 116, 133, 146, 149-50, 184, 224-5, 227-31, 233, 236
Snow castles 239-40
Social construction 30-33, 223
Social tourism 57
Sport city 103-7
Stages of a tourist trip 29-30
Stimulation 56-7
Struve's Geodetic Arc 210
Sustainability 63, 68, 73-4, 80, 116-7, 126-8, 134, 163, 251, 252, 254, 256
Sweden 2, 4, 18, 19, 59, 60, 63-6, 78, 88-90, 92, 94, 97, 98-100, 102, 110, 121, 135, 146-8, 157, 166-7, 169, 175, 176, 178, 181, 189, 192-3, 194, 201, 211, 216-9, 216-7, 223, 224-6, 230, 231-6, 241, 246, 251-63, 264
– Åre 218, 231-6
– Falun 98
– Fredrika 223
– Gotland 212-212203
– Göteberg 96, 167, 213
– Halmstad 169
– Haparanda 100, 239
– Helsingborg 84
– Jokkmokk 216-9
– Jukkasjärvu 241
– Karlskrona 98
– Kvarken 206
– Luleå 98, 232, 233
– Malmö 92
– Piteå 165
– Stockholm 2, 83, 86, 88-9, 100, 192-3
– Sunne 59
– Visby 98

Tourism, definitions of 5-11
Tourism product 11, 13, 25, 29-30, 34-5
Tourism Satellite Accounts 12-15
Tourist city 86, 96-103
Transport 7, 13, 16, 28, 32, 56, 64, 65, 66, 89, 90, 100, 105, 113, 151, 155-63, 166, 175, 246, 248, 251-9

Urban tourism 83-108

Viking Line 155, 157, 159
Vikings 37-8, 46, 49, 95, 201, 205, 206, 213-5
Visual representation 221

Welfare state 52-3, 57, 58, 204-11
Whalewatching 154, 162, 165
Wilderness 1, 130-51, 216, 220, 236, 239, 241, 249
Winter Olympics 104-7, 223
Winter tourism 224-42
Women's Exercise Association 225
World Economic Forum 17-21
World Heritage 21, 87, 97-8, 154, 208-10

For Product Safety Concerns and Information please contact our EU Authorised Representative:

Easy Access System Europe

Mustamäe tee 50

10621 Tallinn

Estonia

gpsr.requests@easproject.com